Multicellularity

Vienna Series in Theoretical Biology

Editor-in-Chief

Gerd B. Müller, University of Vienna and KLI Klosterneuburg

Associate Editors

Johannes Jäger, KLI Klosterneuburg

Thomas Pradeu, CNRS and University of Bordeaux

Katrin Schäfer, University of Vienna

Multicellularity
Origins and Evolution

edited by Karl J. Niklas and Stuart A. Newman

The MIT Press
Cambridge, Massachusetts
London, England

This book was set in Times New Roman by Toppan Best-set Premedia Limited.

Library of Congress Cataloging-in-Publication Data is available.

ISBN: 978-0-262-03415-9 (hc : alk. paper), 978-0-262-54585-3 (pb)

Contents

Werner Callebaut (1952–2014)

The volume editors, the series editors, and the editors at MIT Press jointly dedicate this volume to the memory of Werner Callebaut, who sadly passed away in November 2014. Werner was one of the founding editors of the Vienna Series in Theoretical Biology. He had studied philosophy at Ghent University and subsequently pursued an academic career that led him via the Universities of Brussels, Limburg, and Ghent to Hasselt University, where he became a Professor of Philosophy in 1995. Following two visiting periods at the Konrad Lorenz Institute for Evolution and Cognition Research at Altenberg, he moved to Austria while continuing a part-time affiliation at Hasselt University. In 1999 he became the Scientific Manager and subsequently the Scientific Director of the KLI, and among his tasks were the care of the Vienna Series and the journal *Biological Theory*.

Werner was the perfect editor. His encyclopedic knowledge in vast areas of philosophy of science and his devotion to quality and style made for the formidable success of *Biological Theory*, of which he was the editor-in-chief. He devoted countless hours, day and night, to "his" journal. He would not only run the standard review processes, but would contribute comments, suggestions, and corrections to each manuscript, and oftentimes he engaged in extensive communication with the authors. He also inspired many of the Vienna Series books and was involved in the preparations for the KLI Altenberg Workshop that spawned the present volume. Werner contributed valuable insights during the workshop and gave encouragement to several of the authors joined herein. His remarkable capacity to elicit more substantial results from any intellectual project he came in touch with also contributed to the quality of the present one.

Werner was a remarkable man and a valued colleague, whose strong belief in intellectual exactitude has improved the writings and the thinking of many a scholar. If the present volume contains more flaws than usual, it is because Werner could not edit the manuscripts. He would have loved to see this volume appear. Upon receipt he would have opened it up at a random page, and his swift editorial eye would immediately have detected several mistakes. Now those mistakes will remain hidden for years, but we will remember Werner as our gifted companion who could have detected them. We lift a glass of editor's spirit to his memory!

Gerd B. Müller, Stuart A. Newman, Karl J. Niklas
Klosterneuburg, 17 April 2015

Series Foreword

Biology is a leading science in this century. As in all other sciences, progress in biology depends on the interrelations between empirical research, theory building, modeling, and societal context. But whereas molecular and experimental biology have evolved dramatically in recent years, generating a flood of detailed empirical data, the integration of these results into useful theoretical frameworks has lagged behind. Driven largely by pragmatic and technical considerations, research in biology continues to be less guided by theory than seems indicated. By promoting the formulation and discussion of new theoretical concepts in the biosciences, this series intends to help fill important gaps in our understanding of some of the major open questions of biology, such as the origin and organization of organismal form, the relationship between development and evolution, and the biological bases of cognition and mind. Theoretical biology has important roots in the experimental tradition of early-twentieth-century Vienna. Paul Weiss and Ludwig von Bertalanffy were among the first to use the term *theoretical biology* in its modern sense. In their understanding the subject was not limited to mathematical formalization, as is often the case today, but extended to the conceptual foundations of biology. It is this commitment to a comprehensive and cross-disciplinary integration of theoretical concepts that the Vienna Series intends to emphasize. Today, theoretical biology has genetic, developmental, and evolutionary components, the central connective themes in modern biology, but it also includes relevant aspects of computational or systems biology and extends to the naturalistic philosophy of sciences. The "Vienna Series" grew out of theory-oriented workshops organized by the KLI, an international institute for the advanced study of natural complex systems. The KLI fosters research projects, workshops, book projects, and the journal *Biological Theory*, all devoted to aspects of theoretical biology, with an emphasis on—but not restricted to—integrating the developmental, evolutionary, and cognitive sciences. The series editors welcome suggestions for book projects in these domains.

Gerd B. Müller, Johannes Jäger, Thomas Pradeu, Katrin Schäfer

Foreword: The Evolution of Multicellularity

John Tyler Bonner

A Brief History of Our Interest in Multicellularity

It is generally assumed there was an early period when the only eukaryotic organisms on our planet were unicellular; they were the ancestors of all the multicellular organisms that exist today, including ourselves. The forbearers of this notion did not appear until the eighteenth century; they were the inventors of the microscope and the revealing of a whole world of organisms that were too small to be seen with the naked eye. The pioneer, Antony van Leeuwenhoek, not only invented a microscope but actually used it and gave us our first descriptions of microorganisms, many of which we know today are unicellular. Further, enormous strides forward were made in the next, the nineteenth century, the most important of which was the realization that all organisms were made up of individual cells. It was M. J. Schleiden and T. Swann who formulated this for both plants and animals, respectively. This "cell theory" was completely accepted in the remainder of the nineteenth century, making clear the relation of single-cell organisms to their multicellular descendants. And in this same century the full impact of van Leeuwenhoek's discoveries were felt and botanists had an orgy describing new species of minute algae, while the whole, new field of protozoology came into bloom and large numbers of previously unknown species were described. But curiously the question of how and why multicellularity arose received little attention. That is a subject that has aroused interest only fairly recently, as this symposium attests.

Multicellularity: With and without Photosynthesis

Biologists are generally comfortable with the idea that early in earth history all organisms were unicellular and that multicellularity arose subsequently. The realization that multicellularity was invented more than once by microorganisms is also generally appreciated by biologists. It was not a matter that was, or is, much discussed; it was simply a reasonable assumption. (It should be noted here that many of the so-called unicellular protozoa are in fact multinucleate as, for instance, the foraminifera and the radiolaria, yet they are considered unicellular for they are wrapped in one cell membrane. This is sort of an intermediate point toward multicellularity, but it is a state that has been essentially abandoned in larger multicellular animals and plants; it is only prevalent in what might be called evolutionarily intermediate organisms.)

When I was preparing my first book, *Morphogenesis* (Bonner, 1952), I found a reference to a short note in *Nature* that was original and intriguing and that deserved some thought. It was by J. R. Baker of Oxford University and published in 1948. He noted that there were a greater number of photosynthetic colonial microorganisms than ones that were not capable of photosynthesis; they were particle feeders. Algae could come in a great variety of shapes, each one of which could, with equal ease, gather energy from the sun and channel it into growth and survival; they did not need a special feeding apparatus; they gathered their needed energy directly from the sun. As he pointed out, particle feeders, on the other hand, need a mechanism to capture and digest their prey. In the most primitive forms this consists of the cytoplasmic engulfment of food particles such as we find in amoebae and choanocytes, those flagellated cells with their collar surrounding the flagellum that helps capture the food particles in sponges.

The Multicellularity of Sponges

One of the early successful ventures in multicellularity was having these collar cells come together into chambers, and by connecting the chambers in a suitable fashion a current can be brought in and out in the whole sponge. In this way large sponges can propel the spent water some distance from the sponge so that it is not directly reused and fresh water with food and oxygen can take over, as G. P. Bidder (1923) showed in his famous, lustrous paper. One particularly interesting observation was made by Ankel (1948): in some young, fresh water sponges, before they have enough flagellated chambers to blow the wasted water out of reach so it is not immediately reused, they build a tube which serves as a conduit to carry the wasted water out of reach (see figure 1). This tube disappears only when enough choanocytes grow to provide the necessary pressure to blow the wastewater beyond and out of reach of the growing sponge.

Size Increase

Everyone is content with the idea that finding a system of feeding is required in nonphotosynthetic organisms if they are to increase in size. This is exactly what sponges have done with their connected flagellated chambers. The problem has been solved a number of other ways. Advanced animals and plants begin their development with a provided supply of food: yolk in animals and cotyledons in plants. When those start-up provisions

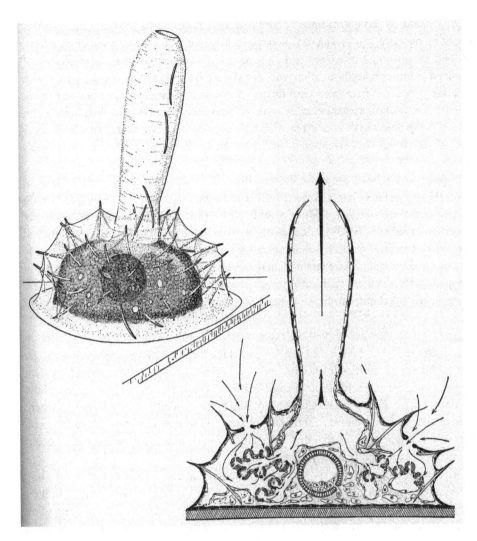

Figure 1
A young fresh water sponge (*Ephydatia fluviatilis*) showing the large osculum chimney. Above: an external view.
Below: a diagrammatic cross-section. (Reprinted from Ankel, 1948.)

are used up, photosynthesis sets in for plants, and an elaborate method of processing solid
food takes over in animals. There are exceptions: in the cellular slime molds, which are
not photosynthetic, the amoebae feed first and then come together to become multicellular.
This may be a clever solution to the problem but is of very limited evolvability. The evolution of slime molds did not lead to any larger forms.

As I have pointed out over the years there is overwhelming evidence that, along with
episodes of trends in size decrease, there has been an overall selection for size increase

over great spans of geological time. "There is always room at the top." For animals this can only be possible if a mechanism is devised to capture and convert the energy from solid food for all living processes. A key invention that made possible the evolution of dinosaurs and blue whales is an alimentary canal—a tube with a mouth at one end and an anus at the other, and populated with digestive glands that produce enzymes that break the food down to small molecules such as amino acids and simple sugars that, in turn, can diffuse into the blood system and reach all of the cells of a huge body. This is a clever invention—or series of inventions—that has made the vast evolution of animals possible.

For some time I have presented figures illustrating that the upper limit of size of both plants and animals has increased over a great time span of the history of life on earth (Bonner, 1965, *et seq.*). More recently my primitive efforts have been greatly extended by Payne et al. (2009) to include the earliest known fossils; they have lengthened the period of scrutiny by three billion years (see figure 2). Nevertheless the principle is maintained: the maximum size attained increases over large periods of time. They find in this new analysis that there is an early period of about a billion years in which there is no increase in the maximum size; the upper limit stays about the same for this extended interval.

There is one very attractive possible explanation. A feature of the ever present selection for size increase is that there have been numerous independent experiments in how to get bigger by becoming multicellular. One need only recall the vast number of independent attempts among the algae that so impressed Baker (1948). It is impossible to estimate how many times multicellularity was reinvented during that extended billion-year period where there was no significant increase in maximum size, but it must have been quite frequent. However, one might presume that they were all dead ends because they, and their descendants, never became significantly larger.

Now for an obvious hypothesis: it is an explanation that could account for the puzzling facts. All the experiments in multicellularity, except for a very few, were limited in their size, and those successful few appeared somewhere near the end of the billion years of no size upper limit change. And they were of such a nature that they allowed the construction of larger and larger organisms. It took a billion years of experimenting to finally have ways of lifting the upper size barrier by finding successful ways of having cells not only stick together but be capable of producing even larger descendants. Plants have cell walls of such a sturdy construction that they can grow massive and thrive, as some large brown algae do, some of which flourish among turbulent waves. Or in the case of land plants the invention of wood and its associated materials gave the strength necessary to support great masses of leaves and fruit, in some cases enormous distances up into the air, as is the norm for large trees. And a similar argument is required for the less formidable fungi; mushrooms also rise into the air—however, not to maximize photosynthesis, which they lack, but to spread their spores more effectively.

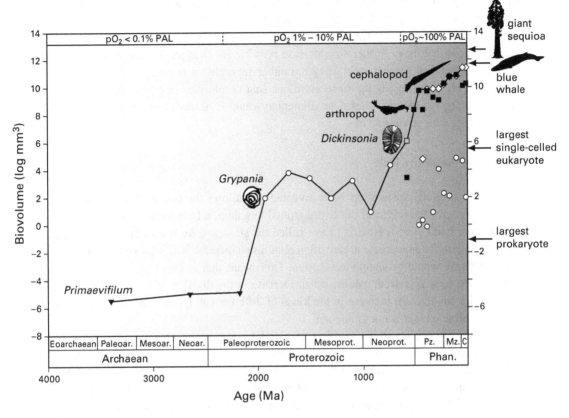

Figure 2
Sizes of the largest fossils through Earth history. Size maxima are illustrated separately for prokaryotes (triangles), single-cell eukaryotes (circles), animals (squares), and plants (diamonds). Also indicated are the estimated oxygen levels. (From Payne et al., 2009.) Reprinted with permission from the authors. PAL, present atmospheric level; pO$_2$, partial pressure of oxygen.

As we have already seen the case for animals goes beyond just having the cells sticking together, for their future success depends on devising a suitable way of taking in energy, of feeding. For the sponges the solution was a canal system in which the flagellated cells both engulfed the food particles and generated a current in the canals that carried in the food from the outside.

The main lesson I want to emphasize is the key role of the selection for size increase. The fact that there have been frequent inventions of multicellularity is a reminder that there is an ever present selection for greater size. Sometimes small size may be adaptive, but over the longest time span there has been an overall increase in the upper limits of size (see figure 2). Becoming multicellular is an easy way of accomplishing a size increase, which no doubt explains why there are numerous independent inventions of multicellular-

ity. And a few of those inventions that made size increase possible turned out to be especially fortuitous in that they had the makings for the further evolution of even larger
organisms, as we find in the large plants and the large animals that are with us today. The
breakthroughs that make further size increase possible are changes in evolvability.

Fungi have taken another tack. They are either saprophytes or parasites; that is, they
get their energy by sopping up those nutritious small molecules (e.g., sugars and amino
acids) directly. No photosynthesis, no alimentary canal is needed; they grab their needs
directly from their environment.

The Size–Complexity Rule

Once a multicellular system has been invented that allows the selection of further size
increase we have the origin of the entire animal kingdom, all plants and fungi, and here
another rule takes over. It is what I have called the *size–complexity rule* (Bonner, 2004),
where selection for an increase in size often requires an increase in complexity, for without
it the organism would be unable to function. This means that as larger organisms evolve,
this increase in size is accompanied by an increase in the division of labor among body
parts, which involves an increase in the kinds of different cell types. Size increase could
not be possible were this not to happen.

This rule also applies to the simplest multicellular inventions. Consider, for instance,
the evolution of *Volvox* and all its relatives, the volvocine algae (see figure 3). When I
was a student we were taught that there was a simple progression of size increase from
the unicellular *Chlamydomonas* to the relatively huge *Volvox,* for indeed this was the
obvious conclusion when morphology was our sole method of devising some sort of
ancestral tree. However, Kirk (1998), using an analysis of their DNA phylogeny, showed
that in their evolution sometimes larger species were ancestral to smaller ones and sometimes it was the reverse and small species were ancestral to larger ones. And there lies an
interesting example of the size–complexity rule. *Volvox* is always the biggest and it, along
with some middle-size species, have two cell types: the cells that remain motile and continue to move the colony and the cells that give rise to the reproductive cells that become
daughter colonies. All the smallest species have only one cell type: all the cells are first
motile, and then after a number of cell divisions each one develops into a daughter colony.
In some of the intermediate-size species the colonies may be of different sizes, and if they
are small, they produce only one cell type, while the larger ones produce two.

It is a perfect example of the size–complexity rule. The transition from one to two cell
types only occurs in the bigger colonies; it is regulated by size. The smaller colonies only
show one cell type while the larger ones show two. And remember *Volvox* is one of the
many, ancient multicellular inventions—one that did not give rise to higher plants, but
nothing more complex than *Volvox* itself. It lacked the necessary ingredients, such as stiff

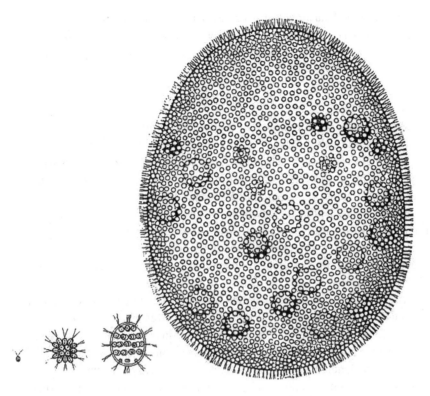

Figure 3
The ranges of size of the volvocine algae. From left to right, the ancestral single cell of *Chlamydomonas*, 16-cell *Gonium, Eudorina*, and *Volvox*. (Selected from W. H. Brown, *The Plant Kingdom,* 1935.)

cell walls, to be able to manage the construction of anything larger. Although *Volvox* failed to give rise to the plant kingdom, it was very successful in its own world. It and all the other volvocine algae that exist today have persisted successfully for millions of years; they have not gone extinct but have settled in stable niches.

Two Multicellular Worlds: The *Haves* and the *Have-Nots*

In those billion years of size stasis we saw in figure 2 most of the experiments in multicellularity reached an upper size limit that is very small, such as *Volvox*. A few, by chance, developed the necessary building materials to construct larger organisms and ultimately managed whales and trees. They are the *haves* compared to the smaller and more ancient *have-nots* that never break through the size barrier. The *haves* have evolvability. The distinction between the two is also aptly described by the terms *micro-* and *macro*organisms. Or the two categories might be labeled *multi*cellular organisms for the simple colonies and *multi-multi*cellular organisms for the elephants and the trees. The division line of the distinction is not a sharp one but nevertheless a clear one. It is one that is breached by the creation of the larger plants and animals.

Natural Selection and Size

One big difference between the micro- and the macroworld is that small organisms are to some degree less affected by the whims of natural selection compared to large organisms (Bonner 2013). This goes against the accepted dogma of evolutionary biologists wherein all organisms should be considered equally affected. However, I think the evidence is overwhelming that with increased size selection becomes more effective. This advantage for macroorganisms is reflected in the enormous success of their evolutionary progress. In contrast, the forms of microorganisms often vary as the direct result of mutation and many of the variants will be undetected by natural selection. This means that becoming multi-multicellular opens up the possibility of evolution almost wholly guided by natural selection—it is a new world.

There is considerable resistance to this notion by many evolutionary biologists today. They view adaptations to be the same for organisms of all sizes and do not see the need for a special rule for microorganisms. My arguments are that there are an inordinately large number of species among some groups of microorganisms (e.g., radiolaria, foraminifera, diatoms), a phenomenon that could most easily be explained by assuming they were relatively immune to natural selection, and random shapes frequently survive. Another argument is that mutations in microorganisms are immediately expressed in the adult while those in larger organisms have to undergo strict editing during an extended development. This means selection can channel changes—such as changes in shape—in larger organisms and produce significant and stable adaptations.

There is another argument that supports the notion that natural selection operates differently at different size levels. With the advent of molecular phylogenies, P. Schaap and S. Baldauf and their coworkers (Schaap et al., 2006) made a tree of the 75 or so species of cellular slime molds known. They found that they fell into four groups which are now known to span at least 400 million years. Many of the earliest species are with us today; the ancient ones are what Darwin (1859) called "living fossils" that have not gone extinct. It is remarkable that so many ancient species of microorganisms still exist today. The same story also applies to the volvocine algae,. Among macroorganisms living fossils, such as the horseshoe crab, are exceedingly rare; the vast majority of larger species have gone extinct. Were this not the case, the great African veldt, or Patagonia, would still be populated by herds of dinosaurs alongside the mammals—an unimaginable state.

Multicellularity is the first major step in the evolution of size increase. When megaorganisms crashed through the next size ceiling, natural selection became central and the billion years of fumbling with many experiments in multicellularity were superseded. In the transition from unicellular forms to simple multicellular colonies to larger animals and plants are steps toward a progressive increase in the power of natural selection. All the evidence points to this conclusion.

Acknowledgments

I would like to thank Vidyanand Nanjundiah and Slawa Lamont for their most helpful suggestions and comments.

References

Ankel, W. E. (1948). Über Fragen der Formbildung und der Zelldetermination bei Susswasserschwämmen. *Verhandlungen der Deutschen Zoologischen Gesellschaft*, 58–66.

Baker, J. R. (1948). The status of the protozoa. *Nature*, 161, 548–551, 587–589.

Bidder, G. P. (1923). The relation of the form of a sponge to its currents. *Quarterly Journal of Microscopical Science*, 67, 293–323.

Bonner, J. T. (1952). *Morphogenesis: An essay on development*. Princeton, NJ: Princeton University Press.

Bonner, J. T. (1965). *Size and cycle*. Princeton, NJ: Princeton University Press.

Bonner, J. T. (2004). Perspective: The size–complexity rule. *Evolution; International Journal of Organic Evolution*, 58, 1883–1890.

Bonner, J. T. (2013). *Randomness in evolution*. Princeton, NJ: Princeton University Press.

Darwin, C. (1859). *On the origin of species*. London: John Murray.

Kirk, D. L. (1998). *Volvox*. Cambridge, UK: Cambridge University Press.

Payne, J. L., Boyer, A. G., Brown, J. H., Finnegan, S., Kowalewski, M., Krause, R. A., Jr., et al. (2009). Two-phase increase in the maximum size of life over 3.3 billion years reflects biological innovation and environmental opportunity. *Proceedings of the National Academy of Sciences of the United States of America*, 106, 24–27.

Schaap, P., Winkler, T., Nelson, M., Alvarez-Curto, E., Elgie, B., Hagiwara, H., et al. (2006). Molecular phylogeny and evolution of morphology in the social amoebas. *Science*, 314, 661–663.

Preface

Multicellularity has evolved independently in ten different major clades and over 25 times in individual lineages within these clades, each of which had a unicellular ancestral condition. Its evolution involved the appearance of physiological mechanisms resulting in cell-to-cell adhesion and sustained intercellular communication among adjoining cells. A comparative approach among extant lineages shows that these two requirements have been achieved in different ways among different plant, animal, and fungal groups; for example, cell-to-cell adhesion and communication in metazoans typically involves membrane-bound glycoproteins and cell-cell junctions whereas land plant cell adhesion and communication results from a pectinaceous middle lamella and plasmodesmata. Critical questions in this context are these: What are the genomic and developmental commonalities among multicellular lineages that permit cell-to-cell adhesion and communication, and what are the unique features of the lineages? What are the selective advantages and disadvantages of these defining features of multicellular organisms? What physical consequences follow from the multicellular state of life, and can these provide insight into the origins of morphological novelties?

In addition, the evolutionary transition from the unicellular to the multicellular condition represents a major change in the definition of individuality since a new kind of organism emerges from the interactions and cooperation among subunits (cells). Until recently, discussions about this transition have been characterized by two divergent schools of thought, one focusing on the so-called "unicellular bottleneck" between alternating generations, and another focusing on "soma–germ" specialization. More recently, the focus has become more synthetic by considering the unicellular bottleneck in terms of an "alignment-of-fitness" phase (wherein genetic similarity among cells prevents internal conflict) and soma–germ specialization in terms of an "export of fitness" (wherein cellular components become interdependent and collaborate in reproductive effort). This perspective raises a number of important questions. For example, does the unicellular bottleneck in the life cycles of multicellular organisms ensure an alignment of fitness? Does multicellular "individuality" evolve as the result of a gain in fitness achieved by cellular specialization?

Finally, recent knowledge of the physical underpinnings of morphogenesis and pattern formation in multicellular organisms raises questions that go beyond the adaptationist framework of traditional evolutionary theory. In particular, to what extent are morphological motifs inevitable physical outcomes of a new scale of cellular life that afford opportunities to explore and create entirely novel niches? What is the implication of such morphological inevitabilities (which may be brought about by nonlinear processes) for phyletic gradualism? Is the evolution of multicellularity typically a "left wall"—that is, is it a generally irreversible event in the history of many, if not the majority of lineages, and if so, why?

In view of the importance of understanding the origins and consequences of the evolution of multicellularity, a group of scholars active in these areas gathered at the Konrad Lorenz Institute, Klosterneuburg, Austria, September 25–28, 2014, to share their views and to explore new avenues of research. The results of these discussions, which continue to the present, along with contributions from several key investigators in this field not present at the workshop, are contained in the 14 chapters of this book. These chapters, which collectively cover a broad spectrum of organisms, perspectives, and approaches, are organized into five sections focusing on selected topics that present particular challenges.

Part I provides insights from the fossil record regarding the paleoecological circumstances in which animal multicellularity evolved and a consideration of the ancient physiological, biochemical, and developmental predispositions for the evolution of multicellularity in general. The first chapter explores the phyletic distribution of the different methods heterotrophs use to acquire food and reviews the sequence of their appearance in the fossil record (Knoll and Lahr). One conclusion emerging from this analysis is that different methods of nutrient acquisition may have driven the evolution of simple and complex multicellularity in some lineages. Next, certain molecular aspects of cell type specification and elaboration are explicated (Niklas and Dunker). It is hypothesized that the functional repertoire of eukaryotic proteomes has been evolutionarily expanded (without increasing the size of genomes) by the evolutionary innovation of proteins containing intrinsically disordered domains, as exemplified by the ubiquitous and metabolically versatile pan-kingdom calmodulin and calmodulin-like proteins.

Part II focuses on plants, here broadly defined to include all photosynthetic eukaryotes, encompassing the algae as well as the land plants. One chapter presents new insights into how the distinction between somatic and germ cell lines may have evolved in the green alga, *Volvox*, by the co-optation of an ancient (premulticellularity) stress-induced transcriptional repressor (König and Nedelcu). A second contribution discusses how the mechanical and chemical coupling of cells and combinations of "dynamical patterning modules" (defined as associations of gene products with the physical forces and effects they mobilize) may lead to the formation of recurrent patterns in plants and aggregates of unicellular entities (Arias, Azpeitia, Benítez, Escalante, Hernández-Hernández, and Mora van Cauwelaert). The final chapter in this section presents a multifaceted view of how the evolution of multicellular plants under constraints of nutrient gathering and escape from predation led to developmental mechanisms that are modular, distributed, self-organizing, integrative, and highly adaptive, without centralized control (Leyser).

Part III is devoted to the Amoebozoa and fungi, and theoretical approaches relevant to these groups. Like the volvocine algae, the cellular slime molds have long provided a robust model system with which to dissect and explore the evolution of multicellularity in large part because of the seminal work of John Tyler Bonner, who has graciously provided a preface to this volume. New insights continue to emerge from their study, such as the extent to which different species, as well as different genotypes of the same species, can cooperate to form fruiting bodies, a phenomenon that calls into question previous conceptualizations of individuality (Nanjundiah). Also called into question is whether multicellularity evolved as an adaptation, an exaptation, or as a nonadaptive happenstance. The available evidence depends on the organisms being examined. Experimental manipulation of yeast under selective regimes rapidly yields "snowflake" aggregates that manifest apoptosis (Conlin and Ratcliff), favoring an adaptationist account (though not necessarily a gradualist one). Whether generalities can be drawn from this remains an open question. Nevertheless, theoretical explorations identify recurrent themes involving a minimal set of generative rules (such as differential adhesion, growth, and cell death) that promote the evolution of multicellularity in simulated systems (Duran-Nebreda, Montañez, Bonforti, and Solé).

Part IV continues with the thematic treatment of multicellularity by focusing on the genomic toolkits of the animals (metazoans) and the mechanisms by which they mediate morphological complexity. For example, are the toolkits required for metazoan multicellularity found in the unicellular ancestors of this vast clade? Although this has been addressed and answered affirmatively, in part, for the choanoflagellates, the closest unicellular relatives, an examination of extant representatives of the more phylogenetically distant Ichthyosporea and Filasterea indicates that this conservation runs even deeper, and that early branching unicellular relatives of the metazoans manifest surprisingly complex genomic repertoires and associated multicellular organizational effects (Ruiz-Trillo). These and other conclusions nevertheless depend on the extent to which phylogenetic relationships are resolved, which lineages are identified as early divergent branches of metazoan evolution and which more closely resemble the colonial ancestors of multicellular animals. Analyses of the characteristics of the choanocytes of marine sponges, for example, and comparisons with choanoflagellates, reveal patterns of gene content and expression that permit a plausible scenario linking the sponges to the ancestral multicellular colonies from which the metazoans evolved (Adamska).

But how do cells achieve the necessary division of labor that permits different functionalities? Mathematical models employing multilevel consistency dynamics shed light on this question. As the number of model cells in an interactive aggregate increases, simulations yield divergent fates among the originally identical systems, a "germ line-soma" separation within the resulting heterogeneous colony from which cells may detach to produce other colonies (Kaneko). Finally, the characteristic animal body plan motifs of tissue multilayering, lumen formation, and tissue elongation are discussed in terms of the ancestry of the genes whose products first mediated these morphogenetic effects in ancient cell clusters. However, such morphologically complex cell aggregates were still not multicellular individuals, and it is proposed that the emergence of an egg stage of development *subsequent* to the origination of the forms was required to produce such entities (Newman).

Part V explores some of the philosophical aspects of what is meant by individuality and how different evolutionary pathways led to the appearance of integrated multicellular phenotypes. The attainment of mutual interdependence between the constitutive and inter-active process of a multicellular organism, which thereby achieves an integrated pheno-type, is characterized in terms of the "constitutive–interactive closure principle" (Arnellos and Moreno). This chapter also discusses the implications and drawbacks of taking the metazoans as the paradigmatic model system for multicellularity, since fully integrated multicellular organisms appear in other lineages, such as the land plants, although in qualitatively different ways, especially with respect to the relation and interdependence between their constitutive and interactive dimensions. The final chapter identifies a key bifurcation in the broad domain of questions ("problem agenda") concerning the origin of multicellularity (Love). By addressing the phenomenon from the contrasting but interre-lated perspectives of evolutionary dynamics (the "when and why" of multicellularity) and developmental biology (the "how" of multicellularity), philosophers of biology, through the activities of reasoning explication and problem clarification, can bring coherence to a scientific effort whose practitioners might otherwise seem at odds with one another.

Collectively, the book's 14 chapters address a broad spectrum of topics, organisms, experimental protocols, and philosophical as well as practical issues.

These chapters show that the evolutionary expansion of preexisting gene families encod-ing regulatory proteins in combination with novel physical and regulatory interactions resulting from such expansions played critical roles and likely contributed to the evolution of multicellular complexity. Future research is required to identify their targets and to determine their participation in developmental processes. However, as noted by many of the authors, the extent to which the details of transcription factor regulation and gene network architecture carry over from one organism to another is an open question since sequence homologies do not necessarily imply the conservation of function. Likewise, functional homologies are not invariably the result of genomic or developmental homol-ogy, as is evident from the broad spectrum of molecules providing cell adhesion, intercel-lular communication, and so forth.

One of the many interesting questions raised in this volume is whether multicellularity confers any selective advantage. The abundance of extant multicellular organisms gives the impression that it does. However, although it is true that major evolutionary innovations are likely not to be retained within a lineage if they are incompatible with survival and reproduction, it is not always the case that every transition requires a large or even measurable advantage, nor is it the case that phenotypic novelties are invariably adaptive. The retention of multicellularity in some lineages may reflect a random and an adaptively neutral exploration of a body plan morphospace that, once achieved, is, on average, irreversible. These and other speculations will continue to impel researchers to explore the origins and consequences of multicellularity, an effort to which this volume is intended to contribute important information and concepts.

I FUNCTIONAL AND MOLECULAR PREDISPOSITIONS TO MULTICELLULARITY

1 Fossils, Feeding, and the Evolution of Complex Multicellularity

Andrew H. Knoll and Daniel J. G. Lahr

The evolution of complex multicellularity is commonly viewed as a series of genomic events with developmental consequences. It is surely that, but a focus on feeding encourages us to view it, as well, in terms of functional events with ecological consequences. And fossils remind us that these events are also historical, with environmental constraints and consequences.

Several definitions of complex multicelluarity are possible; here we adopt the view that complex multicellular organisms are those with tissues or organs that permit bulk nutrient and gas transport, thereby circumventing the limitations of diffusion (Knoll, 2011). Organisms that fit this description occur exclusively in eukaryotic clades, and so we begin by tracing the early evolution of eukaryotes, based on the fossil record but informed by phylogeny and environmental geochemistry. In particular, we ask, what might have facilitated a major increase in eukaryotic diversity, documented in the geologic record ca. 800 million years ago? By analogy to explanations for the Cambrian radiation of animals, we hypothesize that the origin or expansion of *eukaryovory*—protists ingesting other protists—can account, at least in part, for the observed historical record. Insofar as simple multicellularity can provide protection from eukaryovorous predators, molecular clocks that place the origin of animals in this same time frame may also find at least partial explanation in the ecological consequences of novel feeding modes. More broadly, mapping feeding mode onto a modern framework for eukaryotic phylogeny suggests that escape from phagotrophy was a prerequisite for the evolution of complex multicellular organisms. That stated, physical changes in the Earth system also influenced the Ediacaran–Cambrian radiation of animals and have to be taken into account in the construction of historical narratives.

A Brief History of Early Eukaryotic Evolution

The identification of ancient microfossils as eukaryotic can be challenging, as neither DNA nor histological details are preserved in the deep sedimentary record. What is preserved are cell walls, imparting an immediate bias to the microfossil record—protistan fossils will be restricted to those groups that synthesize preservable cell walls at some stage of their life cycles (Knoll, 2014). Under favorable circumstances, lipids can also be preserved, and steranes (the geologically stable derivatives of sterols) provide a molecular account of eukaryotic evolution that complements the conventional record of body fossils (e.g., Summons and Lincoln, 2012).

In combination, large size, complex morphology, and complex ultrastructure as revealed by transmission electron microscopy (TEM) provide strong evidence of eukaryotic origin, and microfossils that meet this criterion occur in rocks as old as 1,600–1,800 Ma (Knoll et al., 2006). None of these microfossils can be assigned with confidence to an extant subclade of the Eukarya, and, indeed, molecular clock estimates suggest that some or all could be stem group eukaryotes (e.g., Parfrey et al., 2011; Eme et al., 2014).

Two examples illustrate what can and cannot be inferred from early eukaryotic microfossils. *Shiuyousphaeridium macroreticulatum* is a large (110–180 μm) ellipsoidal vesicle that bears many symmetrically distributed cylindrical processes, as well as tessellated 2-μm fields that cover the vesicle surface (see figure 1.1A); TEM shows a complex multilayered wall ultrastructure (Xiao et al., 1997; Javaux et al., 2004; interpreted as a vegetative wall within a cyst by Agic et al., 2015). As Cavalier-Smith (2002) remarked, "Cysts with spines or reticulate surface sculpturing would probably have required both an endomembrane system and a cytoskeleton, the most fundamental features of the eukaryotic cell, for their construction." *Shiuyousphaeridium* is thus interpreted with confidence as eukaryotic, but it cannot be assigned to a specific clade within the domain. Functional interpretation is also possible: *Shiuyousphaeridium* is the preserved wall of a resting cell, or cyst. Therefore, in addition to a dynamic cytoskeleton/membrane system, the *Shiuyousphaeridium* organism had a life cycle that included cell differentiation, a key character in the later evolution of complex multicellularity (Mikhailov et al., 2009). How old are these fossils? Their age is constrained by radiometric dating to be older than 1,400 million years but younger than 1,700 million years (Lan et al., 2014).

Other fossils of comparable age add nuance to this picture. *Tappania plana* (see figure 1.1B) is a morphologically complex microfossil found in 1,400- to 1,600-million-year-old shales from Australia, China, India, Siberia, and North America (Yin, 1997; Xiao et al., 1997; Javaux and Knoll, in review). *Tappania* specimens are up to 150 μm in diameter and bear a limited number of cylindrical extensions that emerge asymmetrically from the vesicle surface; the extensions sometimes show evidence of dichotomous branching, suggesting an actively metabolizing cell that grew finger-like extensions, changing its shape in real time. Like *Shiuyousphaeridium*, *Tappania* was eukaryotic but cannot easily be assigned to an extant eukaryotic clade. *Tappania* does, however, document two features that underpin complex multicellularity: cell polarity and, again, the dynamic cytoskeleton/ membrane system (Knoll et al., 2006; Javaux, 2011). Among living eukaryotes, actively metabolizing cells with preservable cell walls tend to be either photosynthetic or osmotrophic, along with mixotrophs such as dinoflagellates. Many protists are capable of taking in organic molecules, and a few clades, notably the fungi and oomycetes, gain nutrition

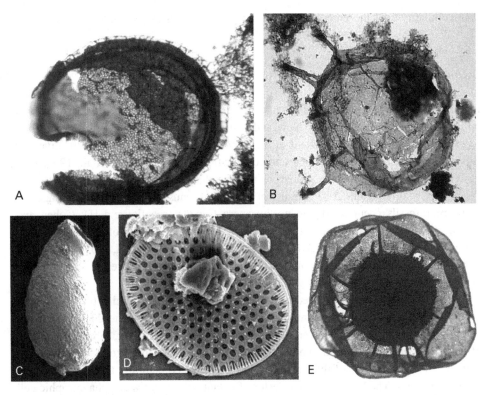

Figure 1.1
Proterozoic eukaryotes. (A) *Shuiyousphaeridium macroreticulatum*, the 1,400- to 1,700-million-year-old Ruyang Group, China. (B) *Tappania plana*, the 1,400- to 1,500-million-year-old Roper Group, Australia. (C) *Bonniea dacruchares*, the ca. 750-million-year-old Chuar Group, Arizona. (D) *Thorakidictyon myriocanthum*, the ca. 800-million-year-old Fifteenmile Group, Canada. (E) *Trachyhystrichosphaera polaris*, 750- to 800-million-year-old Svanbergfjellet Formation, Spitsbergen. Bar in D = 40 microns for A, = 50 microns for B, = 20 microns for C, = 14 microns for D, and = 100 microns for E.

primarily by elaborating cylindrical projections to facilitate feeding by absorption. Oomycetes and fungi are widely separated in terms of phylogeny, and molecular clocks discourage interpretation of *Tappania* as either one, but functionally, these fossils seem to indicate the early differentiation of osmotrophy as a principal feeding mode within the Eukarya.

In general, late Paleoproterozoic and Mesoproterozoic eukaryotes are problematic microfossils that occur in limited diversity within sedimentary rocks deposited beneath coastal oceans. Molecular fossils, in turn, suggest that in these oceans, bacteria were the principal sources of primary production (Brocks et al., 2005). By 1,100–1,200 million years ago, however, large populations of the early-branching red alga *Bangiomorpha* (Butterfield, 2000) simultaneously document the presence of an extant eukaryotic clade, simple multicellularity, and the emergence of photosynthesis as a major feeding mode within the Eukarya.

Feeding Mode and Neoproterozoic Protistan Diversification

Fossils suggest that an important change in eukaryotic diversity occurred around 800 million years ago (Knoll, 2014). New and distinctive organic-walled protists enter the record (see figure 1.1E; Butterfield et al., 1994; Butterfield, 2005a, 2005b), diversity expands among simple multicellular and coenocytic forms (Butterfield et al., 1994; Butterfield, 2004, 2009a), diverse protists with vase-shaped tests proliferate globally (see figure 1.1C; Porter et al., 2003; Strauss et al., 2014), unusually diverse assemblages of scale microfossils occur locally in northwestern Canada (see figure 1.1D; Cohen and Knoll, 2012), and steranes suggest the rising ecological importance of algae as primary producers in marine environments (Summons et al., 1988; Ventura et al., 2005). How might we explain the paleontologically observed diversification of eukaryotes so long after the origin of the group?

A working hypothesis is inspired by another radiation that occurred 250 million years later—the Cambrian explosion of animal diversity (Erwin and Valentine, 2013). For many years, Cambrian radiation has been viewed as driven, at least in part, by the emergence of carnivores that initiated an evolutionary arms race among predators and prey (e.g., Stanley, 1973). The logic of this argument can be extended naturally to the microscopic world of protists. The last common ancestor of extant eukaryotes was a phagotrophic cell that ingested bacteria-sized particles (Cavalier Smith, 2013; see below). In principle, then, the mid-Neoproterozoic origin or expansion of protists that feed on other protists could have set in motion an earlier arms race, resulting in accelerated eukaryotic diversification (Porter, 2011; Knoll, 2014).

Fossils provide both direct and inferential support for the eukaryovory hypothesis. Vase-shaped microfossils, which preserve easily and so have first appearances in the fossil record that at least broadly approximate their evolutionary origins, have never been observed in rocks older than about 800 million years, but are among the most common fossils in 800–740 Ma sedimentary successions observed globally (Strauss et al., 2014; agglutinated tests have also been reported from ca. 660- to 670-million-year-old carbonates in Namibia and Mongolia; Bosak et al., 2011; Dalton et al., 2013). These fossils provide three lines of evidence consistent with the eukaryovory hypothesis. First, at least some Neoproterozoic vase-shaped microfossil populations closely resemble the tests of extant arcellinid amoebozoans, a diverse group of bacterivorous and eukaryorous protists (Porter et al., 2003). Second, the emergence of testate protists in marine environments needs functional explanation, and reasonable conjecture holds that the organic walled (in some cases likely scale-encrusted; Porter and Knoll, 2000) tests provided defense against protistan predators. And third, vase-shaped microfossils from shales exposed deep within the Grand Canyon, Arizona, sometimes have regular half-moon holes cut out of the test wall (Porter et al., 2003). Such features are not easily explained as diagenetic structures; more likely they provide a direct record of predation by vampyrellid or other eukaryovorous protists in the local community (Porter et al., 2003; Hess et al., 2012).

Ca. 800-million-year-old limestones of the Fifteenmile Group, northwestern Canada, contain a remarkable abundance of biogenic scales preserved by phosphate mineralization (Cohen and Knoll, 2012). Some 40 distinct taxa have been identified, and while none can be assigned to an extant taxonomic group with confidence, they generally resemble the organic and siliceous scales that armor the walls of some prymnesiophytes, ciliates, centrohelid heliozoans, rhizarians, chrysophytes, and green algae. Again, defense against protistan predators provides a reasonable, if not unique, interpretation of Neoproterozoic scale synthesis.

Molecular fossils independently document eukaryotic expansion. For example, gammacerane (a geologically stable derivative of tetrahymanol) records the presence of ciliates, a clade rich in eukaryovorous species (Summons et al., 1988), while steranes indicate the growing importance of green and red algae as primary producers along continental shelves (Summons et al., 1988; Knoll et al., 2007). We note that the genes required for tetrahymanol synthesis have spread through the Eukarya by horizontal transfer (Takishita et al., 2012), but gene trees suggest that ciliate genes are sister to all others. Thus, if gammacerane is present in sedimentary rocks, ciliates existed at the time those rocks formed. Limited experiments also support the view that eukaryovory might have facilitated the rise to ecological prominence of eukaryotic algae. For example, when Ratti et al. (2013) grew eukaryotic algae in the presence of ciliate predators, both specific growth rate and protein content increased; cyanobacteria subjected to the same treatment showed either no response or decreased specific growth rates.

At present, then, the hypothesis that eukaryovory spurred Neoproterozoic protistan diversification can be supported from a number of paleontological perspectives. Intriguingly, molecular clocks suggest that animals differentiated as part of this Neoproterozoic diversification (Erwin et al., 2011). Could eukaryovory, then, be relevant to the origins of animal multicellularity? It could, and in a direct way. Experiments (e.g., Boraas et al., 1998; see also Hessen and van Donk, 1993) show that protistan predators can induce multicellularity in previously unicellular algal populations—there is strong selection pressure because the predators are not capable of phagocytosing the larger, multicellular individuals. The late Neoproterozoic–Cambrian expansion of complex multicellularity and complex macroscopic coenocytes in both green and red algae (Butterfield, 2009a; Xiao et al., 2002; Yuan et al., 2011) might also reflect both eukaryovory and the subsequent radiation of animal grazers.

Alegado et al. (2012) have hypothesized that simple multicellularity could actually enhance capture of prey bacteria in choanoflagellates, and other functional advantages of

simple multicellularity are possible (e.g., Knoll, 2011; Parfrey and Lahr, 2013). And Butterfield (2015) has suggested that nascent metazoans themselves provided the ecological
spark for Neoproterozoic eukaryotic diversification. This interpretation relies on the phylogeny and molecular clock estimate of Erwin et al. (2011), which view sponges as both
basal and paraphyletic, but it is far from obvious why the emergence of filter feeding by
sponges would facilitate the patterns of morphological diversification actually observed
in the fossil record. An alternative version of this hypothesis would ascribe protistan
diversification to predation by tiny ancestral bilaterians—which differs from the "straight
eukaryovory" hypothesis more in degree than in kind. A number of distinct feedbacks are
possible between early metazoans and protists, and in all probability most were in play in
Neoproterozoic oceans. Whether by avoiding or enhancing prey capture, feeding mode
provides an illuminating functional perspective on the emergence of complex multicellular
organisms in Neoproterozoic oceans.

You Are *How* You Eat

To place arguments about feeding mode and its evolutionary consequences into phylogenetic context, we performed analyses of trait evolution over samples of available phylogenetic trees. Details of data sets and methodologies are found in Lahr and Knoll (in
review); here we present only a brief summary of methods and results in this paper (see
figure 1.2).

   We began with the files used to generate the dated phylogeny of Parfrey et al. (2011).
We used both the 1,000 bootstrapped trees generated by their maximum-likelihood analysis and the over 1,000 trees sampled in each of seven independent Bayesian searches.
These collections of trees provide a sample of the most probable topology and branch
lengths, given the data, and enable assessment of uncertainty while generating probabilities
for specific patterns of trait evolution. In order to reconstruct the ancestral states for
selected nodes (i.e., hypothetical ancestors), we scanned the literature to determine the
feeding mode in each terminal taxon originally used by Parfrey et al. (2011). Specifically,
we characterized the primary feeding mode of each terminal taxon as osmotrophy, autotrophy, bacteriovory, and/or eukaryovory. Finally, we calculated the probabilities of each
particular feeding mode for nodes within the phylogeny using the program BayesTraits.
Combinations are possible, for example, the mixotrophy found in photosynthetic dinoflagellates (Jeong et al., 2010). The reconstruction of an ancestral state that has equally
shared probabilities for each of the four feeding modes may thus be interpreted either as
an ancestor that could do everything, or, perhaps more likely, as simply uncertainty in the
reconstruction.

   Such an exercise comes with several potential pitfalls. First, the phylogeny itself may
be uncertain (or wrong), which can result in incorrect estimates of ancestral character

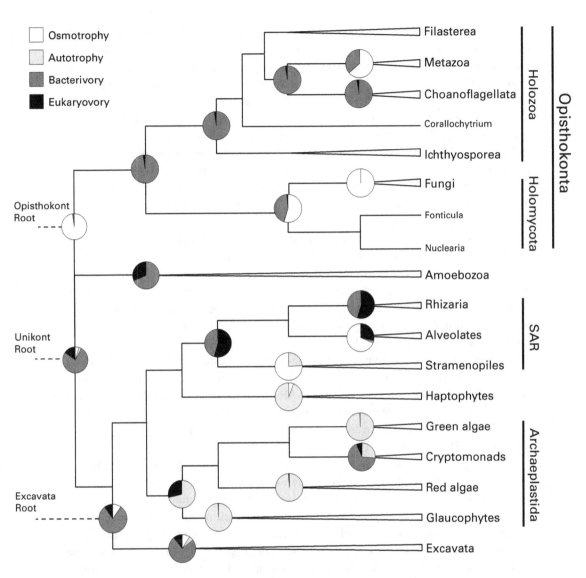

Figure 1.2
A summary of the evolution of carbon uptake mechanisms in eukaryotes, based on character trait analyses. The cartoon was constructed based on several distinct analyses detailed in Lahr and Knoll (in review). The backbone of the tree is based on the Parfrey et al. (2011) reconstruction, except for relationships within the SAR and opisthokont clades, grafted from Sierra et al. (2013) and Paps et al. (2013), respectively. The plotted probabilities are based on maximum-likelihood reconstruction of character trait evolution based on the respective trees, with terminal states obtained from a large body of literature. An opisthokont rooting was used for the traits plotted on this tree, with the exception that we show what the ancestral eukaryote would look like if the rooting were unikont-based. Branch lengths are not to scale; that is, they do not reflect evolutionary or real time.

states. Second, the taxa used to construct the phylogeny may not capture the range of feeding modes found on any particular branch of the tree—that is, fuller sampling of a clade might alter the probability of ancestral feeding mode for a clade. To address these issues, we revisited the major clades of the eukaryotic tree, using other published phylogenies containing several taxa not employed by Parfrey et al. (2011). In particular, we tested the same approach using the phylogenetic framework for amoebozoans published in Lahr et al. (2013), opisthokonts (based on the phylogenetic framework of Paps et al. (2013), and Rhizaria (based on Sierra et al., 2013). Moreover, we tested our conclusions for ancestral feeding mode in the Excavata and Archaeplastida, finding them to be in agreement with results from other analyses (Cavalier Smith, 2013, and Deschamps and Moreira, 2009, respectively). The summary figure shown here provides a composite of results from these multiple analyses (figure 1.2).

Consistent with earlier opinion (e.g., Cavalier Smith, 2013), placement of the root to the eukaryotic tree between opisthokonts + amoebozoans and everything else (the unikont root) or within or near the excavates (excavate root) suggests that that the last common ancestor of extant eukaryotes was a bacterivorous phagotroph (Lahr and Knoll, in review). Interestingly, however, placement of the root between opisthokonts and all other eukaryotes (the ophisthokont root) suggests that the last common ancestor could have been osmotrophic (see the discussion below). In any case, eukaryovory is a derived feeding mode, found principally within a limited number of protistan clades. The SAR clade may have evolved eukaryovory in some of its earliest representatives, but the ecologically dominant clades of eukaryovorus SAR protists—ciliates, dinoflagellates, and foraminiferans—have Neoproterozoic origins (Parfrey et al., 2011; Eme et al., 2014; Groussin et al., 2011). Eukaryovorous amoebozoans also diversified during the Neoproterozoic Eon (Parfrey et al., 2011).

Integrating feeding mode with molecular clocks, then, provides further support for the eukaryovory hypothesis to explain Neoproterozoic eukaryotic diversification. Another interesting result also derives from this analysis. No complex multicellular organisms occur within phagotrophic clades. The paucity of even simple multicellular species among phagotrophic protists was noted earlier by Baker (1948; see Bonner, this volume), who attributed this to the specific feeding requirements of phagotrophs. Whether for this alone or for additional reasons, the cell biological requirements for phagotrophy are inconsistent with the formation of functionally differentiated tissues. Thus, complex multicellularity evolved several times within photosynthetic clades (green algae, florideophyte red algae, and laminarialean brown algae) and in osmotrophs (ascomycetes and basidiomycete fungi, eumetazoan animals). That is, complex multicellularity appears to accompany the functional escape from phagotrophy.

In this regard, the evolution of osmotrophy in eumetazoans stands as a defining moment in animal evolution. While the phylogenetic relationships of early branching animals remain uncertain, there is reason to believe that the earliest animals were bacterivorous, like most sponges and choanoflagellates (Adamska, this volume). Eumetazoans feed primarily by secreting extracellular digestive enzymes and then absorbing metabolizable molecules. Placozoans are little differentiated beyond upper and lower epithelia but are capable of taking up nutrients via both extracellular digestion and the phagocytosis of entire cells (through both epithelia; Wenderoth, 1986; Srivastava et al., 2008). Cnidarians and bilaterian animals, in turn, have a differentiated gut cavity to localize this function. Something approaching consensus now suggests that many of the familiar quilt-like forms of the macroscopic Ediacaran biota (580–541 million years old) reflect an early stage in the evolution of eumetazoan osmotrophy, before body plan reorganizaton led to a differentiated digestive system (e.g., Sperling and Vinther., 2010; Laflamme et al., 2009; Cuthill and Conway Morris, 2014). Filasterea, the clade that is sister to animals + choanoflagellates, contains phagotophs that consume bacteria and small eukaryotes. Thus, osmotrophy is derived within this opisthokont branch. Intriguingly, *Corallochytrium* and the ichythosporeans are osmotrophs.

Fungi are also osmotrophs, but their sister group, containing *Fonticula* and *Nuclearia*, is bacterivorous, and the resulting uncertainty of feeding mode in their common ancestor plays a key role in governing the root dependency of the ancestral feeding mode within eukaryotes as a whole. It remains to be seen whether the probabilities of specific feeding mode changes are all equal, but accepting the last common ancestor of opisthokonts as a bacterivore, osmotrophy must have evolved at least three times within this supergroup. Regardless of the proper history of feeding mode evolution, however, opisthokonts support the general conclusion that complex multicellularity occurs in eukaryotic clades characterized by osmotrophy or photosynthesis.

Physical Environment Also Matters

The preceding arguments suggest that feeding mode played several key roles in the evolution of multicellular organisms in early oceans, providing both selective pressures favoring multicellular organization and cell biological constraints on multicellular organization. This does not, however, mean that physical environment was irrelevant. Historically, the emergence of large animals has been linked mechanistically to rising oxygen tensions in the atmosphere and oceans (summarized by Mills and Canfield, 2014). In recent years, the oxygen facilitation hypothesis has been challenged on several fronts (e.g., Butterfield, 2009b), but stratigraphic evidence still shows a correlation in time between the paleontological record of animal diversification and geochemical indications of change in global redox conditions (Lyons et al., 2014). Both ecological surveys of present day oxygen minimum zones (summarized by Sperling, Frieder, et al., 2013; Sperling, Halverson, et al., 2013) and physiological experiments on sponges (Mills et al., 2014) make it clear that small and anatomically simple animals could have originated when pO_2 was as low as

1%–4% of present atmospheric levels (PAL), consistent with atmospheric compositions traditionally accepted for 800 million years ago. Larger animals with derived feeding modes may, however, have been inhibited until pO_2 rose to higher levels, as inferred from the chemistry of Ediacaran sedimentary rocks. Sperling, Frieder, et al. (2013) reconciled decades of debate about ecological versus environmental drivers of metazoan evolution by showing that carnivory, the feeding mode thought to have initiated an evolutionary arms race in Cambrian oceans (e.g., Stanley, 1973) requires more oxygen than other metazoan feeding strategies. Thus, it appears that increasing oxygen availability in Ediacaran and Cambrian oceans could have set ecological drivers of metazoan diversification in motion.

Oxygen facilitation scenarios depend as much on low pre-Ediacaran oxygen levels as they do on subsequent higher pO_2 (Mills and Canfield, 2014). If pO_2 reached levels of, say, 10% PAL well before the Ediacaran Period, hypotheses of oxygen facilitation would be falsified. If, however, earlier Proterozoic oxygen levels hovered around 1% PAL, then oxygen facilitation would remain a viable hypothesis for both the origin of animals and the later expansion of large animals, including carnivores. Planavsky et al. (2014) have recently argued on the basis of chromium isotope geochemistry that pO_2 was as low as 0.1% PAL until ca. 800 million years ago. This result is new and sure to attract critical attention, but even if it is off by an order of magnitude, the role of oxygen in animal evolution would be supported strongly.

Might the earlier expansion of eukaryovory similarly reflect environmental facilitation? Such a view is consistent with the phylogenetic inference that eukaryovory evolved independently in several clades during the Neoproterozoic Era. It is also consistent with arguments based on oxygen demand. In animals, the oxygen requirements for carnivory are thought to reflect, at least in part, the transiently high oxygen demand of digesting large prey (Sperling, Frieder, et al., 2013). Logically, the oxygen demand of ingesting a 10-μm protistan cell would be three orders of magnitude higher than that incurred by phagocytosing a 1-μm bacterial cell, and so even the modest increase in pO_2 inferred by Planavsky et al. (2014; see also Frei et al., 2009) for the Neoproterozoic Earth could have facilitated the ingestion of larger cells. Without question, experimental insights and improved geochemical resolution of redox history will both be needed to evaluate such hypotheses.

Conclusions

John Tyler Bonner has long underscored the importance of feeding in the evolution of multicellular organisms, not least in his preface to this volume (Bonner, this volume), and our results further support the need for functional perspective on this grand evolutionary issue. The fossil record of early eukaryotic evolution provides insights into the acquisition of cell biological characters thought to play an important role in the evolution of complex multicellularity. It also inspires an ecological hypothesis for Neoproterozoic protistan diversification that focuses on eukaryotic feeding mode and its diversification through time. Phylogenetic perspectives on feeding mode evolution complement paleontological evidence in favor of this hypothesis and also highlight the importance of feeding mode for the probability of evolving complex multicellularity within any given clade. Improved phylogenetic exploration of feeding mode, continuing paleontological discovery, better resolution of geochemical proxies for redox evolution, and novel physiological experimentation will collectively help to test and refine hypotheses that aim to place the evolution of multicellular organisms into its proper functional, ecological, and paleoenvironmental context.

Acknowledgments

We thank Karl J. Niklas and Stuart A. Newman for the invitation to participate in a stimulating workshop at KLI, and we thank Erik Sperling and Pete Girguis for discussion. AHK was supported by the NASA Astrobiology Institute; DJGL is supported by a FAPESP Young Investigator Award (2013/04585–3).

References

Agic, H., Moczydłowska, M., & Yin, L.-M. (2015). Affinity, life cycle, and intracellular complexity of organic-walled microfossils from the Mesoproterozoic of Shanxi, China. *Journal of Paleontology*, 89, 28–50.

Alegado, R. A., Brown, L.W., Cao, S., Dermenjian, R. K., Zuzow, R., Fairclough, S. R., et al. (2012). Bacterial sulfonolipid triggers multicellular development in the closest living relatives of animals. *eLife, 1*, UNSP e00013.

Baker, J. R. (1948). The status of the protozoa. *Nature*, 161, 548–551.

Boraas, M. E., Seale, D. B., & Boxhorn, J. E. (1998). Phagotrophy by a flagellate selects for colonial prey: A possible origin of multicellularity. *Evolutionary Ecology*, 12, 153–164.

Bosak, T., Lahr, D. J. G., Pruss, S. B., Macdonald, F. A., Dalton, L., & Matys, E. (2011). Agglutinated tests in post-Sturtian cap carbonates of Namibia and Mongolia. *Earth and Planetary Science Letters*, 308, 29–40.

Brocks, J. J., Love, G. D., Summons, R. E., Knoll, A. H., Logan, G. A., & Bowden, S. (2005). Biomarker evidence for green and purple sulfur bacteria in an intensely stratified Paleoproterozoic ocean. *Nature*, 437, 866–870.

Butterfield, N. J. (2000). *Bangiomorpha pubescens* n. gen., n. sp.: Implications for the evolution of sex, multicellularity and the Mesoproterozoic/Neoproterozoic radiation of eukaryotes. *Paleobiology*, 26, 386–404.

Butterfield, N. J. (2004). A vaucheriacean alga from the middle Neoproterozoic of Spitsbergen: Implications for the evolution of Proterozoic eukaryotes and the Cambrian explosion. *Paleobiology*, 30, 231–252.

Butterfield, N. J. (2005 a). Reconstructing a complex early Neoproterozoic eukaryote, Wynniatt Formation, arctic Canada. *Lethaia*, 38, 155–169.

Butterfield, N. J. (2005 b). Probable Proterozoic fungi. *Paleobiology*, 31, 165–182.

Butterfield, N. J. (2009 a). Modes of pre-Ediacaran multicellularity. *Precambrian Research*, 173, 201–211.

Butterfield, N. J. (2009 b). Oxygen, animals and oceanic ventilation: An alternative view. *Geobiology*, 7, 1–7.

Butterfield, N. J. (2015). Early evolution of the Eukaryota. *Palaeontology*, *58*, 5–17. doi:1111/pala.12139.

Butterfield, N. J., Knoll, A. H., & Swett, K. (1994). Paleobiology of the Neoproterozoic Svanbergfjellet Formation, Spitsbergen. *Fossils and Strata*, 34, 1–84.

Cavalier Smith, T. (2002). The neomuran origin of Archaebacteria: The Negibacteria root of the universal tree and bacteria megaclassification. *International Journal of Systematic Microbiology*, 52, 7–76.

Cavalier Smith, T. (2013). Early evolution of eukaryote feeding modes, cell structural diversity, and classification of the protozoan phyla Loukozoa, Sulcozoa, and Choanozoa. *European Journal of Protistology*, 49, 115–178.

Cohen, P. A., & Knoll, A. H. (2012). Neoproterozoic scale microfossils from the Fifteen Mile Group, Yukon Territory. *Journal of Paleontology*, 86, 775–800.

Cuthill, J. F. H., & Conway Morris, S. (2014). Fractal branching organizations of Ediacaran rangeomorph fronds reveal a lost Proterozoic body plan. *Proceedings of the National Academy of Sciences of the United States of America*, 111, 13122–13126.

Dalton, L. A., Bosak, T., Macdonald, F. A., Lahr, D. J. G., & Pruss, S. B. (2013). Preservational and morphological variability of assemblages of agglutinated eukaryotes in Cryogenian cap carbonates of northern Namibia. *Palaios*, 28, 67–79.

Deschamps, P., & Moreira, D. (2009). Signal conflicts in the phylogeny of the primary photosynthetic eukaryotes. *Molecular Biology and Evolution*, 26, 2745–2753.

Eme, L., Sharpe, S. C., Brown, M. W., & Roger, A. J. (2014). On the age of eukaryotes: Evaluating evidence from fossils and molecular clocks. *Cold Spring Harbor Perspectives in Biology*. doi:.10.1101/cshperspect.a016139

Erwin, D. H., Laflamme, M., Tweedt, S. M., Sperling, E. A., Pisani, D., & Peterson, K. J. (2011). The Cambrian conundrum: Early divergence and later ecological success in the early history of animals. *Science*, 334, 1091–1097.

Erwin, D. H., & Valentine, J. (2013). *The Cambrian explosion: The construction of animal biodiversity*. Framingham, MA: Roberts.

Frei, R., Gaucher, C., Poulton, S. W., & Canfield, D. E. (2009). Fluctuations in Precambrian atmospheric oxygenation recorded by chromium isotopes. *Nature*, 461, 250–253.

Groussin, M., Pawlowski, J., & Yang, Z. (2011). Bayesian relaxed clock estimation of divergence times in foraminifera. *Molecular Biology and Evolution*, 61, 157–166.

Hess, S., Sausen, N., & Melkonian, M. (2012). Shedding light on vampires: The phylogeny of vampyrellid amoebae revisited. *PLoS One*, 7(2), e31165. doi:.10.1371/journal.pone.0031165

Hessen, D. O., & Van Donk, E. (1993). Morphological changes in *Scenedesmus* induced by substances released by *Daphnia*. *Archiv für Hydrobiologie*, 127, 129–140.

Javaux, E. J. (2011). Evolution of early eukaryotes in Precambrian oceans. In M. Gargaud, P. Lopez-Garcia, & H. Martin (Eds.), *Origins and evolution of life: An astrobiological perspective* (pp. 414–449). Cambridge, UK: Cambridge University Press.

Javaux, E. J., & Knoll, A. H. (in review). Micropaleontology of the Lower Mesoproterozoic Roper Group, Australia, and implications for early eukaryotic evolution.

Javaux, E., Knoll, A. H., & Walter, M. R. (2004). TEM evidence for eukaryotic diversity in mid-Proterozoic oceans. *Geobiology*, 2, 121–132.

Jeong, H. J., Yoo, Y. D., Kim, J. S., Seong, K. A., Kang, N. S., & Ki, T. H. (2010). Growth, feeding and ecological roles of the mixotrophic and heterotrophic dinoflagellates in marine planktonic food webs. *Ocean Science Journal*, 45, 65–91.

Knoll, A. H. (2011). The multiple origins of complex multicellularity. *Annual Review of Earth and Planetary Sciences, 39,* 217–239.

Knoll, A. H. (2014). Paleobiological perspectives on early eukaryotic evolution. *Cold Spring Harbor Perspectives in Biology.* doi:.10.1101/cshperspect.a016121

Knoll, A. H., Javaux, E. J., Hewitt, D., & Cohen, P. (2006). Eukaryotic organisms in Proterozoic oceans. *Philosophical Transactions of the Royal Society of London. Series B, Biological Sciences, 361,* 1023–1038.

Knoll, A. H., Summons, R. E., Waldbauer, J., & Zumberge, J. (2007). The geological succession of primary producers in the oceans. In P. Falkowski & A. H. Knoll (Eds.), *The evolution of primary producers in the sea* (pp. 133–163). Burlington, MA: Elsevier.

Laflamme, M., Xiao, S., & Kowalewski, M. (2009). Osmotrophy in modular Ediacara organisms. *Proceedings of the National Academy of Sciences of the United States of America, 106,* 14438–14443.

Lahr, D., & Knoll, A. H. (in review). Reconstructing the early evolution of feeding mode in eukaryotes.

Lahr, D. J. G., Grant, J. R., & Katz, L. A. (2013). Multigene phylogenetic reconstruction of the Tubulinea (Amoebozoa) corroborates four of the six major lineages, while additionally revealing that shell composition does not predict phylogeny in the Arcellinida. *Protist, 164,* 323–339.

Lan, Z., Li, X., Chen, Z. Q., Li, Q., Ofmann, A., Zhang, Y., et al. (2014). Diagenetic xenotime age constraints on the Sanjiaotang Formation, Luoyu Group, southern margin of the North China Craton: Implications for regional stratigraphic correlation and early evolution of eukaryotes. *Precambrian Research, 251,* 21–32.

Lyons, T. W., Reinhard, C. T., & Planavsky, N. J. (2014). The rise of oxygen in Earth's early ocean and atmosphere. *Nature, 506,* 307–315.

Mikhailov, K. V., Konstantinova, A. V., Nikitin, M. A., Troshin, P. V., Rusin, L. Y., Lyubetsky, V. A., et al. (2009). The origin of Metazoa: A transition from temporal to spatial cell differentiation. *BioEssays, 31,* 758–768.

Mills, D. B., & Canfield, D. E. (2014). Oxygen and animal evolution: Did a rise of atmospheric oxygen trigger the origin of animals? *BioEssays, 36,* 1–11.

Mills, D. B., Ward, L. M., Jones, C., Sweeten, B., Forth, M., Treusch, A. H., et al. (2014). Oxygen requirements of the earliest animals. *Proceedings of the National Academy of Sciences of the United States of America, 111,* 4168–4172.

Paps, J., Medina-Chacón, L. A., Marshall, W., Suga, H., & Ruiz-Trillo, I. (2013). Molecular phylogeny of unikonts: New insights into the position of apusomonads and ancyromonads and the internal relationships of opisthokonts. *Protist, 164,* 2–12.

Parfrey, L. W., & Lahr, D. J. G. (2013). Multicellularity arose several times in the evolution of eukaryotes (Response to DOI 10.1002/bies. 201100187). *BioEssays, 35,* 339–347.

Parfrey, L., Lahr, D., Knoll, A. H., & Katz, L. A. (2011). Estimating the timing of early eukaryotic diversification with multigene molecular clocks. *Proceedings of the National Academy of Sciences of the United States of America, 108,* 13624–13629.

Planavsky, N. J., Reinhard, C. T., Wang, X., Thompson, D., McGoldrick, P., Rainbird, R. H., et al. (2014). Low mid-Proterozoic atmospheric oxygen levels and the delayed rise of animals. *Science, 346,* 635–638.

Porter, S. M. (2011). The rise of predators. *Geology, 39,* 607–608.

Porter, S. M., & Knoll, A. H. (2000). Testate amoebae in the Neoproterozoic Era: Evidence from vase-shaped microfossils in the Chuar Group, Grand Canyon. *Paleobiology, 26,* 360–385.

Porter, S. M., Meisterfeld, R., & Knoll, A. H. (2003). Vase-shaped microfossils from the Neoproterozoic Chuar Group, Grand Canyon: A classification guided by modern testate amoebae. *Journal of Paleontology, 77,* 205–225.

Ratti, S., Knoll, A. H., & Giordano, M. (2013). Grazers and phytoplankton growth in the oceans: An experimental and evolutionary perspective. *PLoS One*, 8(10), e77349. doi:.10.1371/journal.pone.0077349

Sierra, R., Matz, M. V., Aglyamova, G., Pillet, L., Decelle, J., Not, F., et al. (2013). Deep relationships of Rhizaria revealed by phylogenomics: A farewell to Haeckel's Radiolaria. *Molecular Phylogenetics and Evolution*, 67, 53–59.

Sperling, E. A., Frieder, C. A., Girguis, P. R., Raman, A. V., Levin, L. A., & Knoll, A. H. (2013). Oxygen, ecology, and the Cambrian radiation of animals. *Proceedings of the National Academy of Sciences of the United States of America*, 110, 13446–13451.

Sperling, E. A., Halverson, G. P., Knoll, A. H., Macdonald, F. A., & Johnston, D. T. (2013). A basin redox transect at the dawn of animal life. *Earth and Planetary Science Letters*, 371–372, 143–155.

Sperling, E. A., & Vinther, J. (2010). A placozoan affinity for Dickinsonia and the evolution of late Proterozoic metazoan feeding modes. *Evolution & Development*, 12, 201–209.

Srivastava, M., Begovic, E., Chapman, J., Putnam, N. H., Hellsten, U., Kawashima, T., et al. (2008). The *Trichoplax* genome and the nature of placozoans. *Nature*, 454, 955–960.

Stanley, S. M. (1973). An ecological theory for the sudden origin of multicellular life in the late Precambrian. *Proceedings of the National Academy of Sciences of the United States of America*, 70, 1486–1489.

Strauss, J. V., Rooney, A. D., Macdonald, F. A., Brandon, A. D., & Knoll, A. H. (2014). Circa 740 Ma vase-shaped microfossils from the Yukon Territory: Implications for Neoproterozoic biostratigraphy. *Geology*, 42, 659–662.

Summons, R. E., Brassell, S. C., Eglinton, G., Evans, E. J., Horodyski, R. J., Robinson, N., et al. (1988). Distinctive hydrocarbon biomarkers from fossiliferous sediment of the late Proterozoic Walcott Member, Chuar Group, Grand Canyon, U.S.A. *Geochimica et Cosmochimica Acta*, 52, 2625–2673.

Summons, R. E., & Lincoln, S. A. (2012). Biomarkers: Informative molecules for studies in geobiology. In A. H. Knoll, D. E. Canfield, & K. O. Konhauser (Eds.), *Fundamentals of geobiology* (pp. 269–296). Chichester, UK: Wiley-Blackwell.

Takishita, K., Chikaraishi, Y., Leger, M. M., Kim, E., Yabuki, A., Ohkouchi, N., et al. (2012). Lateral transfer of tetrahymanol-synthesizing genes has allowed multiple diverse eukaryote lineages to independently adapt to environments without oxygen. *Biology Direct*, 7(5). http://www.biology-direct.com/content/7/1/5.

Ventura, G. T., Kenig, F., Grosjean, E., & Summons, R. E. (2005). Biomarker analysis of extractable organic matter from the Neoproterozoic Kwagunt Formation, Chuar Group (~800–742 Ma). Fall AGU meeting abstracts.

Wenderoth, H. (1986). Transepithelial cytophagy by *Trichoplax adhaerens* F. E. Schulze (Placozoa) feeding on yeast. *Zeitschrift für Naturforschung. Section C. Biosciences*, 41, 343–347.

Xiao, S., Knoll, A. H., Kaufman, A. J., Yin, L., & Zhang, Y. (1997). Neoproterozoic fossils in Mesoproterozoic rocks? Chemostratigraphic resolution of a biostratigraphic conundrum from the North China Platform. *Precambrian Research*, 84, 197–220.

Xiao, S. H., Yuan, X. L., Steiner, M., & Knoll, A. H. (2002). Macroscopic carbonaceous compressions in a terminal Proterozoic shale: A systematic reassessment of the Miaohe biota, South China. *Journal of Paleontology*, 76, 347–376.

Yin, L. (1997). Acanthomorphic acritarchs from Meso-Neoproterozoic shales of the Ruyang Group, Shanxi, China. *Review of Palaeobotany and Palynology*, 98, 15–25.

Yuan, X., Chen, Z., & Xiao, S. (2011). An early Ediacaran assemblage of macroscopic and morphologically differentiated eukaryotes. *Nature*, 470, 390–393.

2 Alternative Splicing, Intrinsically Disordered Proteins, Calmodulin, and the Evolution of Multicellularity

Karl J. Niklas and A. Keith Dunker

Comprehensive bioinformatics analysis of protein function shows that proteins involved in cell signaling and regulation (Iakoucheva et al., 2002) and proteins underlying cellular differentiation (Xie et al., 2007; Dunker et al., 2015) are much richer in intrinsically disordered protein (IDP) residues compared to proteins involved in virtually all other types of biological functions. Furthermore, both alternative splicing (AS) (Romero et al., 2006) and posttranslational modification (PTM), especially phosphorylation (Iakoucheva et al., 2004; Gao et al., 2010), are strongly associated with IDP domains. Finally, AS, IDPs, and PTMs have been observed to work in concert to modulate the functions of important regulatory proteins (Dunker et al., 2008; Rautureau, Day, and Hinds, 2010).

The goal of this review is to explore the evolutionary origins of multicellularity within the context of how cellular functionalities are expanded by the three aforementioned processes. This focus is driven by an interest in the mechanisms that underlie cell fate specification and the evolutionary transition from organisms composed of one or only a few cell types (e.g., choanoflagellates and charophycean algae) to organisms consisting of many cell types (e.g., insects and vertebrates). It is also motivated by the evidence that AS, IDPs, and PTMs comprise regulatory components that significantly expand the protein and signaling functionalities of cells sharing the same genome without affecting proteome or genome size (Iakoucheva et al., 2004; Romero et al., 2006; Dunker et al., 2008; L. Chen et al., 2014). This evidence shows further that AS, IDPs, and PTMs are extremely ancient and spatially and temporally context dependent. We therefore surmise that AS, IDPs, and PTMs were important participants in the evolutionary transition from unicellular and colonial organisms to ancient and subsequently more derived multicellular eukaryotes. The absence of AS in prokaryotes suggests to us that the evolution of multicellularity involved a simpler IDP and PTM network (Dunker et al., 2015; Niklas et al., 2015).

The validation of these and other claims requires a broad phylogenetic (comparative) approach for at least two reasons. First, multicellularity evolved multiple times in different ways in very different clades, and, second, the evidence that will be reviewed indicates that AS, IDPs, and PTMs participated in the evolution of multicellularity in different ways within these different clades. Estimates of the exact number of times multicellularity has evolved differ depending on how multicellularity is defined and in what phylogenetic context. When described simply as a cellular aggregation, multicellularity is estimated conservatively to have evolved at least 25 times (Grosberg and Strathmann, 2007), which makes it a "minor major" evolutionary transformation. However, when requirements for sustained cell-to-cell interconnection, communication, and cooperation are applied, multicellularity has evolved only once in the Animalia, three times in the Fungi (chytrids, ascomycetes, and basidiomycetes), and six times among the algae (twice each in the rhodophytes, stramenopiles, and chlorobionta; Niklas and Newman, 2013; Niklas et al., 2013; Niklas, 2014).

Regardless of how it is defined or how many times it has evolved, the emergence and maintenance of multicellularity raises a number of important questions that pertain as much to the study of modern organisms as to our understanding of Earth's biological history. Some of these questions are particularly relevant to the thesis advanced here. For example, many workers have speculated on whether a canonical toolkit or "master" gene regulatory network responsible for multicellularity exists and is shared among all or most clades. Is it possible that the multiple origins of multicellularity involved developmental mechanisms that had very different functionalities but that were co-opted from a shared last common (unicellular) ancestor? This last question is particularly intriguing in light of (1) multilevel selection theory that requires the evolution of an alignment of fitness among the cells of a multicellular progenitor (reviewed by Folse and Roughgarden, 2010) and (2) the fact that all eukaryotes shared a last common (unicellular) ancestor that must have achieved an alignment among the various metabolic interests of its endosymbionts to integrate the activities of proto-organelles, for example, the TOC–TIC translocon protein-import system of the land plant plastid envelope incorporates a cyanobacteria-like core (Jarvis and Soll, 2001). Canonical answers to these questions are currently unavailable. However, the literature treating AS, IDPs, and PTMs supports the proposition that these processes comprise a generic and very ancient toolkit that, at the very least, facilitated the evolution of multicellularity.

The following sections review the basic aspects of AS, IDPs, and (to a limited extent) PTMs in the context of cell signaling pathways, metabolic regulation, and cell fate specification. Calmodulin (CaM) and calmodulin-like (CaML) proteins are used as examples (albeit not exclusively) of how AS, IDPs, and PTMs affect cell functionalities because they are phylogenetically pervasive and influence a broad spectrum of cellular processes in fungi, algae, land plants, and animals. This review concludes with speculations concerning the roles played by the AS–IDPs–PTMs motif during the evolution of multicellularity.

Alternative Splicing

AS results in the formation of different protein isoforms from the same precursor messenger RNA (mRNA). This process retains or excludes different exons to produce mRNA variants. AS occurs in all eukaryotic lineages (Black, 2003) and becomes more prevalent as organismic complexity (as estimated by the number of different types of cells an organism can produce) increases (L. Chen et al., 2014). Among the five basic types of AS (i.e., alternative 3′ acceptor site, alternative 5′ donor splice site, intron retention, mutually exclusive exon splicing, and exon skipping; see Black, 2003), exon skipping appears to be the most frequent (Modrek et al., 2001).

Regulation and selection of the splice sites are performed by trans-acting splicing activator and splicing repressor proteins within an RNA–protein complex called the spliceosome, which is canonically composed of five small nuclear RNAs (i.e., U1, U2, U4–U6) and a range of assorted protein factors. Splicing is regulated by trans-acting repressor–activator proteins and their corresponding cis-acting regulatory silencers and enhancers on the precursor mRNA (Matera and Wang, 2014). The effects of splicing factors are often position dependent. For example, a splicing factor that functions as an activator when bound to an intronic enhancer element may function as a repressor when bound to its splicing element in the context of an exon (Lim et al., 2011). It is particularly noteworthy that many spliceosome proteins are highly disordered and therefore adopt multiple 3-D conformations depending on their molecular context (Korneta and Bujnicki, 2012; Coelho Ribeiro et al., 2013).

The secondary structure of the precursor mRNA transcript also plays a role in regulating which exons and introns will be spliced, for example, by bringing together splicing elements or by masking a sequence that would otherwise serve as a binding element for a splicing factor. Collectively, the activators, repressors, and the secondary structure of the precursor mRNA form a splicing "code" that defines the protein isoforms that will be produced under different cellular conditions in response to external as well as internal signals. Additionally, the elements within this code function interdependently in ways that are context dependent, both intra- and extracellularly. For example, (1) cis-acting regulatory silencers and enhancers are influenced by the presence and relative position of other RNA sequence features, (2) the trans-acting context is affected by intracellular conditions that are in turn influenced by external conditions (C. D. Chen et al., 1999; Matlin et al., 2005; Wang and Burge, 2008) and other RNA sequence features, and (3) some cis-acting elements can reverse the effects on splicing if specific proteins are expressed in a cell (Boutz et al., 2007; Spellman et al., 2007).

AS is a highly conserved mechanism that provides adaptive advantages (Fairbrother et al., 2004; Ke et al., 2008). For instance, Chang et al. (2014) report a conserved AS pattern for heat shock transcription factors in the moss *Physcomitrella patens* and the flowering plant *Arabidopsis thaliana*, which indicates that the AS mechanism for heat regulation among the land plants is an ancestral condition. AS is also cell and tissue specific. For example, using mRNA sequence data, Pan et al. (2008) report that transcripts from ≈ 95% of human multi-exon genes undergo AS and that ≈100,000 intermediate- to high-abundance AS events occur in different tissue systems. Similar results are reported by Johnson et al. (2003) based on microarray analyses of human tissues.

Intrinsically Disordered Proteins

The phenotypic domain an individual cell can occupy without increasing its proteome size is significantly expanded because AS produces a disproportionate number of IDPs that lack equilibrium 3-D structures and thus can assume multiple functional roles under normal physiological conditions. Additionally, the majority of eukaryotic transcription factors have IDP domains affected by AS (Liu et al., 2006; Minizawa et al., 2006).

IDPs are characterized as having multispecificity and variable affinity and thus enhanced binding diversity (Uversky and Dunker, 2010; Oldfield and Dunker, 2014). These proteins also have the ability to form both large and small interaction surfaces (Tompa et al. 2009), possess fast association and complex dissociation rates (Shammas et al., 2013; Galea et al., 2008; Dunker and Uversky, 2008; Mitrea et al., 2012), exhibit polymorphisms in the bound state (Dunker et al., 1998; Fuxreiter and Tompa, 2012; Hsu et al., 2013), and have reduced intracellular lifetimes (van der Lee et al., 2014). These traits make IDPs ideal signaling and regulatory molecules (e.g., CaM and CaML proteins). Studies have identified intrinsically disordered domains (IDDs) as enriched in the nonconstitutive exons, indicating that protein isoforms may display functional diversity due to the alteration of functional modules within these regions (e.g., CaM EF-hand motifs and linker regions; Evens et al., 2011). IDDs can exist as molten globules with defined but metastable secondary structure or as unfolded chains that can function through transitions among different folded states. Their functional conformations can change by binding to other proteins and nucleic acids (Uversky, 2002; Oldfield et al., 2008; Hsu et al., 2013). PTMs (e.g., phosphorylation) can also alter IDP functionalities (Iakoucheva et al., 2004; Dyson and Wright, 2005; Oldfield et al., 2008).

Examples of IDPs involved in transcriptional regulation are well-known (Campbell et al., 2000; Haynes and Iakoucheva, 2006; Sun et al., 2013). For example, the C-terminal activation domain of the bZIP proto-oncoprotein c-Fos, which effectively suppresses transcription in vitro, is intrinsically disordered and highly mobile (Campbell et al., 2000) whereas the unbound N-terminal domains of the DELLA proteins, which are central to the integration of plant developmental and environmental signaling by interacting with other transcription factors as corepressors or coactivators to regulate gibberellin (GA)-responsive gene expression patterns, are intrinsically disordered and undergo disorder–order transitions upon binding to interacting proteins (Sun et al., 2010). Significantly, the DELLA proteins are similar in their domain structures to the GRAS protein family whose N-domains are intrinsically disordered (Sun et al., 2011). Like the DELLAs, the GRAS proteins are extensively involved in plant signaling by virtue of their ability to undergo disorder–order transformations with a variety of molecular partners involved in root and shoot development, light signaling, nodulation, and auxin signaling and transcription regulation to biotic and abiotic stresses.

Like the DELLA–GA signaling system in plants, metazoans carry out a significant fraction of their intercellular signaling via a collection of small molecules that bind to their cognate proteins called nuclear hormone receptors (NHRs). Following ligand binding, the NHRs translocate to the nucleus where they act as transcription factors. In addition to the structured ligand and DNA binding domains, like other transcription factors, these NHRs have flanking and linking IDP domains that use their flexibility to bind to large numbers of partners. Simons and Kumar (2013) suggest that these domains may be responsible for the variable responses or context-dependent responses that follow from hormone signaling.

These examples illustrate that intrinsically disordered transcription factors play central roles in animal and plant development and homeostasis. They are not exceptional. Two separate computational biology studies on transcription factors using different disorder predictors identified similar, very large amounts of disorder in eukaryotic transcription factors (Liu et al., 2006; Minezaki et al., 2006). Liu et al. (2006) found that 82.6% to 93.1% of the transcription factors in three databases contain extended regions of intrinsic disorder, in contrast to 18.6% to 54.5% of the proteins in two control data sets. Minezaki et al. (2006) focused on human transcription factors, and, in addition to a disorder predictor, these researchers also used hidden Markov models to search for regions that are homologous to structured protein domains. Supporting the large amounts of predicted disorder in both studies, only 31% of the transcription factor residues aligned with known structured domains, which is only half of the 62% structurally aligned residues for *Escherichia coli* proteins that regulate transcription. Since protein–DNA recognition and protein–protein recognition are central transcription factor functionalities, the data reported by Liu et al. (2006) and by Minezaki et al. (2006) illustrate the extent to which eukaryotic transcription factors manifest extensive flexibility as a consequence of disorder-associated signaling and transcriptional regulation. This flexibility permits transcription factors with IDP domains to bind to a greater array of partners that in turn can induce conformational changes in bound protein and DNA substrates (Oldfield et al., 2005), for example, the high mobility group A (HMGA) protein family associated with a large number of the transcription factors participating in intra- and extracellular signaling (see Reeves, 2001).

Posttranslational Modifications

PTMs regulate transcription at different levels such as chromatin structure and transcription factor interactions. For example, the multiprotein complex called Mediator, which is massively disordered (Tóth-Petróczy et al., 2008) and involved in RNA Pol II-regulated transcription, is positively and negatively regulated by phosphorylation (Gonzalez et al., 2014; Yogesha et al., 2014). Indeed, phosphorylation is a common PTM mechanism that affects different macromolecular recognition/binding events involved in cell- or tissue-specific protein–protein interactions. For example, DNA binding by the transcription factor ETS-1, which is allosterically coupled to a serine-rich region (Lee et al., 2008), is modulated by Ca^{2+} signaling that induces phosphorylation of this region. Phosphorylation of the intrinsically disordered PAGE4 protein as part of the stress-response pathway causes PAGE4 to release the transcription factor c-Jun, enabling its activity in transcription regulation (Mooney et al., 2014). Phosphorylation can also increase interactions among cofactors. For example, the cytokines TNF and IL-1 induce phosphorylation of the p65 subunit of NF-κB, which in turn induces a conformational change that allows p65 ubiquitination and interaction with transcriptional cofactors (Milanovic et al., 2014). Association of Elk-1 and ETS domain transcription factors with Mediator and histone acetyltransferases is dependent on Elk-1 phosphorylation (Galbraith et al., 2013).

As a final example, the *Drosophila* Hox protein Ultrabithorax is multiply phosphorylated (Gavis and Hogness, 1991) and possesses an intrinsically disordered region that regulates DNA binding (Y. Liu et al., 2008) required for interactions with other transcription factors (Bondos et al., 2006). Given that phosphorylation has the potential to regulate as well as coordinate multiple transcription factor functions, it is not surprising that this mechanism is widely used. Indeed, transcription factors are disproportionately phosphorylated compared to other classes of proteins, and their phosphorylation correlates with the evolution of novel transcription factor sequences and functions (Kaganovich and Snyder, 2012).

It is important to note that multiple PTM combinations greatly increase signaling diversity. Consider a protein segment with three phosphorylation sites, each of which can be phosphorylated or unphosphorylated. This condition gives $2^3 = 8$ different PTM states, each of which can lead to different downstream consequences. This scenario is an oversimplification because PTM sites are typically IDP domains that are associated with target binding sites. Consequently, a "PTM code" exists (Lothrop et al., 2013) and operates after the "spliceosome code" (see Pejaver et al., 2014).

Calmodulin and Calmodulin-Like Proteins

One of the fundamental properties of every uni- or multicellular organism is the ability to respond to changes in its external and internal environment. This context-dependent responsiveness ensures survival as well as helps to regulate basic processes such as cell division, cell fate specification, and reproduction. It is not surprising therefore that organisms have evolved the ability to perceive and integrate different external and internal signals in ways that allow them to alter their physiology and development by virtue of signal transduction pathways that modulate adaptive adjustments of gene expression patterns. The foregoing sections show that AS, IDPs, and PTMs play important roles in organismic adaptability. This feature is nowhere better illustrated than by surveying the biological roles played by CaM and CaML proteins.

CaM contains two globular domains connected by a long, helical, and flexible protein linker that is intrinsically disordered (see figure 2.1). Each globular domain consists of a

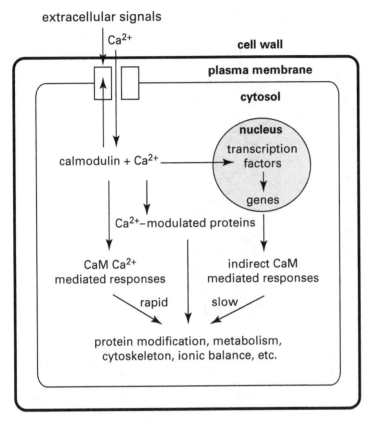

Figure 2.1
Schematic of slow and rapid CAM-Ca^{2+}- and Ca^{2+}-mediated responses to external and internal signals. External signals (which must pass through a cell wall in the case of plants) registered by plasma membrane bound receptors result in transient changes in intracellular Ca^{2+} ion concentrations in the cytosol and organelles, including the nucleus. Some of these ions become bound to calmodulin (CaM) and calmodulin-like (CaML) proteins, which interact with numerous target proteins interacting with metabolic processes and cytoskeletal structure. CaM-Ca^{2+} also affects gene expression patterns directly by binding to transcription factors or indirectly by signal transduction pathways. Rapid changes in cell functionalities result from CaM-Ca^{2+} or CaML-Ca^{2+} protein binding to target proteins; slower responses result from CaM-Ca^{2+}-mediated gene expression. Adapted from Snedden and Fromm (1998).

pair of Ca^{2+}–binding EF-hand helix–loop–helix structural domains or *motifs*. In this manner, Ca^{2+} ions serve as secondary messengers mediating responses to abiotic and biotic external signals by directly influencing cytosolic processes and by modulating gene expression patterns (see figure 2.1). The four EF-hand motifs and the linker region give CaM a dumbbell-like structure (see figure 2.2). Nuclear magnetic resonance studies indicate that the linker is entirely unstructured. The helix observed in its crystalline form is likely an artifact that occurs transiently in solution, if at all. Likewise, in the absence of

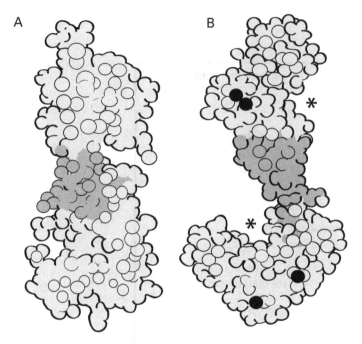

Figure 2.2
Stereochemical diagram of unbound calmodulin (CaM) and CaM bound to four Ca^{2+} ions indicated by black circles (A and B, respectively). When bound to Ca^{2+}, the complex exposes nonpolar surfaces (on EF-hand motifs bound by a linker, shown in light and dark gray, respectively) that then bind to nonpolar regions on target proteins. When bound to Ca^{2+}, the nonpolar amino acids form two grooves (shown with asterisks), which are configured to bind with a target protein. Because these two regions can assume different configurations by virtue of their intrinsically disordered domains, CaM acts as a versatile regulatory protein and its targets are not required to possess any specific amino acid sequence or structural binding motifs.

Ca^{2+}, the four EF-hand motifs are highly nonrigid and perhaps even molten globular. Even upon binding with Ca^{2+}, they retain some flexibility (see figure 2.2).

Ca^{2+} binding causes the helices of the EF-hand motifs to rotate and expose two hydrophobic faces (surrounded by negative charges) on each globular domain. This reconfiguration primes CaM to bind to its target proteins (with an affinity in the nanomolar range), which also tend to have exposed hydrophobic regions. The CaM hydrophobic faces are rich in methionine, an especially long, unbranched, and flexible residue, which allows them to adapt their conformations to other differently shaped hydrophobic surfaces. CaM is known to bind to over 300 targets (Yap et al., 2000), and a bioinformatics approach indicates that CaM target-protein binding domains are themselves intrinsically disordered (Radivojac et al., 2006), which adds yet another dimension to CaM's functional versatility.

When bound to different ligands, the helices in the EF-hands undergo slightly different amounts of rotation relative to each other, another feature that enables CaM to bind to different partner sequences, which typically contain IDDs that form helices when bound to CaM, for example, calcineurin (Dunlap et al., 2013). CaM wraps up its target helix, with the linker and two domains almost completely surrounding it. When bound to a partner, the linker is not the straight helix observed in the dumbbell-like form, but rather curves around the helical rod of the target. The multifunctional diversity resulting from the IDDs in CaM and its target proteins is impressive. Among animals, over 25 CaM targets have been identified, including kinases, receptors, G-proteins, and ion channels. Among plants, CaM has been less well studied, but it has been implicated in gravitropic, photomorphogenetic, and pathogenic responses (for reviews, see Snedden and Fromm, 1998; Yang and Poovaiah, 2003). For example, CaM–Ca^{2+}–dependent pathways for light-mediated gene activation and chloroplast development have been identified using the *aurea* tomato mutant whereas transgenic tobacco plants expressing glutamate decarboxylase lacking its CaM-binding domain become excessively stunted as a result of the failure of stem cortical parenchyma cells to elongate (Baum et al., 1996). CaM also functions in the Ca^{2+}-free state. For example, chromosome segregation in yeast is dependent on Ca^{2+}-free CaM. Likewise, neuromodulin, which participates in axonal and dendritic filopodia induction, binds to CaM with a higher affinity in the absence of Ca^{2+}.

Fungal, plant, and animal cells respond to a large number of external signals by means of a rapid and transient increase in the concentration of cytosolic calcium ions (Ca^{2+}) that have a high affinity for CaM and CaML proteins that regulate the downstream activities of target proteins such as CaM-activated kinases (e.g., CaM kinase kinase and myosin light chain kinase).

Baum et al. (1996) provided the first study documenting in vivo CaM binding to a plant target protein (i.e., glutamate decarboxylase), which is ubiquitously expressed in pro- and eukaryotes. Subsequent studies show that CaM is a highly conserved molecule (see figure 2.3) and that many aspects of plant and animal development involve CaM–Ca^{2+} and CaML protein downstream regulation. For example, the pollen-specific CaM binding protein NPG1 directly interacts with pectate lyase-like proteins that modify the pollen cell wall and regulate the emergence of the pollen tube (Shin et al., 2014). Likewise, the CaML protein CML39 serves as a Ca^{2+} sensor in the transduction of light signals that promote early seedling establishment in *Arabidopsis* (Bender et al., 2013). Overall cell growth rates are also often correlated with CaM gene transcript levels as illustrated by studies of unicellular organisms. Wen et al. (2014) examined the abundance of the *cam* gene during the growth of the dinoflagellate *Alexandrium catenella* and found that the transcript level increased eightfold from the lag growth phase to the exponential growth phase, and decreased from the latter to the stationary phase. These authors also found that the response of the *cam* gene to heat stress affected Ca^{2+} signal transduction in ways that correlated with cell growth (Wen et al., 2014).

AS and PTMs of CaM and CaML proteins extend their functionalities further. For example, Zheng et al. (2014) have shown that the CaM–Ca^{2+}-dependent protein kinase II

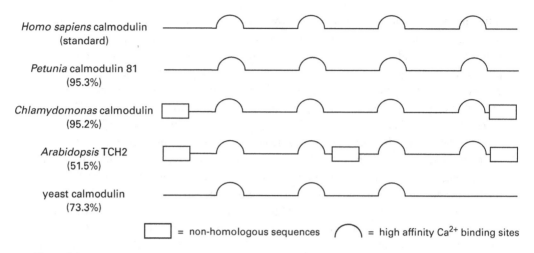

Figure 2.3
Schematic comparisons of calmodulin (CaM) and calmodulin-like (CaML) proteins from selected phyletically diverse eukaryotic species. Species are listed in descending order based on the percentage similarity of their aligned CaM or CaML protein sequences to a standard sequence (here, taken as human CaM). Data taken from GeneBank.

(CaMKII) β has four isoforms (CaMKII β, β∋, βe, and β∋e) resulting from the AS of three exons (E1, E3, and E4). Two of these exons are subject to AS (E1 and E4) and produce the only two isoforms involved with the reorganization of the actin cytoskeleton in developing neurons (CaMKII β and β∋). These authors speculate that the AS of CaMKII β in developing neurons serves as a developmental switch for actin cytoskeleton–associated isoforms correlated with dendritic branching and synapse formation (Zheng et al., 2014). Similarly, Taillebois et al. (2013) examined five splice variants of the CaM–Ca^{2+}-dependent protein kinase II in the cockroach *Periplaneta americana* and found that two variants are preferentially expressed in the nerve cord and in muscle tissue. Taillebois et al. (2013) suggest that AS is a common mechanism to control isoform expression patterns during tissue differentiation in insects. Another example is the CaM–Ca^{2+}-dependent protein kinase (CaMK) II delta. This protein is predominantly expressed in the heart and has three isoforms resulting from the AS of exons 14, 15, and 16. Examples of CaM AS are rare but nevertheless occur. For example, an alternate splice variant of a CaM-2 protein, called CaM-2-extended (CaM-2-ext), interacts with the C-terminal tail of human $Ca_V2.3$ voltage-gated calcium channels (Kamp et al., 2012), which are expressed in excitable cells and trigger neurotransmitter and peptide-hormone release. Only the CaM-2-ext splice variant is functional in this process.

PTM of CaM and CaML proteins is illustrated by CaM N-methyltransferase (CaM KMT), a highly conserved enzyme in eukaryotes that transfers three methyl groups to a conserved lysyl residue at position 115 in CaM. Banerjee et al. (2013) have shown that the methylation status of CaM plays a pivotal role in *Arabidopsis* root development. Plants with suppressed CaM methylation had longer roots whereas CaM KMT knockout plants had shorter roots. Banerjee et al. (2013) conclude that CaM functionalities are fine-tuned by CaM KMT in cell-, tissue-, and organ-specific patterns corresponding to normal development.

AS–IDPs–PTMs and Multicellularity

Multicellularity has evolved numerous times in prokaryotes and at least once in every major eukaryotic clade (and in every ploidy-level). However, in each case, it involved four developmental modules: (1) the ability to produce cell-to-cell adhesives, (2) control over the orientation of the plane of cell division, (3) the establishment of intercellular lines of communication for spatial-dependent patterns of differentiation, and (4) a mechanism to establish axial and lateral polarity (see Hernández-Hernández et al., 2012; Niklas, 2000, 2014). According to multilevel selection (MLS) theory, the transition from a unicellular or colonial body plan to a multicellular level of organization also required an *alignment-of-fitness* phase to eliminate cell–cell conflict and an *export-of-fitness* phase to establish a reproductively integrated phenotype (Michod and Anderson, 1979; Damuth and Heisler, 1988; Michod and Nedelcu, 2003; for a general review, see Folse and Roughgarden, 2010). According to this standard evolutionary scenario, the integration of these four modules while passing through the two MLS phases resulted in a "unicellular-to-colonial-to-multicellular" transformation series of body plans, although a "coenocytic-to-multicellular" transformation series is more consistent with some aspects of the evolution of certain fungal and algal groups (Niklas et al., 2013; Niklas, 2014).

Space precludes a detailed exposition of this standard model. However, there is ample evidence that AS and IDPs played critical roles in the evolution of multicellular organisms. For example, a *condicio sine qua non* for the transition from a unicellular to a colonial or multicellular body plan is the ability to produce, deliver, and export cell-to-cell adhesives to the plasma membrane. Numerous studies show that AS plays a critical role in this developmental module. This is illustrated by the modulation of the homophilic glycoprotein called neural cell adhesion molecule (NCAM), which appears on the surface of neurons, skeletal muscle, and natural killer cells during early embryonic development. AS produces NCAM variants differing in their cytoplasmic domains that differentially interact with the cell membrane (Cunningham et al., 1987). A similar regulatory mechanism links NCAM binding function with other primary cellular processes during embryonic patterning.

IDPs also play critical roles in cell adhesion as, for example, the transport of cytoplasmic soluble protein adhesives, such as interleukin 1β and galectin-1. Studies show that several mechanisms for transporting and targeting these and other soluble proteins to their delivery points involve exosomes, contractile vacuoles, and plasma membrane transporters, such as those found in amoebae and protozoa. CaM plays an important role in many of these mechanisms. For example, Sriskanthadevan et al. (2013) report that CaM forms a complex with the Ca^{2+}-dependent cell–cell adhesion molecule DdCAP-1 in the social amoeba *Dictyostelium discoideum*. CaM–Ca^{2+} complex binds to DdCAP-1 on the surface of contractile vacuoles and facilitates its delivery to the plasma membrane. Co-immunoprecipitation and pull-down studies show that only CaM–Ca^{2+} can bind DdCAP-1 and promote its import into the vacuoles targeting the plasma membrane (Sriskanthadevan et al., 2013). A similar phenomenology occurs in yeast (Peters and Mayer, 1998). Additional examples of IDP cell adhesives are E-cadherin and desmoglein 1. Both of these calcium-dependent adhesion proteins have been experimentally classified as having cytosolic IDP domains (Huber et al., 2001; Kami et al., 2009).

Another important component to the evolutionary acquisition of multicellularity is a module to regulate the plane of cell division since this is required for the formation of simple multicellular body plans such as unbranched and branched filaments. Once again, it is clear that AS and IDPs play critical roles in this context. For example, during *Drosophila* development the planar cell polarity protein called Prickle becomes concentrated along the proximal surfaces of cells and is present in multiple forms as a result of AS (Bastock et al., 2003). The participation of IDPs in cell division and polarity is also clear since Ca^{2+} signaling, CaM, and CaML proteins are involved in critical aspects of cell division and cell orientation. For example, CaM–Ca^{2+} undergoes demonstrably significant spatiotemporal reconfigurations during mitosis that prefigure the plane of cell furrowing. Likewise, during HeLa cell division, one CaM–Ca^{2+} cellular fraction is concentrated at the spindle poles and another is aggregated at the equatorial region where it becomes selectively activated (Li et al., 1999).

The participation of IDPs during cell division is ancient in its origins. Conformational analyses of the Sic1 protein, which plays a key regulatory role in *Saccharomyces cerevisiae* cell division, indicate that the Sic1 kinase-inhibitor domain manifests a dynamical helical structure characteristic of an IDP (Brocca et al., 2011). Especially interesting is the interaction between Sic1 and cell division control protein 4 (Cdc4), which leads to Sic1 digestion, which in turn leads to entry into the S-phase of the cell cycle. Multiple phosphorylated motifs located in an IDP domain in Sic1 compete for a single binding site on Cdc4. As more of the unbound motifs become activated for binding by phosphorylation, the binding constant increases due to more available targets for the single binding site on Cdc4. This IDP-PTM-based mechanism makes entry into S-phase ultrasensitive to the degree of Sic1 phosphorylation (Mittag et al., 2008).

The behavior of the full-length ZipA protein, which is an essential component in the cell division machinery of *E. coli*, is also indicative of an IDP (Lopez-Montero et al., 2013) as is the C-terminal linker of the tubulin homolog FtsZ, which provides the cytoskeletal framework and constriction force for *E. coli* cell division (Gardner et al., 2013). Cell fate specification, as well as cell division, is also influenced by disorder–order transitions. For example, p21 and p27 are two cyclin-dependent kinase (CdK) regulators that participate in cell division and cell fate specification. Studies show that both kinases undergo folding upon binding but retain flexibility within their functional protein complexes (Mitrea et al., 2012).

Number of Cell Types and IDPs

Our hypothesis that AS–IDPs–PTMs enhance cell functionality and thus participated in the evolution of multicellularity is consistent with the observation that a disproportionate number of transcription factors contain IDP domains. As noted, Liu et al. (2006) found that a disproportionate number of transcription factors in three databases contain extended regions of intrinsic disorder, in contrast to the proteins in two control data sets. Similar results are reported by Minezaki et al. (2006) for human transcription factors. The hypothesis that AS–IDPs–PTMs have significantly participated in the evolution of multicellularity is also consistent with data gathered from the primary literature reporting the number of cell types (NCTs), proteome size, the number of intrinsically disordered protein residues (IDResidues) per proteome, and genome size for 19 species of algae ($n = 8$), nonvascular and vascular land plants ($n = 6$), and invertebrates and vertebrates ($n = 5$) whose genomes have been completely sequenced (see figure 2.4). Bivariate regression of these variables reveals a statistically significant and positive correlation exists as gauged by ordinary least squares protocols (which stipulate Y and X as the dependent and independent variables, respectively) using $\log_{10}$-transformed data (see table 2.1 and figure 2.4). However, the numerical values of the slopes of each of the bivariate regression curves is far more informative (see table 2.1) because these slopes are scaling exponents (denoted here by α) that indicate numerically the extent to which NCT proportionally increases or decreases with respect to increasing values of each of the other variables on a log scale (Niklas, 1994). Inspection of these slopes indicates that NCT scales roughly as the 2.18 power of the number of IDResidues and as the 2.36 power of proteome size. Thus, the degree of cellular specialization increases dramatically as each of these two measures of organismic information content increases. Although IDResidues scale nearly one-to-one (isometrically) with respect to proteome size (i.e., $\alpha = 0.97$), NCTs fails to scale one-to-one with respect to increases in genome size (i.e., $\alpha = 0.88$), which helps in part to explain the so-called G paradox (for a discussion of this paradox in the context of AS, see Hahn and Wray, 2002). An information-centered perspective on these same data is provided by Niklas et al. (2014).

Clearly, these scaling relationships are consistent with the thesis that the proportion of IDResidues in a proteome influences cell fate specification. However, the strong positive and disproportionate scaling of NCTs with respect to IDResidues does not provide proof

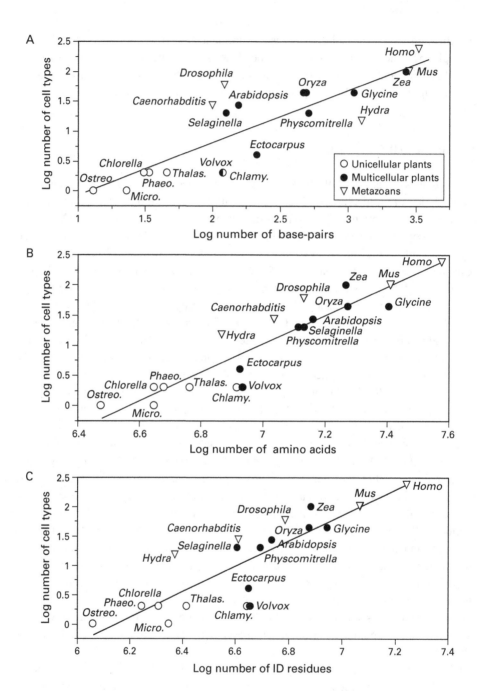

Figure 2.4
Log_{10}-transformed data for the number of different cell types (NCTs) plotted against log_{10}-transformed data for genome size (in 10^6 base-pairs) (A), proteome size (number of amino acids) (B), and protein functional diversity (number of intrinsically disordered protein residues, IDResidues, per proteome) (C) reported for unicellular algae, multicellular algae, land plants, invertebrates, and vertebrates. Straight lines are ordinary least squares regression curves. Chlamy., *Chlamydomonas reinhardii*; Micro., *Micromonas pusilla* NOUM 17; Ostreo., *Ostreococcus tauri*; Phaeo., *Phaeodactylum tricornutum*; Thalas., *Thalassiosira pseudonana*. For statistical regression parameters, see table 2.1. Data for NCTs were taken from Bell and Mooers (1997). Data for IDResidues and AA were taken from Oates et al. (2013). Data for IDResidues are based on predictions made by PONDR VLS2b (K. Peng et al., 2006), which ranked as the best for predicting segments of disorder and gave the highest per-residue accuracy for long regions of disorder (Z.-L. Peng and Kurgan, 2012). Data for genome size were taken from Internet sources.

Table 2.1
Bivariate regression parameters (see figure 2.4) for log_{10}-transformed data on the number of different cell types (NCT), genome size (G, in mbp), the number of amino acids (AA), the number of intrinsically disordered residues (IDResidues) reported for 19 species of algae, land plants, and animal species

Log Y vs. Log X	Slope	r^2	p	F
	(α value)			
NCT vs. IDResidues	2.18	0.721	<0.0001	44.0
NCT vs. AA	2.36	0.894	<0.0001	110
IDResidues vs. AA	0.97	0.940	<0.0001	46.7
NCT vs. G	0.88	0.709	<0.0001	43.9

Note: Species composition: brown algae (*n* = 1 species; *Ectocarpus siliculosus*), green algae (*n* = 3 species: *Chlorella* sp., *Chlamydomonas reinhardtii*, and *Volvox carteri*), diatoms (*n* = 4 species: *Micromonas pusilla* NOUM 17, *Ostreococcus tauri*, *Phaeodactylum tricornutum*, and *Thalassiosira pseudonana*), mosses (*n* = 1 species: *Physcomitrella patens*), lycophytes (*n* = 1 species: *Selaginella moellendorfii*), flowering plants (*n* = 4 species: *Arabidopsis thaliana*, *Glycine max*, *Oryza sativa*, and *Zea mays*), and metazoans (*n* = 5 species: *Hydra attenuate*, *Caenorhabditis elegans*, *Drosophila melanogaster*, *Mus musculus*, and *Homo sapiens*). Data for NCT were taken from Bell and Mooers (1997). Data for IDResidues and AA were taken from Oates et al. (2013). IDResidues values are based on PONDR VLS2b (see K. Peng et al., 2006) predictions, which is ranked as the best protocol for predicting segments of disorder and gives the highest per-residue accuracy for long regions of disorder (Z.-L. Peng and Kurgan, 2012). Data for G were taken from Internet sources.

for a "cause-and-effect" relationship. Such proof requires detailed experimental evidence explicating how IDPs actively participate in cell phenotypic diversification in model lineages (e.g., volvocine algae).

Concluding Remarks

The evolution of multicellularity evokes three obvious, but very important questions. When did it evolve? How did it evolve? And why did it evolve? This review attempts to shed some light primarily on the second of these three questions by illustrating how AS, IDPs, and PTMs (particularly multiple mPTMs) amplify metabolic and developmental functionalities without increasing proteome or genome size, thereby allowing cells sharing the same genome to achieve different phenotypes. The examples presented here are consistent with the hypothesis that the network established by AS, IDPs, and PTMs modulates the temporal- and spatial-specific modifications of protein functionalities required for context-dependent cell signaling, regulation, and differentiation. In another study of a collection of different development-associated proteins, this same motif is also implicated in the evolution of multicellular organisms (Dunker et al., 2015). The regulatory control exercised by this motif can evolve because it has the capacity to buffer a cell's genome against the immediate negative effects of mutations affecting protein functions, and, by so doing, AS, IDPs, and PTMs can provide sufficient time for an organism to adapt to the downstream effects of mutations until such time as these mutations can be adaptively assimilated or expunged.

These features lead us to conclude that gene regulatory networks function in a post-translation milieu that is responsive to external cues (e.g., the state of neighboring cells and physical factors) and internal conditions (e.g., intracellular physiological conditions and cell cycle status). According to this paradigm, cell fate specification emerges from a developmental program operating within well-defined boundary conditions set by regulatory networks, but in a manner that remains free to adaptively respond within these boundary conditions to changes in intra- and extracellular conditions. The classical perspective on gene regulatory networks dynamics therefore is untenable because it assumes that if a network is initiated with a specific initial combination of states (i.e., gene expression patterns), the subsequent trajectory of its behavior is strictly determined by those initial states. According to this perspective, cell fate specification cannot deviate once the initial conditions of gene regulatory networks are prescribed. This deterministic model clearly does not comply with the effects of the AS–IDP motif or with the effects of PTMs. It is therefore insufficient to explain cell fate specification. The development of a new paradigm for gene regulatory networks is beyond the scope of this chapter, but a profitable beginning to the development of such a model requires deciphering the "spliceosome code" (Wang et al., 2004; Wang and Burge, 2008; Barash et al., 2010) and the "post-translational modification code" (Lothrop et al., 2013; Pejaver et al., 2014).

In summary, AS and IDPs occur in all eukaryotic lineages. Unlike promoter activity, which primarily regulates the amount of transcripts, AS changes the structure of transcripts and their encoded proteins. Unlike most structured proteins, IDPs are polymorphic in their bound states (thus, each can have many target proteins). It is worth noting that proteins generally involved with cell signaling and regulation and proteins associated with differentiation are strikingly richer in IDP domains than are virtually all other functional classes of proteins. In tandem, AS and IDPs increase protein functionalities without increasing proteome or genome size, particularly since many important signaling molecules (e.g., CaM and CaML proteins) and the majority of transcription factors contain intrinsically disordered domains. Further, the operation of AS and the functionality of IDPs are dependent on external and internal cellular conditions. These features enlighten our understanding of the diversification of cell types during the evolution of multicellularity because they help to explain how cells sharing the same genome can achieve different physiological and morphological states. AS and IDPs together with PTMs provide a core interactive motif that mediates the time- and cell-specific modifications of protein functionalities required for context-dependent cell signaling, regulation, and differentiation. A review of these features indicates that an ancient AS, IDPs, and PTMs network, especially with multiple forms of PTMs, played a central role in the evolution of eukaryotic multicellularity.

Acknowledgments

The authors thank the anonymous reviewers for their insightful suggestions. Funding is gratefully acknowledged from the College of Agriculture and Life Sciences, Cornell University (to KJN) and Molecular Kinetics, Inc. (to AKD).

References

Banerjee, J., Magnani, R., Nair, M., Dirk, L. M., DeBolt, S., Maiti, I. B., et al. (2013). Calmodulin-mediated signal transduction pathways in *Arabidopsis* are fine-tuned by methylation. *Plant Journal*, 25, 4493–4511.

Barash, Y., Calarco, J. A., Gao, W., Pan, Q., Wang, X., Shai, O., et al. (2010). Deciphering the splicing code. *Nature*, 465, 53–59.

Bastock, R., Strutt, H., & Strutt, D. (2003). Strabismus is asymmetrically localised and binds to Prickle and Dishevelled during *Drosophila* planar polarity patterning. *Development*, 130, 3007–3014.

Baum, G., Lev-Yadun, S., Fridmann, Y., Arazi, T., Katsnelson, H., Zik, M., et al. (1996). Calmodulin binding to glutamate decarboxylase is required for the regulation of glutamate and GABA metabolism and normal development in plants. *EMBO Journal*, 15, 2988–2996.

Bell, G., & Mooers, A. O. (1997). Size and complexity among multicellular organisms. *Biological Journal of the Linnean Society. Linnean Society of London*, 60, 345–363.

Bender, K. W., Rosenbaum, D. M., Vanderbeld, B., Ubaid, M., & Snedden, W. A. (2013). The *Arabidopsis* calmodulin-like protein, CML39, functions during early seedling development. *Plant Journal*, 76, 634–647.

Black, D. L. (2003). Mechanisms of alternative pre-messenger RNA splicing. *Annual Review of Biochemistry*, 72, 291–336.

Bondos, S. E., Tan, X.-X., & Matthews, K. S. (2006). Physical and genetic interactions link hox function with diverse transcription factors and cell signaling proteins. *Molecular & Cellular Proteomics: MCP*, 5, 824–834.

Boutz, P. L., Stoilov, P., Li, Q., Lin, C. H., Chawla, G., Ostrow, K., et al. (2007). A post-transcriptional regulatory switch in polypyrimidine tract-binding proteins reprograms alternative splicing in developing neurons. *Genes & Development*, 21, 1636–1652.

Brocca, S., Testa, L., Sobott, F., Samalikova, M., Natalello, A., Papaleo, E., et al. (2011). Compaction properties of an intrinsically disordered protein: Sic1 and its kinase-inhibitor domain. *Biophysical Journal*, 100, 2243–2252.

Campbell, K. M., Terrell, A. R., Laybourn, P. J., & Lumb, K. J. (2000). Intrinsic structural disorder of the C-terminal activation domain from the bZIP transcription factor Fos. *Biochemistry*, 39, 2708–2713.

Chang, C. Y., Lin, W. D., & Tu, S. L. (2014). Genome-wide analysis of heat-sensitive alternative splicing in *Physcomitrella patens. Plant Physiology*, 165, 826–840.

Chen, C. D., Kobayashi, R., & Helfman, D. M. (1999). Binding of hnRNP H to an exonic splicing silencer is involved in the regulation of alternative splicing of the rat α tropomyosin gene. *Genes & Development*, 13, 593–606.

Chen, L., Bush, S. L., Tovar-Corona, J. M., Castillo-Morales, A., & Urrutia, A. O. (2014). Correcting for differential transcript coverage reveals a strong relationship between alternative splicing and organism complexity. *Molecular Biology and Evolution*, 31, 1402–1413.

Coelho Ribeiro, M. de L., Espinosa, J., Islam, S., Martinez, O., Thanki, J. L., Mazariegos, S., et al. (2013). Malleable ribonucleoprotein machine: Protein intrinsic disorder in the *Saccharomyces cerevisiae* spliceosome. *PeerJ*, 1, e2.

Cunningham, B. A., Hemperly, J. J., Murray, B. A., Prediger, E. A., Brackenbury, R., & Edelman, G. M. (1987). Neural adhesion molecule: Structure, immunoglobulin-like domains, cell surface modulation, and alternative RNA splicing. *Science*, 236, 799–806.

Damuth, J., & Heisler, I. J. (1988). Alternative formulations of multilevel selection. *Biology & Philosophy*, 3, 407–430.

Dunker, A. K., & Uversky, V. N. (2008). Signal transduction via unstructured protein conduits. *Nature Chemical Biology*, 4, 229–230.

Dunker, A. K., Bondos, S. H., Huang, F., & Oldfield, C. J. (2015). Intrinsically disordered proteins and multicellular organisms. *Seminars in Cell & Developmental Biology*, 37, 44–55.

Dunker, A. K., Garner, E., Guilliot, S., Romero, P., Albrecht, K., Hart, J., et al. (1998). Protein disorder and the evolution of molecular recognition: theory, predictions and observations. *Pacific Symposium on Biocomputation, 1998*, 473–484.

Dunker, A. K., Silman, I., Uversky, V. N., & Sussman, J. L. (2008). Function and structure of inherently disordered proteins. *Current Opinion in Structural Biology*, 18, 756–764.

Dunlap, T. B., Cook, E. C., Rumi-Masante, J., Arvin, H. G., Lester, T. E., & Creamer, T. P. (2013). The distal helix in the regulatory domain of calcineurin is important for domain stability and enzyme function. *Biochemistry*, 52, 8643–8651.

Dyson, H. J., & Wright, P. E. (2005). Intrinsically unstructured proteins and their functions. *Nature Reviews. Molecular Cell Biology*, 6, 197–208.

Evens, T. I. A., Hell, J. W., & Shea, M. A. (2011). Thermodynamic linkage between calmodulin binding calcium and contiguous sites in the C-terminal tail of Ca(v)1.2. *Biophysical Chemistry*, 159, 172–187.

Fairbrother, W. G., Holste, D., Burge, C. B., & Sharp, P. A. (2004). Single nucleotide polymorphism–based validation of exonic splicing enhancers. *PLoS Biology*, 2, e268.

Folse, H. J., Jr., & Roughgarden, J. (2010). What is an individual organism? A multilevel selection perspective. *Quarterly Review of Biology*, 85, 447–472.

Fuxreiter, M., & Tompa, P. (2012). Fuzzy complexes: A more stochastic view of protein function. *Advances in Experimental Medicine and Biology*, 725, 1–14.

Galbraith, M. D., Saxton, J., Li, L., Shelton, S. J., Zhang, H., Espinosa, J. M., et al. (2013). ERK phosphorylation of MED14 in promoter complexes during mitogen-induced gene activation by Ekk-1. *Nucleic Acids Research*, 41, 10241–10253.

Galea, C. A., Wang, Y., Sivakolundu, S. G., & Kriwacki, R. W. (2008). Regulation of cell division by intrinsically unstructured proteins: Intrinsic flexibility, modularity, and signaling conduits. *Biochemistry*, 47, 7598–7609.

Gao, J., Thelen, J. J., Dunker, A. K., & Xu, D. (2010). Musite, a tool for global prediction of general and kinase specific phosphorylation sites. *Molecular & Cellular Proteomics*, 9, 2586–2600.

Gardner, K. A., Moore, D. A., & Erickson, H. P. (2013). The C-terminal linker in *Escherichia coli* FtsZ functions as an intrinsically disordered peptide. *Molecular Microbiology*, 89, 264–275.

Gavis, E. R., & Hogness, D. S. (1991). Phosphorylation, expression and function of the Ultrabithorax protein family in *Drosophila melanogaster*. *Development*, 112, 1077–1093.

Gonzalez, D., Hamidi, N., Del Sol, R., Benschop, J. J., Nancy, T., Li, C., et al. (2014). Suppression of Mediator is regulated by Cdk8-dependent Grr1 turnover of the Med3 coactivator. *Proceedings of the National Academy of Sciences of the United States of America*, 111, 2500–2505.

Grosberg, R. K., & Strathmann, R. R. (2007). The evolution of multicellularity: A minor major transition? *Annual Review of Ecology Evolution and Systematics*, 38, 621–654.

Hahn, M. W., & Wray, G. A. (2002). The g-value paradox. *Evolution & Development*, 4, 73–75.

Haynes, C., & Iakoucheva, L. M. (2006). Serine/arginine-rich splicing factors belong to a class of intrinsically disordered proteins. *Nucleic Acids Research*, 34, 305–312.

Hernández-Hernández, V., Niklas, K. J., Newman, S. A., & Benítez, M. (2012). Dynamical patterning modules in plant development and evolution. *International Journal of Developmental Biology*, 56, 661–674.

Hsu, W.-L., Oldfield, C. J., Xue, B., Meng, J., Huang, F., Romero, P., et al. (2013). Exploring the binding diversity of intrinsically disordered proteins involved in one-to-many binding. *Protein Science*, 22, 258–273.

Huber, A. H., Stewart, D. B., Laurents, D. V., Nelson, W. J., & Weis, W. I. (2001). The cadherin cytoplasmic domain is unstructured in the absence of β-catenin: A possible mechanism for regulating cadherin turnover. *Journal of Biological Chemistry*, 276, 12301–12309.

Iakoucheva, L. M., Brown, C. J., Lawson, J. D., Obradović, Z., & Dunker, A. K. (2002). Intrinsic disorder in cell-signaling and cancer-associated proteins. *Journal of Molecular Biology*, 323, 573–584.

Iakoucheva, L. M., Radivojac, P., Brown, C. J., O'Connor, T. R., Sikes, J. G., Obradovic, Z., et al. (2004). The importance of intrinsic disorder for protein phosphorylation. *Nucleic Acids Research*, 32, 1037–1049.

Jarvis, P., & Soll, M. (2001). Toc, Tic, and chloroplast protein import. *Biochimica et Biophysica Acta*, 1541, 64–79.

Johnson, J. M., Castle, J., Garrett-Engele, P., Kan, Z., Loerch, P. M., Armour, C. D., et al. (2003). Genome-wide survey of human alternative pre-mRNA splicing with exon junction microarrays. *Science*, 302, 2141–2144.

Kaganovich, M., & Snyder, M. (2012). Phosphorylation of yeast transcription factors correlates with the evolution of novel sequence and function. *Journal of Proteome Research*, 11, 261–268.

Kami, K., Chidgey, M., Dafforn, T., & Overduin, M. (2009). The desmoglein-specific cytoplasmic region is intrinsically disordered in solution and interacts with multiple desmosomal protein partners. *Journal of Molecular Biology*, 386, 531–543.

Kamp, M. A., Shakeri, B., Tevoufouet, E. E., Krieger, A., Henry, M., Behnke, K., et al. (2012). The C-terminus of human $Ca_V2.3$ voltage-gated calcium channel interacts with alternatively spliced calmodulin-2 expressed in two human cell lines. *Biochimica et Biophysica Acta*, 1824, 1045–1057.

Ke, S., Zhang, X. H.-F., & Chasin, L. A. (2008). Positive selection acting on splicing motifs reflects compensatory evolution. *Genome Research*, 18, 533–543.

Korneta, I., & Bujnicki, J. M. (2012). Intrinsic disorder in the human spliceosome proteome. *PLoS Computational Biology*, 8, e1002641.

Lee, G. M., Pufall, M. A., Meeker, C. A., Kang, H. S., Graves, B. J., & McIntosh, L. P. (2008). The affinity of Ets-1 for DNA is modulated by phosphorylation through transient interactions of an unstructured region. *Journal of Molecular Biology*, 382, 1014–1030.

Li, C.-J., Heim, R., Lu, P., Pu, Y., Tsien, R. Y., & Chang, D. C. (1999). Dynamic redistribution of calmodulin in HeLa cells during cell division revealed by a GFP–calmodulin fusion protein technique. *Journal of Cell Science*, 112, 1567–1577.

Lim, K. H., Ferraris, L., Filloux, M. E., Raphael, B. J., & Fairbrother, W. G. (2011). Using positional distribution to identify splicing elements and predict pre-mRNA processing defects in human genes. *Proceedings of the National Academy of Sciences of the United States of America*, 108, 11093–11098.

Liu, J., Perumal, N. B., Oldfield, C. J., Su, E. W., Uversky, V. N., & Dunker, A. K. (2006). Intrinsic disorder in transcription factors. *Biochemistry*, 45, 6873–6888.

Liu, Y., Matthews, K. S., & Bondos, S. E. (2008). Multiple intrinsically disordered sequences alter DNA binding by the homeodomain of the *Drosophila* Hox protein Ultrabithorax. *Journal of Biological Chemistry*, 283, 20874–20887.

Lopez-Montero, I., Lopez-Navajas, P., Mingorance, J., Rivas, G., Vélez, M., Vicente, M., et al. (2013). Intrinsic disorder of the bacterial cell division protein ZipA: Coil-to-brush conformational transition. *FASEB Journal*, 27, 3363–3375.

Lothrop, A. P., Torres, M. P., & Fuchs, S. M. (2013). Deciphering post-translational modification codes. *FEBS Letters*, 587, 1247–1257.

Matera, A. G., & Wang, Z. (2014). A day in the life of the spliceosome. *Nature Reviews. Molecular Cell Biology*, 15, 108–121.

Matlin, A. J., Clark, F., & Smith, C. W. J. (2005). Understanding alternative splicing: Towards a cellular code. *Nature Reviews. Molecular Cell Biology*, 6, 386–398.

Michod, R. E., & Anderson, W. W. (1979). Measures of genetic relationship and the concept of inclusive fitness. *American Naturalist*, 114, 637–647.

Michod, R. E., & Nedelcu, A. M. (2003). On the reorganization of fitness during evolutionary transitions in individuality. *Integrative and Comparative Biology*, 43, 64–73.

Milanovic, M., Kracht, M., & Schmitz, M. L. (2014). The cytokine-induced conformational switch of nuclear factor κB p65 is mediated by p65 phosphorylation. *Biochemical Journal*, 457, 401–413.

Minezaki, Y., Homma, K., Kinjo, A. R., & Nishikawa, K. (2006). Human transcription factors contain a high fraction of intrinsically disordered regions essential for transcriptional regulation. *Journal of Molecular Biology*, 359, 1137–1149.

Mitrea, D. A., Yoon, M.-K., Ou, L., & Kriwacki, R. W. (2012). Disorder-function relationships for the cell cycle regulatory proteins p21 and p27. *Biological Chemistry*, 393, 259–274.

Mittag, T., Orlicky, S., Choy, W. Y., Tang, X., Lin, H., Sicheri, F., et al. (2008). Dynamic equilibrium engagement of a polyvalent ligand with a single-site receptor. *Proceedings of the National Academy of Sciences of the United States of America*, 105, 17772–17777.

Modrek, B., Resch, A., Grasso, C., & Lee, C. (2001). Genome-wide detection of alternative splicing in expressed sequences of human genes. *Nucleic Acids Research*, 29, 2850–2859.

Mooney, S. M., Qiu, R., Kim, J. J., Sacho, E. J., Rajagopalan, K., Johng, D., et al. (2014). Cancer/testis antigen PAGE4, a regulator of c-Jun transactivation, is phosphorylated by homeodomain-interacting protein kinase 1, a component of the stress-response pathway. *Biochemistry*, 53, 1670–1679.

Niklas, K. J. (1994). *Plant allometry: The scaling of form and process*. Chicago: University of Chicago Press.

Niklas, K. J. (2000). The evolution of plant body plans—A biomechanical perspective. *Annals of Botany*, 85, 411–438.

Niklas, K. J. (2014). The evolutionary-developmental origins of multicellularity. *American Journal of Botany*, 101, 6–25.

Niklas, K. J., Bondos, S. E., Dunker, A. K., & Newman, S. A. (2015). Rethinking gene regulatory networks in light of alternative splicing, intrinsically disordered protein domains, and post-translational modifications. *Frontiers in Cell and Developmental Biology*. doi:.10.3389/fcell.2015.00008

Niklas, K. J., Cobb, E. D., & Crawford, D. R. (2013). The evo-devo of multicellular cells, tissues, and organisms, and an alternative route to multicellularity. *Evolution & Development*, 15, 466–474.

Niklas, K. J., Cobb, E. D., & Dunker, A. K. (2014). The number of cell types, information content, and the evolution of multicellularity. *Acta Societatis Botanicorum Poloniae*, 83, 337–347.

Niklas, K. J., & Newman, S. A. (2013). The origins of multicellular organisms. *Evolution & Development*, 15, 41–52.

Oates, M. E., Romero, P., Ishida, T., Ghalwash, M., Mizianty, M. J., Xue, B., et al. (2013). D^2P^2: Database of disordered protein predictions. *Nucleic Acids Research*, 41, D508–D516.

Oldfield, C. J., Cheng, Y., Cortese, M. S., Romero, P., Uversky, V. N., & Dunker, A. K. (2005). Coupled folding and binding with α-helix-forming molecular recognition elements. *Biochemistry*, 44, 12454–12470.

Oldfield, C. J., & Dunker, A. K. (2014). Intrinsically disordered proteins and intrinsically disordered regions. *Annual Review of Biochemistry*, 83, 553–584.

Oldfield, C. J., Meng, J., Yang, J. Y., Yang, M. Q., Uversky, V. N., & Dunker, A. K. (2008). Flexible nets: Disorder and induced fit in the associations of p53 and 14–3–3 with their partners. *BMC Genomics*, 9(Suppl 1), S1.

Pan, Q., Shai, O., Lee, L. J., Frey, B. J., & Blencowe, B. J. (2008). Deep surveying of alternative splicing complexity in the human transcriptome by high-throughput sequencing. *Nature Genetics*, 40, 1413–1415.

Pejaver, V., Hsu, W. L., Xin, F., Dunker, A. K., Uversky, V. N., & Radivojac, P. (2014). The structural and functional signatures of proteins that undergo multiple events of post-translational modification. *Protein Science*, 23, 1077–1093.

Peng, K., Radivojac, P., Vucetic, S., Dunker, A. K., & Obradovic, Z. (2006). Length-dependent prediction of protein intrinsic disorder. *BMC Bioinformatics*, 7, 208.

Peng, Z.-L., & Kurgan, L. (2012). Comprehensive comparative assessment of in-silico predictors of disordered regions. *Current Protein & Peptide Science*, 13, 6–18.

Peters, C., & Mayer, A. (1998). Ca^{2+}-calmodulin signals the completion of docking and triggers a late step of vacuole fusion. *Nature*, 396, 575–580.

Radivojac, P., Vucetic, S., O'Conner, T. R., Uversky, V. N., Obradovic, Z., & Dunker, A. K. (2006). Calmodulin signaling: Analysis and prediction of a disorder-dependent molecular recognition. *Proteins*, 63, 398–410.

Rautureau, G. J., Day, C. L., & Hinds, M. G. (2010). Intrinsically disordered proteins in bcl-2 regulated apoptosis. *International Journal of Molecular Sciences*, 11, 1808–1824.

Reeves, R. (2001). Molecular biology of HMGA proteins: Hubs of nuclear function. *Gene*, 277, 63–81.

Romero, P. R., Zaidi, S., Fang, Y. Y., Uversky, V. N., Radivojac, P., Oldfield, C. J., et al. (2006). Alternative splicing in concert with protein intrinsic disorder enables increased functional diversity in multicellular organisms. *Proceedings of the National Academy of Sciences of the United States of America*, 103, 8390–8395.

Shammas, S. L., Travis, A. J., & Clarke, J. (2013). Remarkably fast coupled folding and binding of the intrinsically disordered transactivation domain of cMyb to CBP KIX. *Journal of Physical Chemistry*, 117, 13346–13356.

Shin, S. B., Golovkin, M., & Reddy, A. S. N. (2014). A pollen-specific calmodulin-binding protein, NPG1, interacts with putative pectate lyases. *Scientific Reports, 4*, srep05263.

Simons, S. S., & Kumar, R. (2013). Variable steroid receptor responses: Intrinsically disordered AF1 is the key. *Molecular and Cellular Endocrinology, 376*, 81–84.

Snedden, W. A., & Fromm, H. (1998). Calmodulin, calmodulin-related proteins and plant responses to the environment. *Trends in Plant Science, 3*, 299–304.

Spellman, R., Llorian, M., & Smith, C. W. J. (2007). Crossregulation and functional redundancy between the splicing regulator PTB and its paralogs nPTB and ROD1. *Molecular Cell, 27*, 420–434.

Sriskanthadevan, S., Brar, S. K., Manoharan, K., & Siu, C. H. (2013). Ca^{2+}-calmodulin interacts with DdCAD-1 and promotes DdCAD-1 transport by contractile vacuoles in *Dictyostelium* cells. *FEBS Journal, 280*, 1795–1806.

Sun, X., Jones, W. T., Harvey, D., Edwards, P. J., Pascal, S. M., Kirk, C., et al. (2010). N-terminal domains of DELLA proteins are intrinsically unstructured in the absence of interaction with GID1/gibberellic acid receptors. *Journal of Biological Chemistry, 285*, 11557–11571.

Sun, X., Rikkerink, E. H. A., Jones, W. T., & Uversky, V. N. (2013). Multifarious roles of intrinsic disorder in proteins illustrate its broad impact on plant biology. *Plant Cell, 25*, 38–55.

Sun, X., Xue, B., Jones, W. T., Rikkerink, E., Dunker, A. K., & Uversky, V. N. (2011). A functionally required unfoldome from the plant kingdom: Intrinsically disordered N-terminal domains of GRAS proteins are involved in molecular recognition during plant development. *Plant Molecular Biology, 77*, 205–223.

Taillebois, E., Heuland, E., Bourdin, C. M., Griveau, A., Quinchard, S., Tricoire-Leignel, H., et al. (2013). Ca^{2+}/calmodulin-dependent protein kinase II in the cockroach *Periplaneta americana*: Identification of five isoforms and their tissue distribution. *Archives of Insect Biochemistry and Physiology, 83*, 138–150.

Tompa, P., Fuxreiter, M., Oldfield, C. J., Simon, I., Dunker, A. K., & Uversky, V. N. (2009). Close encounters of the third kind: Disordered domains and the interactions of proteins. *BioEssays, 31*, 328–335.

Tóth-Petróczy, A., Oldfield, C. J., Simon, I., Takagi, Y., Dunker, A. K., Uversky, V. N., et al. (2008). Malleable machines in transcription regulation: The mediator complex. *PLoS Computational Biology, 4*, e1000243.

Uversky, V. N. (2002). Natively unfolded proteins: A point where biology waits for a physics. *Protein Science: A Publication of the Protein Society, 11*, 739–756.

Uversky, V. N., & Dunker, A. K. (2010). Understanding protein non-folding. *Biochim Biophys Acta, 1804*, 1231-64.e.

van der Lee, R., Lang, B., Kruse, K., Gsponer, J., Sánchez de Groot, N., Huynen, M. A., et al. (2014). Intrinsically disordered segments affect protein half-life in the cell and during evolution. *Cell Reports, 8*, 1832–1844.

Wang, Z., & Burge, C. B. (2008). Splicing regulation: From a parts list of regulatory elements to an integrated splicing code. *RNA, 14*, 802–813.

Wang, Z., Rolish, M. E., Yeo, G., Tung, V., Mawson, M., & Burge, C. B. (2004). Systematic identification and analysis of exonic splicing silencers. *Cell, 119*, 831–845.

Wen, R., Sui, Z., Bao, Z., Zhou, W., & Wang, C. (2014). Isolation and characterization of calmodulin gene of *Alexandrium catenella* (Dinoflagellate) and its performance in cell growth and heat stress. *Journal of Ocean University of China, 13*, 290–296.

Xie, H., Vucetic, S., Iakoucheva, L. M., Oldfield, C. J., Dunker, A. K., Uversky, V. N., et al. (2007). Functional anthology of intrinsic disorder: I. Biological processes and functions of proteins with long disordered regions. *Journal of Proteome Research, 6*, 1882–1898.

Yang, T., & Poovaiah, B. W. (2003). Calcium/calmodulin-mediated signaling networks in plants. *Trends in Plant Science*, 8, 505–512.

Yap, K., Kim, J., Truong, K., Sherman, M., Yuan, T., & Ikura, M. (2000). Calmodulin target database. *Journal of Structural and Functional Genomics*, 1, 8–14.

Yogesha, S. D., Mayfield, J. E., & Zhang, Y. (2014). Cross-talk of phosphorylation and prolyl isomerization of the C-terminal domain of RNA polymerase II. *Molecules*, 19, 1481–1511.

Zheng, J., Redmund, L., Xu, C., Kuang, J., & Liao, W. (2014). Alternative splicing in the variable domain of CaMKII affects the level of F-actin association in developing neurons. *International Journal of Clinical and Experimental Pathology*, 7, 2963–2975.

II   PLANTS AND RELATED THEORY

3 The Mechanistic Basis for the Evolution of Soma during the Transition to Multicellularity in the Volvocine Algae

Stephan G. König and Aurora M. Nedelcu

Overview

The most dramatic increases in complexity levels during the evolution of life have involved a series of evolutionary transitions in individuality during which previously independent individuals formed groups that ultimately became integrated into higher-level individuals. The newly emerged individuals possessed variable and heritable traits at this new level of organization on which natural selection could act (Michod, 2007). One such major evolutionary transition led single-cell individuals to form highly integrated multicellular organisms with new levels of fitness emerging at the higher level. The transition to multicellularity required the reorganization of the two fitness components (survival and reproduction) between soma (specialized in survival) and germ (specialized in reproduction; Michod and Nedelcu, 2003). In fact, the hallmark of this transition is the rise of sterile somatic cells, which was also responsible for the further increases in complexity that took place in some lineages.

Although somatic cell differentiation is a basic process in the development of most multicellular organisms, the mechanisms involved in the separation of germ and soma are well understood only in a handful of lineages. For instance, in most animals, the asymmetrical distribution of germ granules determines germ fate, as a set of cells is separated during early embryogenesis to form the germline (Voronina et al., 2011). The differentiation of germ cells involves three general phases (Cinalli, Rangan, and Lehmann, 2008): (1) transcriptional repression of somatic cell programs in progenitor germ cells (PGC); (2) migration of PGCs to the somatic gonad, which forms a microenvironment for the maintenance of germ cells; and (3) posttranscriptional regulation of gene regulatory networks, which represses somatic functions on the one hand and might promote germ cell–specific functions (including meiosis and gametogenesis) on the other hand. In stark contrast, in flowering plants, germ cells differentiate late from somatic cells of the mature plants (Yang, Shi, and Chen, 2010). Specifically, the germline develops from the so-called archesporial cells, which will enter meiosis to form gametophytes and ultimately gametes, a process that also depends on complex intercellular signaling between somatic and germline cells (Feng, Zilberman, and Dickinson, 2013). Even less is known about the mechanistic basis for the evolution of soma. Recently, we have argued that the evolution of germ–soma separation in multicellular individuals involved the co-option of molecular mechanisms underlying fitness and life history trade-offs in unicellular lineages, by changing their expression from a temporal into a spatial context (Nedelcu and Michod, 2006; Nedelcu, 2009b; Nedelcu and Michod, 2011).

Multicellularity evolved at least 25 times independently (Grosberg and Strathmann, 2007). In addition to the lineages leading to animals and land plants, multicellular forms evolved among bacteria (e.g., cyanobacteria, myxobacteria); green, red, and brown algae; slime molds; fungi; and several other groups (Grosberg and Strathmann, 2007). In this context, the chlorophyte green algae have a special place as 9 out of 11 orders contain multicellular forms that each evolved multicellularity independently (Kirk, 2005). Furthermore, among chlorophytes, the order Volvocales has been intensively studied as it includes a series of genera, known as the volvocine algae, that exhibit several distinct morphologies associated with different levels of complexity, from simple unicellular individuals to multicellular forms with distinct soma and germ lines (Herron and Nedelcu, 2015). Below, we discuss the volvocine algae as a model system with which to investigate the mechanistic basis for the evolution of multicellularity and present our new insights into the evolutionary history of a master regulator of somatic cell differentiation in this group.

The Volvocine Algae as a Model System

Volvocine algae are flagellated algae that inhabit freshwater environments (from small ephemeral puddles to ponds and large lakes) around the world (Kirk, 1998). Some are unicellular (e.g., *Chlamydomonas*), while others consist of either 4–32 uniform unspecialized cells (e.g., *Gonium, Eudorina*), 32–64 cells with differentiated somatic cells but no specialized reproductive cells (*Pleodorina*), or up to 10,000 cells with terminally differentiated somatic cells and specialized reproductive cells known as gonidia (e.g., *Volvox* species; see figure 3.1). Complexity increased several times independently in this group, and this is reflected in the fact that nominal genera such as *Eudorina*, *Pleodorina*, and *Volvox* are actually polyphyletic groups (see figure 3.2).

Multicellular volvocine algae without specialized cells are characterized by a biphasic life cycle (Kirk, 2003). During the vegetative phase, they are motile (each cell possesses a pair of flagella) and grow, with each cell growing 2^n in size; then, during the reproduction phase, all cells lose their flagella as they enter a series of n mitotic divisions in quick succession, a process called multiple fission (Kirk, 2003). In algae with germ–soma differentiation, this temporal biphasic pattern (which is the ancestral developmental program in this group) is instead realized in a spatial context, resulting in division of labor among

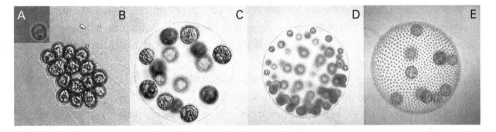

Figure 3.1
A subset of volvocine green algae that show a progressive increase in cell number, volume of extracellular matrix per cell, division of labor between somatic and reproductive cells, and proportion of vegetative cells. (A) *Chlamydomonas reinhardtii*; (B) *Gonium pectorale*; (C) *Eudorina elegans*; (D) *Pleodorina californica*; (E) *Volvox carteri*. Where two cell types are present, the smaller cells are the vegetative (somatic) cells, whereas the larger cells are the reproductive cells (gonidia).

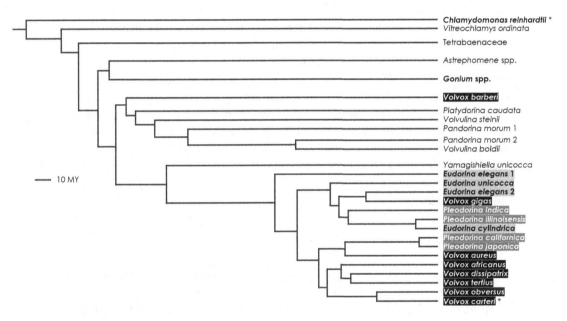

Figure 3.2
Schematic representation of our current understanding of phylogenetic relationships among the main volvocine lineages (adapted from Herron and Michod, 2008, and Herron et al., 2009). Genera mentioned in the main text are in bold, and the two main species discussed here are marked with asterisks. Species in the same genus are indicated with the same background and font color; note that *Eudorina, Pleodorina,* and *Volvox* are actually polyphyletic genera.

cells; that is, the somatic cells are specialized in vegetative functions and retain motility throughout the entire life cycle while the gonidia are not involved in motility and are specialized in reproduction (Kirk, 2003). Notably, all species with a differentiated soma contain only one type of somatic cells.

In this group, the presence of different organizational complexity levels and the availability of developmental mutants for the most studied multicellular species, *Volvox carteri*, facilitate the testing of various hypotheses concerning the selective pressures associated with the evolution of germ–soma differentiation (Kirk, 2003; Solari et al., 2006a, b, 2013, 2015). In addition, soma has evolved several times independently (Herron and Michod 2008; figure 3.2), and this is thought to be an indication that only a small number of changes was necessary to facilitate the increase in complexity from a simpler ancestor (Kirk, 2003). Furthermore, in the volvocine algae, multicellularity is thought to have evolved more recently than in many other lineages, allowing for the possibility that the origin of the genetic program for multicellular development can be deciphered (Kirk, 2003). Recent estimates indicate that, in contrast to other multicellular lineages (such as red algae and metazoans, which evolved around 1,200 and 600 million years ago, respectively; King et al., 2008; Herron et al., 2009), the lineages leading to the multicellular *V. carteri* and the unicellular *Chlamydomonas reinhardtii* diverged from each other about 235 million years ago (Herron et al., 2009). Lastly, both *C. reinhardtii* and *V. carteri* are well-developed experimental model systems, and their genomes have been sequenced (Merchant et al., 2007; Prochnik et al., 2010); the sequencing of several other volvocine genomes is also under way. Altogether, these unique aspects make volvocine algae an extremely suitable model system with which to study the transition to multicellularity and the evolution of somatic cell differentiation from both an ultimate (why) and proximate (how) perspective.

Somatic Cell Differentiation in *Volvox carteri*

Volvox carteri is a spherical multicellular green alga consisting of two different cell types: about 2,000 biflagellated somatic cells with no cell division or redifferentiation potential, and up to 16 nonflagellated reproductive cells called gonidia (Kirk, 1998). During the asexual life cycle (see figure 3.3), each mature gonidium develops into an embryo (by a rapid succession of cell divisions known as cleavage) that forms a new individual (aka spheroid). In *V. carteri* asexual spheroids, cell fate is established in three steps: (1) formation of large and small cells by asymmetrical divisions during embryonic development; (2) cytodifferentiation of small cells and large cells into somatic cells and gonidia, respectively; and (3) terminal differentiation of somatic cells.

Specifically, during the cleavage phase of embryogenesis, each gonidium first undergoes 5 symmetrical divisions resulting in 32 cells (blastomeres) identical in size and dif-

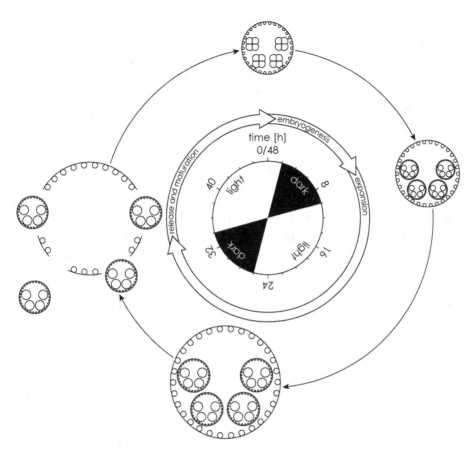

Figure 3.3
The asexual life cycle of *Volvox carteri* (adapted from Hallmann et al., 1998). The life cycle of *V. carteri* takes 48 h when synchronized using a 16 h light:8 h dark regime. The first cleavage division of embryogenesis is defined as time point 0 h and occurs ca. 2 h before the next dark period. Embryogenesis takes about 8 h, and after transition back into the light, both the parent and the newly formed juveniles expand in size as somatic cells synthesize the extracellular matrix. After a full day of growth, juveniles hatch from their parents, and the parental somatic cells undergo senescence and die. The gonidia of the released juveniles mature and enter the next round of embryogenesis toward the end of the light period, concluding the 48-h life cycle.

ferentiation potential. Then, the anterior 16 cells start dividing asymmetrically, each producing a large and a small cell; the large cells continue to divide asymmetrically twice more to produce 16 large cells (presumptive gonidia) ca. 10.5 μm in diameter (Kirk et al., 1993). All other cells (i.e., the small cells formed by asymmetrical divisions and all cells of the posterior hemisphere) divide symmetrically up to a total of 11–12 times, to produce cells ca. 3.3 μm in diameter. At the end of cleavage, the embryo turns inside out (a process called inversion) and the newly formed spheroid is called a juvenile.

In the presence of light, cells differentiate (a process known as cytodifferentiation) into gonidia and somatic cells (Kirk and Kirk, 1985); large cells differentiate as gonidia by resorbing their flagella, becoming spherical and forming large vacuoles, whereas small cells become somatic by flagellar elongation and formation of the eyespot (important for light perception; Coggin and Kochert, 1986; Kirk, 2001). Gonidia start out about 32-fold larger in volume than somatic cells and increase about 140-fold during the life cycle; in contrast, somatic cells only increase ca. 9-fold (Kirk, 1998).

Cell Size Determines Cell Fate in *V. carteri*

As in most animals, the germline in *V. carteri* is set apart early during embryogenesis by a series of asymmetrical divisions. However, in contrast to animals, the different fates undertaken by the two cell types at the end of cleavage do not involve the unequal distribution of a cytoplasmic factor. A series of studies (discussed below) indicated that cell size alone (and not cytoplasmic composition) determines cell fate in *V. carteri*; that is, cells that at the end of cleavage are below 8 μm in diameter (regardless of whether cells originated from symmetrical or asymmetrical cell divisions) develop into somatic cells, whereas cells above 8 μm differentiate as gonidia (Kirk et al., 1993). For instance, when a heat shock was applied during embryogenesis, cells continued to divide for 1–2 times and then stopped dividing (Kirk et al., 1993). When the heat shock was applied early, spheroids with fewer but larger cells were formed, whereas spheroids with more but smaller cells were produced when the shock was applied later. In spheroids with fewer but larger cells, more cells differentiated as gonidia. Similarly, a temperature-sensitive mutant with a cleavage defect divided fewer times when shifted to the restrictive temperature generating cells of various sizes; the percentage of cells above 8 μm directly correlated with the percentage of cells differentiating as gonidia (Kirk et al., 1993). Also, two mutant strains randomly stopped cleavage after an average of 5–6 divisions, forming a wide range of cell sizes and numbers for each spheroid; tracking the cell fate of cells of various sizes showed that cells below 6 μm cell diameter always differentiated as somatic cells and above 9 μm always as gonidia (Kirk et al., 1993).

Microsurgery experiments corroborate the studies mentioned above. The posterior hemisphere never develops gonidia in an undisturbed, whole spheroid. When an embryo was separated into the posterior and anterior hemispheres at the 16-cell stage, 1–2 large cells were formed both in the anterior and posterior half by early cessation of cleavage or by a number of asymmetrical divisions (Kirk et al., 1993). This cannot be explained by assuming an asymmetrical distribution of a cytoplasmic factor, because the posterior hemisphere should not contain this factor since gonidia are usually not formed there. Furthermore, when cells larger than 8 μm were generated by microsurgery (i.e., cell fusions) in the posterior hemisphere, they always differentiated as gonidia (Kirk et al.,

1993). Overall, regardless of how cells with a diameter above 8 μm were formed or of their position within the embryos, they always differentiated as gonidia, whereas cells below 8 μm always differentiated as somatic cells. Therefore, although the germline is segregated early during embryogenesis (as in animal systems), cell size—and not cytoplasmic composition—is responsible for determining cell fate in *V. carteri*.

Somatic Cells Escape Terminal Differentiation in *V. carteri* Regenerator Mutants

The terminal differentiation of nonreplicative/sterile somatic cells is disrupted in the so-called regenerator mutants in which somatic cells regain their reproductive potential (Starr, 1970; Kirk et al., 1999). Small cells initially differentiate as somatic cells, but after one day they redifferentiate and resemble gonidia: they lose their flagella and eyespot, start growing in size, and become vacuolated; by the 3rd day they start dividing to form new spheroids.

Crossing experiments among more than a hundred regenerator (reg) mutants always mapped the mutation to the same locus, and it was concluded that they all are affected by the same gene (Huskey and Griffin, 1979); the gene was later identified and named *regA* (Kirk et al., 1999). More recently, three *regA*-like genes have been found in close proximity to *regA* (forming the so-called *regA* cluster), which allows for the possibility that the early crossing experiments might not have resolved mutations among these clustered genes (Duncan et al., 2006, 2007). However, all genetically characterized regenerator mutants (1 bleomycin-induced deletion mutant and 28 transposon-mediated insertion mutants) mapped specifically to *regA* (Kirk et al., 1999; Duncan et al., 2006), suggesting that *regA* is likely the only gene associated with this phenotype.

A number of regenerator mutants are available in culture collections. In addition to mutants exhibiting single mutations mapping to *regA*, a series of mutant strains that carry additional mutations in other cell differentiation–related genes have been isolated and described, including the so-called *regA⁻*/gonidialess mutants in which all cells start small and resemble somatic cells for one day but start growing and redifferentiate into gonidia later (Tam and Kirk, 1991b; Miller and Kirk 1999; see figure 3.4).

RegA Is a Putative Transcription Repressor

regA is thought to encode a transcriptional repressor (Kirk et al., 1999). Several lines of evidence are consistent with this functional assignment. The RegA protein contains three glutamine-, two alanine-, and two proline-rich regions, often in the form of homopolymeric runs (Kirk et al., 1999). Such regions are found in known or predicted transcriptional repressors that directly prevent transcription initiation as opposed to modulating the activity of a transcriptional activator or chromatin structure (Hanna-Rose and Hansen, 1996;

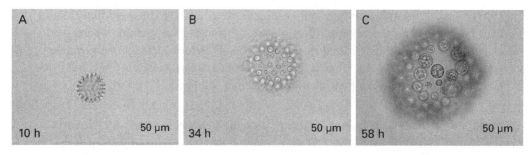

Figure 3.4
The dmAMN *regA⁻*/gonidialess mutant during the first (10 h), second (34 h), and third (58 h) day of its life
cycle. (A, 10 h) A juvenile spheroid composed of small flagellated cells. (B, 34 h) After one day of growth most
of the cells have lost their flagella. (C, 58 h) After two days of growth the cells resemble gonidia and are clearly
vacuolated.

Kirk et al., 1999). When these regions are fused to DNA binding domains, they act as a
repressor domain, that is, they downregulate transcription of a gene next to the respective
DNA binding site (Cowell, 1994; Hanna-Rose and Hansen, 1996). The proposed DNA
binding domain of RegA is a SAND domain—found in transcription factors such as
ULTRAPETALA in *Arabidopsis thaliana* and SP100 in humans (Bottomley et al., 2001;
Carles and Fletcher, 2010)—and is contained in the ca. 100 AA long VARL (Volvocine
Algal *RegA* Like) domain shared by the members of the so-called *VARL* gene family
specific to volvocine algae (Duncan et al., 2006, 2007). The only fully characterized and
published *regA* mutation in a regenerator strain results in the loss of the SAND domain.
Specifically, in the mutant strain HB11A, the *regA* locus has a 284 bp deletion upstream
of the coding region corresponding to the SAND domain; this deletion causes a frameshift
that introduces a premature stop codon 12 bp downstream of the deletion site, and this
results in a short 291 AA gene product (relative to the 1049 AA wild-type product) lacking
a SAND domain (Kirk et al., 1999).

regA is specifically expressed in somatic cells, and its product localizes to their
nucleus; no *regA* transcripts have been identified in gonidia (Kirk et al., 1999). Likewise,
gonidia express at least 18 genes (gonidial genes) that are not expressed in somatic cells
(Tam and Kirk, 1991a). In contrast, in a *regA⁻*/gonidialess mutant gonidial genes are
also expressed in the small somatic-like cells, and their transcript levels exhibit a pattern
that is similar to that of gonidial genes in wild-type gonidia, suggesting that in wild-
type somatic cells these genes are under the control of the RegA protein (Tam and Kirk,
1991b).

Interestingly, all of the 18 gonidial genes code for nuclear-encoded chloroplast proteins—15 of which are directly involved in photosynthesis (Choi, Przybylska, and Straus,
1996; Kirk et al., 1999; Meissner et al., 1999). Taken together, RegA is thought to establish
a nonreproductive state in the small somatic precursor cells by repressing the expression

of nuclear-encoded chloroplast proteins. The repression of such genes prevents the further development of the small embryonic chloroplast in the somatic cell precursors, which—since *V. carteri* is an obligate photoautotroph—prevents cell growth, and consequently cell division (Kirk et al., 1999). However, the exact mechanism that converts information about cell size into cell fate and differential *regA* expression is unknown; specifically, the signal transduction pathway that induces *regA* expression in cells that fall under the 8-µm threshold has not been identified.

Overall, *V. carteri* provides a great model system with which to investigate the mechanistic basis for somatic cell differentiation because of its simple development and the fact that *regA* alone is necessary to establish the somatic cell fate in this species. Furthermore, the evolutionary origin of the genetic basis for somatic cell differentiation can be addressed as a homolog of *regA* was identified in the unicellular relative *C. reinhardtii*, and its expression has been investigated (Nedelcu and Michod, 2006; Nedelcu, 2009b).

The Evolutionary History of *regA*—The Gene Responsible for Somatic Cell Differentiation in *V. carteri*

regA Belongs to a Diverse Volvocine-Specific Gene Family That Predates the Evolution of Multicellularity

regA belongs to a large and diverse gene family known as the *VARL* family with members in both unicellular and multicellular species (Duncan et al., 2006, 2007; Nedelcu and Michod, 2006; Hanschen, Ferris, and Michod, 2014). As mentioned above, *VARL* genes are characterized by the presence of the VARL domain, a ~100 AA region including a DNA binding motif called the SAND domain (Bottomley et al., 2001; Duncan et al., 2007; Carles and Fletcher, 2010). With the exception of the SAND domain, *VARL* genes are quite diverged in gene structure (exon–intron organization) and predicted amino acid sequences (Duncan et al., 2006, 2007; Hanschen, Ferris, and Michod, 2014), suggesting that the SAND domain is the only region under strong negative/purifying selection and functional constraint.

The finding of *VARL* genes in both *V. carteri* (14 genes) and *C. reinhardtii* (12 genes) indicates that *VARL* genes must have already been present in the unicellular ancestor of *C. reinhardtii* and *V. carteri* (Nedelcu and Michod, 2006; Duncan et al., 2007). Interestingly, although a *regA* ortholog has not been found in *C. reinhardtii*, orthologs of *regA* have been identified in several distant *Volvox* species that evolved somatic cell differentiation independently, suggesting that *regA* originated in multicellular volvocine algae before the evolution of somatic cell differentiation (Hanschen, Ferris, and Michod, 2014). Under this scenario, an ancestral *regA*-like gene was later co-opted for a new function in the lineage leading to *V. carteri* (note that whether *regA* is involved in somatic cell differentiation in the other *Volvox* species that have a *regA* ortholog is not known).

The Closest *regA* Homolog in *C. reinhardtii*, *RLS1*, Is Expressed in Response to Stress

The closest homolog of *regA* in *C. reinhardtii* is known as *RLS1*, and its expression appears to be associated with changes in the environment. Specifically, *RLS1* is induced under nutrient limitation (including phosphorus and sulfur deprivation), under light deprivation, and during stationary phase (Nedelcu and Michod, 2006; Nedelcu, 2009b). Furthermore, the induction of this gene coincides with the decline in the reproduction potential of the population grown under limiting conditions (Nedelcu, 2009b). In unicellular algae such as *C. reinhardtii*, this temporary cessation of reproduction in response to nutrient deprivation is part of a general response known as photoacclimation, which is thought to increase survival by avoiding photooxidative damage.

Indeed, compared to heterotrophic organisms, photosynthetic organisms are faced with additional challenges that are related to the presence of a second electron transport chain—the photosynthetic electron transport chain (pETC)—whose imbalance can result in oxida-

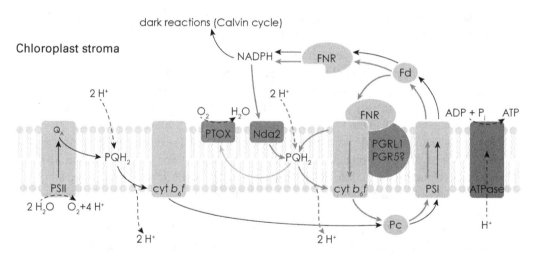

Figure 3.5
Linear electron flow (LEF, black solid arrows) and cyclic electron flow (CEF, gray solid arrows) in the photosynthetic electron transport chain; adapted from Joliot and Johnson (2011) and Peltier et al. (2010). Proton translocation associated with LEF and CEF is indicated by dashed black and gray arrows, respectively. PS, photosystem; cyt b_6f: cytochrome b_6f complex; Pc, plastocyanin; Fd, ferredoxin; FNR, ferredoxin NADP$^+$ reductase; PQH$_2$, plastoquinol; PTOX, plastid terminal oxidase; Nda2, type II NAD(P)H dehydrogenase.

tive stress. Light causes the flow of electrons through the pETC (see figure 3.5), and its energy is ultimately converted into ATP and NADPH (i.e., light reactions); ATP and NADPH are then used to reduce inorganic carbon (i.e., carbon-fixation reactions). Under normal conditions, the electrons flow through both photosystem (PS) II and I and are eventually used to reduce $NADP^+$ to NADPH; this is known as the linear electron flow (LEF). However, when imbalances (such as during nutrient stress—discussed below) lead to excess excitation energy in the pETC and the potential production of reactive oxygen species (ROS), electrons are redirected via the cyclic electron flow (CEF). CEF describes the electron flow around PSI from stromal electron carriers to plastoquinone (PQ) forming plastoquinol (PQH_2). PQH_2 is then oxidized by the cytochrome b_6f complex (cyt b_6f), building a proton gradient across the thylakoid membrane that drives the chloroplastic ATPase; this results in ATP production but no net production of NADPH (see figure 3.5; Alric, 2010).

In *C. reinhardtii*, LEF does not provide sufficient ATP for the carbon-fixation reactions and ATP is supplemented by CEF, which might constitute up to 50% of the total electron flow (Forti et al., 2003). In this alga, the components for CEF are recruited into a super-complex consisting of PSI and its own light-harvesting complex (LHCI), the PSII light-harvesting complex (LHCII), cyt b_6f, ferredoxin-$NADP^+$ reductase, and the integral membrane protein PGR5-Like 1 (PGRL1; Iwai et al., 2010). The formation of the super-complex depends on the reducing power of the chloroplast (i.e., the chloroplast redox poise). That is, when availability of ATP limits the carbon-fixation reactions, NADPH is not depleted, which leads to the overreduction of the stroma; in this way the amount of CEF is modulated to metabolic demands (Takahashi et al., 2013).

This energy balance, called photostasis, can be disrupted by environmental changes; consequently, photosynthetic organisms evolved mechanism (i.e., short-term and long-term acclimation responses) to maintain an intracellular redox balance and avoid oxidative damage (Scheibe et al., 2005). For instance, during sulfur or phosphorus deprivation ATP and NADPH consumption is reduced, leading to excess excitation energy in the pETC and the formation of ROS (Wykoff et al., 1998); these changes to the redox poise of the chloroplast that cause short- and long-term acclimation responses reducing photosynthetic efficiency likely include a switch to CEF and culminate in cessation of cell growth (Wykoff et al., 1998; Pfannschmidt et al., 2009). Because the conditions that induce *RLS1* in *C. reinhardtii* likely affect photostasis, we suggested that *RLS1* is induced by changes in redox status to promote survival (through avoiding the production of damaging reactive oxygen species) at a cost to immediate growth and reproduction (Nedelcu and Michod, 2006; Nedelcu, 2009b).

Gene Co-option and the Evolution of *regA*

Furthermore, since *RLS1* in *C. reinhardtii* is expressed in response to environmental stress, we proposed a hypothesis for the evolution of somatic cells in *V. carteri* involving the co-option of an ancestral environmentally induced *RLS1*-like gene, by switching its regulation from a temporal (environmental) into a spatial (developmental) context (Nedelcu and Michod, 2006; Nedelcu, 2009b). Three potential scenarios (with distinct predictions) can be envisioned for such a change in regulation (see figure 3.6): (A) no new regulatory elements evolved; rather, the same ancestral environmentally induced signaling pathway was also induced during development in early multicellular volvocine algae with somatic cells—that is, the developmental signal simulated the environmental signal (Nedelcu, 2009b) (prediction: *regA* in *V. carteri* can be induced both environmentally and developmentally via the same signaling pathway); (B) an additional layer of regulation evolved as part of a new (developmentally induced) signaling pathway, and both mechanisms have been maintained in *V. carteri* (prediction: *regA* can be induced both environmentally and developmentally, but the signaling pathways are different); or (C) new regulatory elements evolved, and the ancestral regulation was replaced or lost in *V. carteri* (prediction: *regA* can be only be induced developmentally). Notably, the former two scenarios predict that *regA* can still be induced in an environmental context. However, scenario A also predicts that the unknown intracellular signal triggered by environmental stimuli (e.g., reactive oxygen species, Ca^{2+}, ATP/ADP, NADPH/$NADP^+$, ATP/NADPH) is also induced in small cells at the end of embryogenesis. If the ancestral signaling pathway that is now associated with *RLS1* induction in response to various environmental stresses in *C. reinhardtii* is still present in *V. carteri*, conditions that will induce the same signal in *V. carteri* should result in *regA* expression. Thus, learning about the factors that can induce *regA* expression outside its developmental context could provide clues as to the signal that triggers *regA* induction at the end of embryogenesis.

regA Is Induced by a Specific Combination of Environmental Cues, Cell Size, and RegA Status

Recently, we showed that *regA* can, in fact, be induced outside its developmental context (König, 2015; König and Nedelcu, in preparation). Specifically, *regA* was induced in a *regA⁻*/gonidialess mutant (dmAMN) exposed to light following extended darkness. As mentioned earlier, gonidialess mutants do not undergo asymmetrical cell divisions, and at the end of embryogenesis all cells are of the same small size; however, if they also have nonfunctional RegA proteins (such as in the *regA⁻*/gonidialess mutants), the initially small cells are able to grow and later differentiate into gonidia (see figure 3.4). Interestingly, in the dmAMN mutant, *regA* is still induced at the end of embryogenesis (though its product is a truncated nonfunctional RegA protein) and its expression is dependent on both cell size and length of the darkness period, being highest in small dmAMN cells exposed to 4 h of light following 31 h of darkness. On the other hand, *regA* was only slightly induced in the somatic cells of a wild-type *V. carteri*, and its expression levels were independent of duration of darkness, light exposure, or age. Since at the end of

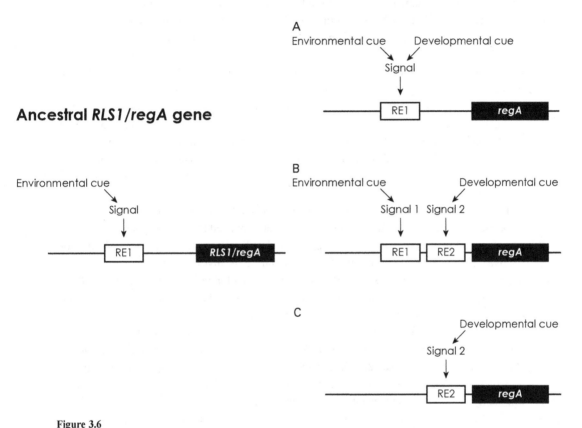

Potential models for *regA* regulation in *Volvox carteri*

Ancestral *RLS1/regA* gene

Figure 3.6
Different models for the evolution of *regA* regulation. Three different models (A, B, C) for the evolution of *regA* via changes in the regulation of an ancestral environmentally induced *RLS1/regA*-like gene (left panel). (A) No new regulatory elements (RE) evolved; rather, the same ancestral environmentally induced signaling pathway was also induced during development in early multicellular volvocine algae with somatic cells (*regA* is induced both environmentally and developmentally by the same signal). (B) In addition to the environmentally induced regulation, a new layer of regulation evolved as part of a new (developmentally induced) signaling pathway (both environmental and developmental cues induce *regA* expression acting via distinct signaling pathways and REs). (C) New REs evolved and the ancestral regulation was replaced or lost (*regA* is only induced by a developmental cue via a signaling pathway and REs unrelated to the ancestral environmental regulation).

embryogenesis wild-type somatic cells and RegA mutant cells are equivalent in size, it is likely that the only difference between the two cell types is related to the status of their RegA protein at the time of exposure to light; namely, a functional protein is likely present in wild-type somatic cells but not in the dmAMN mutant cells. Taken together, these data suggest that this specific combination of factors—exposure to 4 h of light following 31 h of dark; small cell size; and the absence of a functional RegA protein—creates a particular change in the intracellular state of dmAMN mutant cells that induces the expression of *regA*.

Overall, our recent data show that *regA* can be induced in response to environmental changes, suggesting that *regA* has maintained its postulated ancestral environmental regulation and has been later co-opted into a developmental pathway. However, the data are consistent with both A and B scenarios in figure 3.6—that is, the developmental and environmental regulation could involve either the same or different signaling pathways.

The Jun N-terminal kinase (JNK) pathway involved in both development and stress response in *Drosophila melanogaster* has also been proposed to have originated as a stress response pathway that was later co-opted for development (Ríos-Barrera and Riesgo-Escovar 2013). Several important developmental steps involve the JNK pathway and depend on the activation of basket (Bsk) regulating cell elongation: dorsal closure, follicle cell morphogenesis and thorax closure during embryogenesis as well as male genitalia disc rotation/closure during metamorphosis (Ríos-Barrera and Riesgo-Escovar 2013). But Bsk is not only activated in a developmental context but also as stress response, such as UV exposure, irradiation, wound healing, oxidative stress, and immune challenges (Ríos-Barrera and Riesgo-Escovar 2013). One hypothesis regarding the JNK pathway is that it is perpetually "on" but silenced by negative regulators of Bsk; this enables the pathway to quickly respond to unpredictable stimuli like stresses.

Conditions that Trigger *regA* Expression Also Induce Programmed Cell Death

Interestingly, we also found that the conditions that triggered *regA* induction affected cell viability in a manner similar to the *regA* induction pattern, and cell death correlated with the level of *regA* induction (König, 2015; König and Nedelcu, in preparation). Specifically, exposure to light following extended darkness caused extensive cell death in small RegA-deficient dmAMN cells, but less, very little, or no cell death was observed in large dmAMN cells, wild-type somatic cells (small but possessing a functional RegA protein), and gonidia (large and not expressing *regA*), respectively (König, 2015; König and Nedelcu, in preparation). However, in contrast to *regA* induction, the death of small dmAMN cells was influenced by the intensity of light, with lower levels having less negative effect on cell viability. Furthermore, the fragmentation of the DNA in small RegA-

Table 3.1
Summary of conditions that induce *regA* expression and/or programmed cell death (PCD) as a function of cell size (small or large) and the presence of a functional RegA protein

	Wild type		*regA⁻*/gonidialess	
(A) Developmental cue				
Cell size	small	large	small	
Functional RegA protein	no	no	no	
regA *expression*	yes	no	yes	
(B) Environmental cue				
Cell size	small	large	small	large
Functional RegA protein	**yes?**	no	**no**	no
regA *expression*	**no**	no	**yes**	no
(C) Environmental cue				
Cell size	small	large	small	large
Functional RegA protein	**yes?**	no	**no**	no
PCD	**no**	no	**yes**	no
(D) Heat shock				
Cell size	small	large	small	large
Functional RegA protein	yes	no	ND	ND
PCD	no	yes	ND	ND

Note: regA expression in response to developmental (A) and environmental (B) cues; the same environmental cues in B also trigger PCD (C); PCD in response to heat-shock (D) (Nedelcu, 2006; König, 2015). Differences between wild type and *regA⁻*/gonidialess strains are in **bold**.
ND, not determined

deficient cells indicated that cell death was executed via the activation of a programmed cell death (PCD) pathway known to be present in *V. carteri* (Nedelcu, 2006).

Table 3.1 provides a summary of what is currently known about *regA* induction (both developmental and environmental) and PCD in both the wild-type strain and the dmAMN gonidialess mutant. The fact that the viability of both wild-type gonidia and the large dmAMN cells (both not expressing *regA*) is not, or is just minimally, affected by exposure to light suggests that the light-induced signal is dependent on cell size. On the other hand, the difference in response between the small dmAMN cells and the wild-type somatic cells (of similar sizes but differing in the status of their RegA protein) is likely attributed (directly or indirectly) to the absence/presence of a functional RegA protein (assuming that the RegA protein is actually active in the wild-type somatic cell during extended darkness; see discussion below).

Taken together, these observations suggest that a functional RegA protein can confer protection to small cells exposed to light after long periods of dark. To further argue for a role for RegA in stress protection is the previous observation that in *V. carteri* exposed to heat stress, gonidia (not expressing *regA*) undergo PCD, while somatic cells remain alive and continue to provide the colony with motility (Nedelcu, 2006; table 3.1).

A Role for RegA Protein in Mitigating the Effects of Stress

The findings summarized in table 3.1 suggest that the RegA protein might confer resistance to various types of stress, whether mediated by cell size or not. Two possibilities can be envisioned to explain RegA's postulated stress protection: RegA-containing cells either (1) are less sensitive to stress or (2) have a better ability to respond to the effects of stress. Below we discuss these two scenarios and the potential mechanistic basis for the observed differences in response between the wild-type and the RegA-deficient cells (see figure 3.7).

In the first scenario, the two strains might differ in their chloroplast protein compositions (due to the presence/absence of RegA), and these differences reflect in their different sensitivities to stress. RegA is thought to act as a transcription factor that directly or indirectly suppresses the transcription of nuclear genes coding for chloroplast proteins, often referred to as "gonidial genes" (Tam and Kirk, 1991a; Choi, Przybylska, and Straus, 1996; Meissner et al., 1999). At the end of embryogenesis, small cells in gonidialess mutants and wild-type somatic cells share the same morphology and cell size, and both express early somatic genes (Tam and Kirk, 1991b). However, starting on the first day and likely before they encounter light, gonidialess strains express at least eight gonidial genes at much higher levels than wild-type somatic cells (Tam and Kirk, 1991b). Thus, it is possible that the chloroplast composition is different between wild-type somatic cells and small dmAMN cells at the time when they are exposed to light. These differences can affect differently the overall metabolic/redox state of the two cell lines maintained in dark for extended periods and then exposed to light. Specifically, light might induce a metabolic imbalance in the RegA-deficient dmAMN cells, and this imbalance could trigger a signal that induces *regA* expression in an attempt to adjust the chloroplast composition to light. However, in the absence of a functional RegA protein, these cells would be unable to adjust their chloroplast composition and mitigate the redox imbalance, which would result in the accumulation of ROS and ultimately PCD (see figure 3.7A). Importantly, for this scenario to be true, the RegA protein in wild-type *V. carteri* must already be active (and repress the expression of nuclear genes coding for chloroplast proteins) during extended darkness.

Alternatively, if there are no differences in chloroplast composition at the end of the dark period (that is, if RegA in the wild type did not repress gonidial genes in the dark), light will affect both cell lines in a similar manner and induce the same signal; however, cells with a functional RegA protein might respond differently to that signal—that is, in the wild type, RegA will be activated and will repress the expression of gonidial genes, which will in turn mitigate the impending redox imbalance (see figure 3.7B). In this scenario, the RegA protein might already be synthesized during extended darkness but become active only after cells are exposed to light.

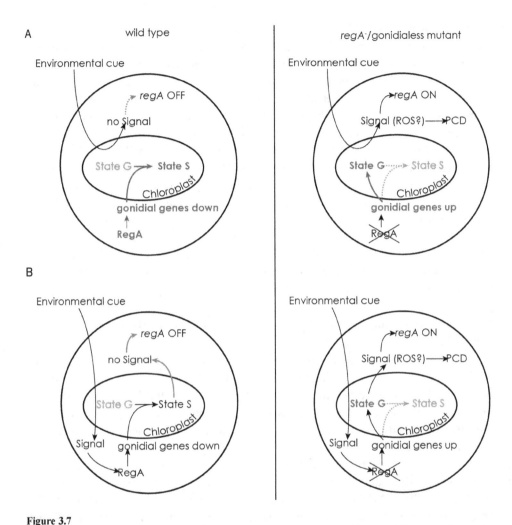

Figure 3.7
Two different models for the role of RegA in stress resistance, with RegA either in a (A) preventative or (B)
responsive role. (A) During extended darkness in wild-type somatic cells, the RegA protein represses gonidial
genes, which encode chloroplast proteins. As a result, the chloroplast composition changes from the inherited
state (from the parental gonidium, State G) to a state specific to somatic cells (State S). In contrast, dmAMN
mutant does not contain a functional RegA protein, and the chloroplast is maintained in State G. Consequently,
the environmental cue does elicit a signal only in the dmAMN chloroplast, which results in *regA* induction and
ultimately programmed cell death (PCD); in the wild type, the preventive action of RegA averts the induction
of such signal. (B) Both wild-type and dmAMN chloroplasts are in State G. However, in wild type, the
environmental cue activates RegA (by posttranscriptional or posttranslational regulation) and initiates the change
of the chloroplast to State S. In contrast, in dmAMN since RegA is not functional, the chloroplast remains in
State G and a secondary signal is triggered initiating additional *regA* induction and ultimately PCD. State G,
chloroplast in gonidial state; State S, chloroplast in somatic state. (See the main text for further discussion.)

In both scenarios, the presence of a functional RegA protein appears to confer stress resistance to wild-type somatic cells, but the role of RegA would be different. In the first scenario, the role of RegA protein is indirect (and "preventative"); its early expression at the end of embryogenesis would affect chloroplast composition in a way that renders cells able to withstand the environmental change. In the second scenario, the RegA protein has a direct (and "responsive") role by activating an acclimation response in response to the environmental change. Furthermore, in the first scenario, the signal that is triggered by the environmental change would be different between dmAMN and wild type—or missing altogether in the wild-type somatic cells. On the other hand, in the second scenario the same signal will be triggered in both cell types. In wild-type cells, this initial signal might trigger a RegA-mediated acclimation response that would alter the chloroplast composition, while in cells lacking the RegA protein, chloroplast composition cannot be adjusted, which might elicit a secondary signal that will induce *regA* expression. However, in the absence of a functional RegA, this signal will ultimately, directly or indirectly, induce PCD. In both scenarios, the signal that induces both *regA* expression and PCD is related to the chloroplast status.

The Environmental Induction of *regA* Likely Involves a Chloroplast-Related Redox Signal

The observation that the combination of small cell size and light exposure following extended darkness triggers both *regA* induction and PCD in the dmAMN mutant (similarly, enlarged cells or presence of a functional RegA protein in small cells fail to induce both *regA* and PCD) allows for the possibility that a common signaling pathway is involved (directly or indirectly) in triggering *regA* expression and PCD. Reactive oxygen species might be involved in this pathway as they are known to act as transduction signals at low levels (e.g., as retrograde signals from the chloroplast regulating nuclear genes) but can also trigger oxidative damage and PCD or even necrosis at higher levels (Reape, Molony, and McCabe, 2008; Mullineaux and Baker, 2010). In support of this possibility is the observation that light intensity did not influence the level of *regA* induction but affected the level of PCD (König, 2015; König and Nedelcu, in preparation). It is likely that up to a specific level and exposure time ROS act as transduction signals, while higher levels of ROS or the accumulation of ROS over time can induce levels of oxidative damage that will correlate with light intensity.

In photosynthetic organisms, ROS can originate in the chloroplast since environmental stresses affect the redox state of the pETC and stromal electron acceptors (Wykoff et al., 1998; Pfannschmidt et al., 2009; Pfannschmidt and Yang, 2012; Suzuki et al., 2012). As discussed earlier, when imbalances lead to excess excitation energy in the pETC and the formation of ROS (such as during phosphate and sulfate deprivation; Wykoff et al., 1998), electrons are redirected via the CEF, which adjusts the ATP:NADPH ratio to metabolic demands. CEF pathways lead to the reduction of the PQ pool (see figure 3.5), one of which involves *PGR5* and *PGRL1* (homologs have been identified in *V. carteri*; Peltier et al., 2010). The *pgr5* mutant of *A. thaliana* showed less efficient CEF than wild type (Nandha et al., 2007; Joliot and Johnson, 2011). A more reduced PQ pool due to CEF might cause more frequent charge recombination events at PSII in the light thus increasing $^1O_2$ evolution involved in retrograde signaling (Krieger-Liszkay, Fufezan, and Trebst, 2008). Alternatively, ROS in form of H_2O_2 may evolve from the nonenzymatic oxidation of PQH_2 (Mubarakshina, Khorobrykh, and Ivanov, 2006). Overall, it is possible that the environmental conditions that resulted in *regA* induction in dmAMN small cells (i.e., long periods of dark followed by light exposure) have affected CEF in the chloroplast.

Interestingly, the duration of darkness influenced the extent of *regA* induction in small RegA-deficient cells (König, 2015). Noteworthy, in barley, prior dark adaptation of 24 h increased CEF in response to a flash of saturating white light, possibly because of a shift in ATP:ADP and/or NADPH:NADP$^+$ ratios (Golding, Finazzi, and Johnson, 2004). In the dark, NADPH (an electron donor for CEF) might be produced by starch metabolism in *C. reinhardtii*, and glycolytic enzymes localized in the chloroplast are considered to cause a more reduced redox poise in comparison to plants (Johnson and Alric, 2012). Thus, NADPH produced in the dark might feed into CEF; also the more negative redox poise might favor the formation of the supercomplex involved in CEF.

Another effect of extended darkness might be an increase in stress. In peach, catalase is upregulated as a response to oxidative stress (Bagnoli et al., 2004), and Giannino et al. (2004) showed that 36 h of darkness also induces the upregulation of catalase. Accordingly, it is possible that 31 h of darkness increased oxidative stress in small RegA mutant cells, and light exposure further increased stress levels, resulting in *regA* induction and ultimately PCD (König, 2015).

The Potential Effects of Cell Size on *regA* Induction

In the dmAMN mutant, in addition to the specific combination of dark and light exposure, the size of cells also influences the level of *regA* induction and extent of PCD. If ROS are actually the signal for *regA* induction and PCD, then ROS seem to be produced more readily in small than in large cells. In the light, imbalances in the pETC might also occur if the ratio between membrane-bound electron transport carriers and soluble cofactors is disrupted. Specifically, since most of the cell is occupied by a cup-shaped chloroplast, small cells could experience an excess of membrane-bound pETC molecules that could lead to excess excitation energy and formation of ROS (Nedelcu, 2009b). Relative to large cells, small cells have about 16-fold higher surface to volume (S:V) ratio and, accordingly, a higher ratio of membrane-bound pETC proteins to electron carriers in the stroma (NADP$^+$, Fd). Thus, small cells are likely more prone to overreduction of the electron acceptors and excess excitation energy in the pETC; the ATP:NADPH ratio might also be affected. We have initially proposed that such a redox change exclusive to small cells could be the signal (that is ultimately translated in the form of ROS) inducing *regA* expression in response to light (Nedelcu, 2009b). Although now we know this is not the case in the somatic cells of the wild type, a similar scenario can account for (or contribute to) *regA* induction in small (but not large) dmAMN cells exposed to light following an extended dark period.

Alternatively, or additionally, the metabolic balance between light and carbon-fixation reactions might be disrupted in small cells since the carbon-fixation reactions are carried out by enzymes in the stroma; in this scenario, NADPH and ATP consumption by the carbon-fixation reactions would be limited, which will lead to excess excitation energy in the pETC including overreduction of the PQ pool (Pfannschmidt and Yang, 2012).

Lastly, the S:V ratio might affect ferredoxin-NADP$^+$ reductase (FNR). FNR can be found either bound to cyt b_6f (in a supercomplex) or soluble in the stroma (see figure 3.5), and Joliot and Johnson (2011) proposed that FNR allocation might regulate electron flow between LEF and CEF. Similar to the *pgr5* mutant of *A. thaliana*, an antisense mutant of tobacco depleted in FNR showed also a reduction of CEF, likely due to less FNR cyt b_6f complexes (Joliot and Johnson, 2011). In the dmAMN mutant, the S:V ratio specific for small or large cells might affect the allocation of FNR (soluble vs. in supercomplex) and thus LEF and CEF distribution and ROS production.

Expanding on the hypothesis that ROS are the signal for *regA* induction in small dmAMN cells exposed to light, the RegA protein would confer stress resistance in the wild-type somatic cells by affecting the chloroplast composition (either before or right after exposure to light). Interestingly, two of the genes thought to be repressed by RegA encode FNR and PGR5 (Tam and Kirk, 1991a; Meissner et al., 1999) whose importance for CEF has been described above, so ultimately RegA might influence the LEF and CEF distribution to abolish ROS production. In small dmAMN cells exposed to light after extended dark, LEF and CEF distribution cannot be regulated and ROS will be produced, and this increase in ROS levels might be the signal triggering the induction of *regA* in small dmAMN cells.

A Model for *regA* Induction

If the developmental *regA* induction depends on a signal similar to that generated environmentally, such signal must be independent of light excitation as the developmental induction of *regA* is independent of light (König, 2015). Could ROS still be the signal for the developmental induction of *regA* (scenario A in figure 3.7)? It is possible that the drastic change in cell size at the end of embryogenesis and the shift in the S:V ratio between large and small cells might affect the distribution of specific components of the LEF and CEF (including the allocation of FNR) or the ratio between other stromal and membrane-bound proteins in a light-independent manner. Such changes could induce redox imbalances and cause a transient production of a ROS signal in somatic cells. Although the model presented below considers this possibility, further investigations are needed to address this suggestion.

Building on our initial hypothesis (Nedelcu, 2009b), we propose the following mechanism for *regA* induction. The drastic changes in cell and chloroplast size at the end of embryogenesis result in imbalances in the ratio of membrane-bound pETC to stromal proteins or a redistribution of LEF and CEF components (including the possible formation of CEF supercomplexes promoted by FNR allocation). These imbalances generate a transient signal that induces *regA* expression as part of an acclimation-like response. In the wild-type strain, the functional RegA counteracts this initial imbalance via downregulation of genes coding for chloroplast proteins, which changes the composition of the chloroplast to adjust to the small cell size. As a result, light exposure following extended darkness does not act as a stressor on wild-type somatic cells. In contrast, the mutant strain is RegA deficient and continues to express gonidial genes, failing to adopt an acclimated-like state in small cells; in this case, light following extended darkness results in an excess formation of ROS that initially induces *regA* expression (in an attempt to trigger an acclimation-like response), but the lack of a functional RegA protein culminates with the accumulation of damaging levels of ROS and the activation of PCD. Under standard growth conditions (16 h of light:8 h of dark), cells in the dmAMN mutant are able to grow in size (as RegA is nonfunctional), which results in the rebalancing of their S:V ratio and the dissipation of the imbalance(s) associated with the initial small cell size.

regA Might Have Evolved to Deal with Drastic Changes in Chloroplast Size due to Multiple Fission

Regardless of the mechanism of induction, we have shown that the postulated ancestral environmental regulation of *regA* is still present in *V. carteri*. Interestingly, *regA* can be induced only in small cells—either developmentally (at the end of embryogenesis in both wild type and dmAMN mutant) or environmentally (in the dmAMN mutant only). Thus, it is likely that the ancestral *regA* might have initially played a role in dealing with the drastic changes in chloroplast size due to the specific type of cell division in this lineage (i.e., multiple fission) and its effect on photosynthesis. Under the selective pressure to increase the number of cells and flagella (which would increase fitness) and the constraint of multiple fission, the only available "solution" would have been a decrease in the size of the cells at the end of embryogenesis. However, such a decrease might have affected the redox status of the chloroplast, especially during the first hours of light. Species that already possessed a good acclimation response might have been able to deal with this initial imbalance due to a decrease in S:V ratio (which would later be counteracted by an increase in cell size as cells start to grow) and would have been selected for. With the evolution of terminally differentiated somatic cells in the larger volvocine species, *regA* could have been co-opted into "locking" the smallest cells into a permanent "acclimated" state to prevent them from further growing and reproducing.

Summary

The evolution of cell differentiation is responsible for the increase in complexity and diversity observed among the extant multicellular lineages. Nevertheless, little is known about the early steps in the transition to multicellular forms with different cell types, especially with respect to the separation of vegetative and reproductive functions between a somatic and a germ line. The volvocine algae represent a great model system with which to study the transition to multicellularity as its members represent different steps toward complex multicellularity, and have been used to demonstrate the benefit of increased group size and germ–soma differentiation (Koufopanou and Bell, 1993; Koufopanou, 1994; Solari et al., 2006b; Solari, Kessler, and Michod, 2006a).

Volvox carteri is a simple multicellular alga that consists of only two cell types (somatic and reproductive). Its low level of morphological complexity and its simple genotype–phenotype map make this alga an ideal system with which to address the mechanistic and evolutionary basis for the evolution of germ–soma separation. Cell fate in this species is determined by cell size at the end of embryogenesis; cells above the threshold size of 8 μm differentiate as germ cells, whereas cells below the threshold will enter a somatic cell fate (Kirk et al., 1993). Furthermore, the terminal differentiation of somatic cells involves the expression of *regA*—a gene that is only expressed in presumptive somatic cells (Kirk et al., 1999). The correlation between cell size and *regA* expression suggests that *regA* induction depends on cell size, but the signaling pathway involved in the cell size-dependent differential regulation of *regA* is not known.

Using a newly isolated *regA* and gonidialess mutant in which all cells start as somatic-like but then differentiate as gonidia, we were recently able to show that *regA* can be induced not only developmentally but also in response to environmental stimuli, and that this induction is also strictly dependent on cell size; that is, the environmental induction of *regA* was observed in small but not in large cells (König and Nedelcu in preparation). In addition, our data suggest a role for *regA* in mitigating the effects of ROS-induced damage.

The finding that *regA* could also be induced by environmental stress argues that *regA*'s proposed ancestral environmental regulation has been conserved and is still present in *V. carteri*. It remains to be seen if the developmental and environmental regulation of *regA* involve the same (or different) signals and regulatory elements. Our data suggest that the environmental signal might involve (directly or indirectly) a change in the cellular/chloroplast redox status. Nevertheless, other signaling pathways could be involved, and further studies are needed to elucidate the exact mechanism that converts environmental signals into *regA* expression. Identifying the environmentally induced signal will allow the direct testing of the models proposed here; that is, if the developmental and environmental regulation are both based on the same signaling pathway, interfering with components of the environmentally induced pathway (e.g., by altering redox status, inhibiting kinases) should also affect the developmental differentiation of somatic cells.

Furthermore, since *RLS1*—the closest homolog of *regA* in the single-celled species *C. reinhardtii*—is also expressed in response to environmental stress, our recent findings provide support to the hypothesis that the evolution of somatic cells involved a change in the regulation of an ancestral life history trade-off gene (which increases survival with a cost to reproduction), from a temporal/environmental context into a spatial/developmental context (Nedelcu and Michod, 2006; Nedelcu, 2009b). Understanding the pathways involved in the developmental and environmental regulation of *regA* will provide insights into the mechanisms responsible for the co-option of an environmentally induced gene into a developmental regulator.

General Remarks

The transition to multicellularity requires a series of specific mechanisms to ensure (1) the physical unity/stability of the multicellular individual, (2) communication and recognition among cells, and (3) regulation of cell growth, proliferation, and differentiation. Components of many of these mechanisms have been co-opted from genes already present in the unicellular ancestors of multicellular lineages. For instance, genes that code for proteins associated with adhesion (e.g., integrins, cadherins), cell signaling, and cell–cell communication (e.g., tyrosine kinases) predate the evolution of Metazoa (e.g., King et al., 2008; Abedin and King, 2010; Sebé-Pedrós et al., 2010; Suga et al., 2013). Similarly, in volvocine algae, genes coding for components of the extracellular matrix have evolved from genes already present in their unicellular ancestors (Prochnik et al., 2010).

Some multicellular development and cell differentiation pathways have also evolved from pathways present in unicellular lineages. In addition to the case of *V. carteri* discussed here, examples include the developmental cyclic AMP signaling in the cellular slime mold *Dictyostelium discoideum* that evolved from a stress response in single-celled amoebae (Schaap, 2011), and genes involved in moss development that have been found in the closest unicellular relatives of land plants (Nedelcu et al., 2006). Similarly, PCD— thought to be an important developmental mechanism in multicellular lineages—is widespread in the unicellular world (e.g., Nedelcu et al., 2011), and many PCD genes have been found in the genomes of single-celled species, including *C. reinhardtii* (e.g., Nedelcu, 2009a).

Gene co-option is known to be an important evolutionary mechanism involved in the evolution of both physiological and morphological traits (True and Carroll, 2002). Genes can be co-opted by changing their pattern of regulation (via mutations in their regulatory elements), by altering the function of the encoded protein (involving structural mutations), or by both. The relative contribution of structural and regulatory mutations to the evolution of new traits is still a controversial matter (Hoekstra and Coyne, 2007), but changes in gene regulation are considered the main contributors to morphological evolution (Carroll, 2000; Ohta, 2003; Carroll, 2008; Wittkopp and Kalay, 2012).

The volvocine system and the *VARL* gene family discussed here provide a great opportunity to address such questions in the context of the evolution of multicellularity and cell differentiation. Understanding how an environmentally induced *RLS1*-like gene was co-opted into a developmental gene will provide new insights into how major morphological innovations can evolve. In addition, understanding how a pathway involved in responses to environmental changes has been co-opted into a developmental program will contribute to the growing interest in reevaluating the role of environment in developmental evolution (Moczek, 2015).

References

Abedin, M., & King, N. (2010). Diverse evolutionary paths to cell adhesion. *Trends in Cell Biology*, 20, 734–742.

Alric, J. (2010). Cyclic electron flow around photosystem I in unicellular green algae. *Photosynthesis Research*, 106, 47–56.

Bagnoli, F., Danti, S., Magherini, V., Cozza, R., Innocenti, A. M., & Racchi, M. L. (2004). Molecular cloning, characterisation and expression of two catalase genes from peach. *Functional Plant Biology*, 31, 349–357.

Bottomley, M. J., Collard, M. W., Huggenvik, J. I., Liu, Z., Gibson, T. J., & Sattler, M. (2001). The SAND domain structure defines a novel DNA-binding fold in transcriptional regulation. *Nature Structural Biology*, 8, 626–633.

Carles, C. C., & Fletcher, J. C. (2010). Missing links between histones and RNA Pol II arising from SAND? *Epigenetics*, 5, 381–385.

Carroll, S. B. (2000). Endless forms: The evolution of gene regulation and morphological diversity. *Cell*, 101, 577–580.

Carroll, S. B. (2008). Evo-devo and an expanding evolutionary synthesis: A genetic theory of morphological evolution. *Cell*, 134, 25–36.

Choi, G., Przybylska, M., & Straus, D. (1996). Three abundant germ line-specific transcripts in *Volvox carteri* encode photosynthetic proteins. *Current Genetics*, 30, 347–355.

Cinalli, R. M., Rangan, P., & Lehmann, R. (2008). Germ cells are forever. *Cell*, 132, 559–562.

Coggin, S., & Kochert, G. D. (1986). Flagellar development and regeneration in *Volvox carteri* (Chlorophyta). *Journal of Phycology*, 22, 370–381.

Cowell, I. G. (1994). Repression versus activation in the control of gene transcription. *Trends in Biochemical Sciences*, 19, 38–42.

Duncan, L., Nishii, I., Harryman, A., Buckley, S., Howard, A., Friedman, N. R., (2007). The *VARL* gene family and the evolutionary origins of the master cell-type regulatory gene, *regA*, in *Volvox carteri. Journal of Molecular Evolution*, 65, 1–11.

Duncan, L., Nishii, I., Howard, A., Kirk, D. L., & Miller, S. M. (2006). Orthologs and paralogs of *regA*, a master cell-type regulatory gene in *Volvox carteri. Current Genetics*, 50, 61–72.

Feng, X., Zilberman, D., & Dickinson, H. (2013). A conversation across generations: Soma–germ cell crosstalk in plants. *Developmental Cell*, 24, 215–225.

Forti, G., Furia, A., Bombelli, P., & Finazzi, G. (2003). In vivo changes of the oxidation-reduction state of NADP and of the ATP/ADP cellular ratio linked to the photosynthetic activity in *Chlamydomonas reinhardtii. Plant Physiology*, 132, 1464–1474.

Giannino, D., Condello, E., Bruno, L., Testone, G., Tartarini, A., Cozza, R., (2004). The gene geranylgeranyl reductase of peach (*Prunus persica* [L.] Batsch) is regulated during leaf development and responds differentially to distinct stress factors. *Journal of Experimental Botany*, 55, 2063–2073.

Golding, A. J., Finazzi, G., & Johnson, G. N. (2004). Reduction of the thylakoid electron transport chain by stromal reductants—Evidence for activation of cyclic electron transport upon dark adaptation or under drought. *Planta*, 220, 356–363.

Grosberg, R. K., & Strathmann, R. R. (2007). The evolution of multicellularity: A minor major transition? *Annual Review of Ecology Evolution and Systematics*, 38, 621–654.

Hallmann, A., Godl, K., Wenzl, S., & Sumper, M. (1998). The highly efficient sex-inducing pheromone system of *Volvox. Trends in Microbiology*, 6, 185–189.

Hanna-Rose, W., & Hansen, U. (1996). Active repression mechanisms of eukaryotic transcription repressors. *Trends in Genetics*, 12, 229–234.

Hanschen, E. R., Ferris, P. J., & Michod, R. E. (2014). Early evolution of the genetic basis for soma in the Volvocaceae. *Evolution*, 68, 2014–2025.

Herron, M. D., Hackett, J. D., Aylward, F. O., & Michod, R. E. (2009). Triassic origin and early radiation of multicellular volvocine algae. *Proceedings of the National Academy of Sciences of the United States of America*, 106, 3254–3258.

Herron, M. D., & Michod, R. E. (2008). Evolution of complexity in the volvocine algae: Transitions in individuality through Darwin's eye. *Evolution*, 62, 436–451.

Herron, M. D., & Nedelcu, A. M. (2015). Volvocine algae: From simple to complex multicellularity. In I. Ruiz-Trillo & A. M. Nedelcu (Eds.), *Evolutionary transitions to multicellular life: Principles and mechanisms* (pp. 129–152). Dordrecht, the Netherlands: Springer.

Hoekstra, H. E., & Coyne, J. A. (2007). The locus of evolution: Evo devo and the genetics of adaptation. *Evolution; International Journal of Organic Evolution*, 61, 995–1016.

Huskey, R. J., & Griffin, B. E. (1979). Genetic control of somatic cell differentiation in *Volvox* analysis of somatic regenerator mutants. *Developmental Biology*, 72, 226–235.

Iwai, M., Takizawa, K., Tokutsu, R., Okamuro, A., Takahashi, Y., & Minagawa, J. (2010). Isolation of the elusive supercomplex that drives cyclic electron flow in photosynthesis. *Nature*, 464, 1210–1213.

Johnson, X., & Alric, J. (2012). Interaction between starch breakdown, acetate assimilation, and photosynthetic cyclic electron flow in *Chlamydomonas reinhardtii. Journal of Biological Chemistry*, 287, 26445–26452.

Joliot, P., & Johnson, G. N. (2011). Regulation of cyclic and linear electron flow in higher plants. *Proceedings of the National Academy of Sciences of the United States of America*, 108, 13317–13322.

King, N., Westbrook, M. J., Young, S. L., Kuo, A., Abedin, M., Chapman, J., (2008). The genome of the choanoflagellate *Monosiga brevicollis* and the origin of metazoans. *Nature*, 451, 783–788.

Kirk, D. L. (1998). *Volvox: The molecular genetic origins of multicellularity and cellular differentiation*. Cambridge, UK: Cambridge University Press.

Kirk, D. L. (2001). Germ–soma differentiation in *Volvox*. *Developmental Biology*, 238, 213–223.

Kirk, D. L. (2003). Seeking the ultimate and proximate causes of *Volvox* multicellularity and cellular differentiation. *Integrative and Comparative Biology*, 43, 247–253.

Kirk, D. L. (2005). A twelve-step program for evolving multicellularity and a division of labor. *BioEssays*, 27, 299–310.

Kirk, M. M., & Kirk, D. L. (1985). Translational regulation of protein synthesis, in response to light, at a critical stage of *Volvox* development. *Cell*, 41, 419–428.

Kirk, M. M., Ransick, A., McRae, S. E., & Kirk, D. L. (1993). The relationship between cell size and cell fate in *Volvox carteri*. *Journal of Cell Biology*, 123, 191–208.

Kirk, M. M., Stark, K., Miller, S. M., Müller, W., Taillon, B. E., Gruber, H., (1999). *regA*, a *Volvox* gene thatplays a central role in germ–soma differentiation, encodes a novel regulatory protein. *Development*, 126, 639–647.

König, S. G. (2015). In A. M. Nedelcu (Ed.), *The genetic and evolutionary basis for somatic cell differentiation in the multicellular alga Volvox carteri: Investigations into the regulation of regA expression (Doctoral disseration*. Fredericton, Canada: University of New Brunswick.

Koufopanou, V. (1994). The evolution of soma in the Volvocales. *American Naturalist*, 143, 907–931.

Koufopanou, V., & Bell, G. (1993). Soma and germ: An experimental approach using *Volvox*. *Proceedings. Biological Sciences*, 254, 107–113.

Krieger-Liszkay, A., Fufezan, C., & Trebst, A. (2008). Singlet oxygen production in photosystem II and related protection mechanism. *Photosynthesis Research*, 98, 551–564.

Meissner, M., Stark, K., Cresnar, B., Kirk, D. L., & Schmitt, R. (1999). *Volvox* germline-specific genes that are putative targets of RegA repression encode chloroplast proteins. *Current Genetics*, 36, 363–370.

Merchant, S. S., Prochnik, S. E., Vallon, O., Harris, E. H., Karpowicz, S. J., Witman, G. B., (2007). The *Chlamydomonas* genome reveals the evolution of key animal and plant functions. *Science*, 318, 245–250.

Michod, R. E. (2007). Evolution of individuality during the transition from unicellular to multicellular life. *Proceedings of the National Academy of Sciences of the United States of America*, 104(Suppl 1), 8613–8618.

Michod, R. E., & Nedelcu, A. M. (2003). On the reorganization of fitness during evolutionary transitions in individuality. *Integrative and Comparative Biology*, 43, 64–73.

Miller, S. M., & Kirk, D. L. (1999). *glsA*, a *Volvox* gene required for asymmetric division and germ cell specification, encodes a chaperone-like protein. *Development*, 126, 649–658.

Moczek, A. P. (2015). Re-evaluating the environment in developmental evolution. *Frontiers in Ecology and Evolution*, 3, 00007.

Mubarakshina, M., Khorobrykh, S., & Ivanov, B. (2006). Oxygen reduction in chloroplast thylakoids results in production of hydrogen peroxide inside the membrane. *Biochimica et Biophysica Acta*, 1757, 1496–1503.

Mullineaux, P. M., & Baker, N. R. (2010). Oxidative stress: Antagonistic signaling for acclimation or cell death? *Plant Physiology*, 154, 521–525.

Nandha, B., Finazzi, G., Joliot, P., Hald, S., & Johnson, G. N. (2007). The role of PGR5 in the redox poising of photosynthetic electron transport. *Biochimica et Biophysica Acta*, 1767, 1252–1259.

Nedelcu, A. M. (2006). Evidence for p53-like-mediated stress responses in green algae. *FEBS Letters*, 580, 3013–3017.

Nedelcu, A. M. (2009 a). Comparative genomics of phylogenetically diverse unicellular eukaryotes provide new insights into the genetic basis for the evolution of the programmed cell death machinery. *Journal of Molecular Evolution*, 68, 256–268.

Nedelcu, A. M. (2009 b). Environmentally induced responses co-opted for reproductive altruism. *Biology Letters*, 5, 805–808.

Nedelcu, A. M., Borza, T., & Lee, R. W. (2006). A land plant-specific multigene family in the unicellular *Mesostigma* argues for its close relationship to Streptophyta. *Molecular Biology and Evolution*, 23, 1011–1105.

Nedelcu, A. M., Driscoll, W. W., Durand, P. M., Herron, M. D., & Rashidi, A. (2011). On the paradigm of altruistic suicide in the unicellular world. *Evolution*, 65, 3–20.

Nedelcu, A. M., & Michod, R. E. (2006). The evolutionary origin of an altruistic gene. *Molecular Biology and Evolution*, 23, 1460–1464.

Nedelcu, A. M., & Michod, R. E. (2011). Molecular mechanisms of life history trade-offs and the evolution of multicellular complexity in volvocalean green algae. In T. Flatt & A. Heyland (Eds.), *Mechanisms of life history evolution* (pp. 271–283). Oxford, UK: Oxford University Press.

Ohta, T. (2003). Evolution by gene duplication revisited: Differentiation of regulatory elements versus proteins. *Genetica*, 118, 209–216.

Pfannschmidt, T., Brautigam, K., Wagner, R., Dietzel, L., Schroter, Y., Steiner, S., (2009). Potential regulation of gene expression in photosynthetic cells by redox and energy state: Approaches towards better understanding. *Annals of Botany*, 103, 599–607.

Pfannschmidt, T., & Yang, C. (2012). The hidden function of photosynthesis: A sensing system for environmental conditions that regulates plant acclimation responses. *Protoplasma*, 249(Suppl 2), S125–S136.

Peltier, G., Tolleter, D., Billon, E., & Cournac, L. (2010). Auxiliary electron transport pathways in chloroplasts of microalgae. *Photosynthesis Research*, 106, 19–31.

Prochnik, S. E., Umen, J. G., Nedelcu, A. M., Hallmann, A., Miller, S. M., Nishii, I., (2010). Genomic analysis of organismal complexity in the multicellular green alga *Volvox carteri*. *Science*, 329, 223–226.

Reape, T. J., Molony, E. M., & McCabe, P. F. (2008). Programmed cell death in plants: Distinguishing between different modes. *Journal of Experimental Botany*, 59, 435–444.

Ríos-Barrera, L. D., & Riesgo-Escovar, J. R. (2013). Regulating cell morphogenesis: The *Drosophila* Jun N-terminal kinase pathway. *Genesis (New York, N.Y.)*, 51, 147–162.

Schaap, P. (2011). Evolution of developmental cyclic AMP signalling in the Dictyostelia from an amoebozoan stress response. *Development, Growth & Differentiation*, 53, 452–462.

Scheibe, R., Backhausen, J. E., Emmerlich, V., & Holtgrefe, S. (2005). Strategies to maintain redox homeostasis during photosynthesis under changing conditions. *Journal of Experimental Botany*, 56, 1481–1489.

Sebé-Pedrós, A., Roger, A. J., Lang, F. B., King, N., & Ruiz-trillo, I. (2010). Ancient origin of the integrin-mediated adhesion and signaling machinery. *Proceedings of the National Academy of Sciences of the United States of America*, 107, 10142–10147.

Solari, C. A., Galzenati, V. J., & Kessler, J. O. (2015). The evolutionary ecology of multicellularity: The volvocine green algae as a case study. In I. Ruiz-Trillo & A. M. Nedelcu (Eds.), *Evolutionary transitions to multicellular life: Principles and mechanisms* (pp. 201–223). Dordrecht, the Netherlands: Springer.

Solari, C. A., Kessler, J. O., & Goldstein, R. E. (2013). A general allometric and life-history model for cellular differentiation in the transition to multicellularity. *American Naturalist*, 181, 369–380.

Solari, C. A., Kessler, J. O., & Michod, R. E. (2006 a). A hydrodynamics approach to the evolution of multicellularity: Flagellar motility and germ–soma differentiation in volvocalean green algae. *American Naturalist*, 167, 537–554.

Solari, C. A., Ganguly, S., Kessler, J. O., Michod, R. E., & Goldstein, R. E. (2006 b). Multicellularity and the functional interdependence of motility and molecular transport. *Proceedings of the National Academy of Sciences of the United States of America*, 103, 1353–1358.

Starr, R. C. (1970). Control of differentiation in *Volvox*. *Symposium of the Society for Developmental Biology*, 29, 59–100.

Suga, H., Chen, Z., de Mendoza, A., Sebé-Pedrós, A., Brown, M. W., Kramer, E., (2013). The *Capsaspora* genome reveals a complex unicellular prehistory of animals. *Nature Communications*, 4, 2325.

Suzuki, N., Koussevitzky, S., Mittler, R., & Miller, G. (2012). ROS and redox signalling in the response of plants to abiotic stress. *Plant, Cell & Environment*, 35, 259–270.

Takahashi, H., Clowez, S., Wollman, F.-A., Vallon, O., & Rappaport, F. (2013). Cyclic electron flow is redox-controlled but independent of state transition. *Nature Communications*, 4, 1954.

Tam, L.-W., & Kirk, D. L. (1991 a). Identification of cell-type-specific genes of *Volvox carteri* and characterization of their expression during the asexual life cycle. *Developmental Biology*, 145, 51–66.

Tam, L.-W., & Kirk, D. L. (1991 b). The program for cellular differentiation in *Volvox carteri* as revealed by molecular analysis of development in a gonidialess/somatic regenerator mutant. *Development*, 112, 571–580.

True, J. R., & Carroll, S. B. (2002). Gene co-option in physiological and morphological evolution. *Annual Review of Cell and Developmental Biology*, 18, 53–80.

Voronina, E., Seydoux, G., Sassone-Corsi, P., & Nagamori, I. (2011). RNA granules in germ cells. *Cold Spring Harbor Perspectives in Biology*, 3(12), a002774.

Wittkopp, P. J., & Kalay, G. (2012). *Cis*-regulatory elements: Molecular mechanisms and evolutionary processes underlying divergence. *Nature Reviews. Genetics*, 13, 59–69.

Wykoff, D. D., Davies, J. P., Melis, A., & Grossman, A. R. (1998). The regulation of photosynthetic electron transport during nutrient deprivation in *Chlamydomonas reinhardtii*. *Plant Physiology*, 117, 129–139.

Yang, W.-C., Shi, D.-Q., & Chen, Y.-H. (2010). Female gametophyte development in flowering plants. *Annual Review of Plant Biology*, 61, 89–108.

4 Physicochemical Factors in the Organization of Multicellular Aggregates and Plants

Juan Antonio Arias del Angel, Eugenio Azpeitia, Mariana Benítez, Ana E. Escalante, Valeria Hernández-Hernández and Emilio Mora Van Cauwelaert

What we are seeking are the first principles of development. As I have pointed out, there are two ways of achieving this: one is by mathematical modeling, and the other is by looking at the beginning of multicellular development.

—John Tyler Bonner, 2001

Summary

Since the formulation of the Modern Synthesis, the causes of phenotypic variation and innovation, as sources of adaptive evolution, have been mainly associated with changes in the DNA sequence. However, diverse avenues of theoretical and empirical research are showing that our view of the possible causes and processes by which phenotypic diversity originates needs to be expanded. Indeed, it has been acknowledged that ecological, social, and developmental factors, as well as generic physicochemical processes, may contribute to phenotypic variation and innovation, just as the genome and changes therein do. One of the major evolutionary innovations regarding phenotypes is the appearance of multicellularity in different lineages. Multicellular organisms are not only the aggregation or incomplete separation of cells but also involve some sort of cell differentiation, metabolic integration, and the appearance of new systemic properties and levels of selection. The generation of these features is now being studied with a diversity of causes in consideration. We focus on the potential role of physicochemical properties on the formation and patterning of multicellular arrangements in lineages that, while evolutionarily very distant, might exhibit suggestive commonalities. In particular, we discuss how mechanical and chemical coupling of cells and the combination of the so-called dynamical patterning modules (DPMs) may lead to the formation of recurrent patterns in plants and aggregates of unicellular entities. We illustrate these ideas in specific model systems and discuss some aspects of plant and microbial development that might help identify principles in the origin, organization, and evolution of multicellular organisms.

Diversity in the Causes of Origin, Variation, and Evolution of Phenotypes

What are the processes and factors underlying the formation of complex multicellular structures? And how do changes in these processes and factors affect the phenotypes at the developmental and evolutionary scales? are among the central questions in developmental biology. While all-embracing answers to these questions are still elusive even in model organisms, it is now clear that a variety of processes and factors are involved in the origin, variation and evolution of phenotypes. Indeed it has been suggested that diverse causal components, besides those postulated by the Modern Synthesis in evolutionary theory, need to be considered in order to fully understand the development and evolution of organisms (see Waddington, 1957; Lewontin, 2000; Jablonka and Lamb, 2005; Pigliucci and Muller, 2010, among many others).

In the context of development, spatiotemporal changes in gene expression are certainly associated with differences in cell type and activity, as well as with other important developmental processes. However, the role of genes in the generation of form and patterns is more complex than once thought. In contrast to a view in which gene products unidirectionally "instruct" cells or tissues to reach a particular cell type or to form a particular structure (see figure 4.1), developmental studies have shown that patterns of differential gene expression both affect and are affected by cellular activity. Thus, rather than being the cause of cell differentiation, changes in gene activity are part of a set of interlinked processes that dynamically generate patterns leading to position-dependent cell differentiation (Newman and Müller, 2000; Jaeger and Reinitz, 2006). The latter scenario is reminiscent of the reaction–diffusion (RD) system proposed by A. Turing in the context of morphogenesis, as it relies on a couple of chemicals that diffuse at the same time as they chemically react,

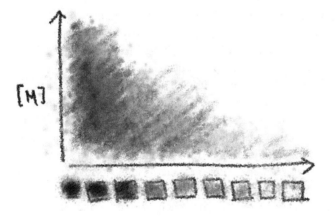

Figure 4.1
French flag model for cell fate determination.

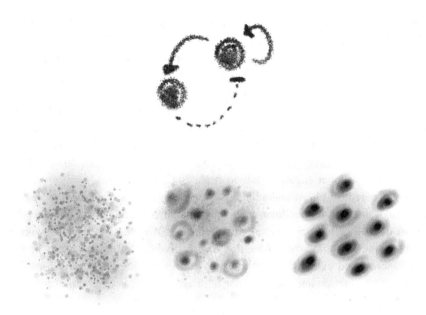

Figure 4.2
Reaction–diffusion model for spontaneous generation of steady patterns in physicochemical systems.

spontaneously giving rise to steady patterns of chemical concentration (Turing, 1952; figure 4.2). The principles of this self-organizing patterning system have been found in a variety of study systems, although often as part of more complex regulatory networks (Pesch and Hülskamp, 2004; Newman and Bhat, 2007; Kondo and Miura, 2010).

In addition to the activity of biochemical regulatory networks, the emergence and change of patterns and structures during development has been shown to depend on other processes and causal agents, such as organism–environment interactions and generic physicochemical aspects. For instance, regarding ecological interactions, plants can vary in fruit size and seed germination rate depending on whether their parent flowers are pollinated manually or by insects (Chautá-Mellizo et al., 2012). Leaf size, shape, and hair density can change dramatically depending on the light, temperature, humidity, plant ages, and other climatic and developmental conditions to which plants are subjected (Tsukaya, 2005; Tsukaya et al., 2000). In another scale of organism–environment interactions, it has been shown that the potato moth is more abundant in plants grown among large crop fields than those grown in complex landscapes (with patches of natural vegetation and cropped land), which affects traits such as plant weight and yield (Poveda et al., 2012). These changes can be understood as an effect of phenotypic plasticity, a long-known phenomenon that has often been considered a nuisance in the study of character development and evolution but that has now been acknowledged as an important cause in the generation and variation, as well as in the evolution, of phenotypes (West-Eberhard, 2003; Robert, 2006; Pigliucci, Murren, and Schlichting, 2006).

Physicochemical processes that are generic to living and some types of nonliving matter have also been recognized as important factors in morphogenesis. Actually, the study of mechanical forces acting during the development of plants and animals has a long-standing tradition (see Thompson, 1917; Green, 1962; Lynch and Lintilhac, 1997; Niklas, 1992, among others) but has recently boomed into an abundance of studies showing that mechanical fields are central in the regulation of cell fate determination, pattern formation, organ shaping, and growth, among other developmental processes (Mammoto et al., 2012; Caballero et al., 2012; Nakayama et al., 2012; Barrio et al., 2013).

As in the development of plants and animals, a variety of factors could have operated in the generation of the first multicellular organisms and may continue to act in the recurrent generation of multicellular aggregates. In any case, it is necessary to count with a theoretical framework that allows consideration of this diversity of causal agents in the study of development and evolution of phenotypes and that enables comparison of the dynamics of pattern formation and multicellular organization among diverse lineages.

Dynamical Patterning Modules as a Framework for Comparative Studies of Multicellular Development

Comparative studies among modern organisms have been central to understanding the evolution of biological forms, for example, through the comparison of gene and protein sequences, function, and expression patterns. These comparisons have provided valuable insights into the conservation and diversification of phenotypes, and even of developmental processes. However, as briefly illustrated above, evidence reveals that a variety of epigenetic factors (i.e., complex regulatory dynamics among molecules, physicochemical processes, ecological interactions, etc.) also contribute to the generation and variation of biological forms. Therefore, to fully understand the generation and evolution of biological forms, it is necessary to extend the current comparative strategies and compare nonlinear relationships among different types of developmental processes.

The DPMs framework proposed by Stuart A. Newman and collaborators (Newman and Bhat, 2008, 2009) allows characterization and comparison of the physicochemical and molecular aspects of key processes in multicellular development. DPMs are conformed by feedbacks between sets of well-conserved genes that mobilize generic physicochemical effects, which in turn can affect gene expression. The physicochemical forces and processes associated with DPMs are common to living and nonliving chemically and mechanically excitable systems and are specific to the mesoscopic scale in which muticellular development takes place. Moreover, the physicochemical processes involved in DPMs are those modified by genes and other molecules acting during development. Then, for example, although gravity certainly affects some aspects of development, it is not regulated by developmental processes and seems to apply in scales larger than collections of only a few cells, so it is not considered in the DPMs (although some nonlinear interactions between organ growth and gravistimulation have been identified in plants, see, e.g., Mason et al., 2002).

It is postulated that such well-conserved molecules were present in unicellular ancestors but acquired a new function when interacting with physical processes correspondent to the multicellular scale (e.g., cohesion, viscoelasticity, diffusion, activator–inhibitor interactions, oscillatory dynamics, etc.). Because this framework goes beyond the comparison of conserved genes and takes into account regulatory interactions and the physical processes of cellular aggregates, it can better capture the dynamical richness of multicellular development.

Newman and collaborators first proposed a set of DPMs that can, in principle, generate the basic features of animal development (formation of cellular masses, segmentation, pattern formation, periodic patterning, appendage formation, etc.). Although the evolutionary history and the origin of plants are not as well described as in animals, a preliminary DPM catalog was later proposed for plants (Hernández-Hernández et al., 2012; see table 4.1). In order to postulate the plant DPMs, some of the key and characteristic features of plant organization were considered: the presence of a relatively rigid cell wall; absence of cellular migration; presence of plasmodesmata; the possibility of an open indeterminate development that is largely influenced by environmental conditions. Despite important differences among plants, animals, fungi, and other multicellular organisms, we suggest that the DPMs concept could underlie comparative studies that look beyond gene homologies, which may not be informative about the mechanisms of morphogenesis that could be shared (or not) at the physical level. In the following section we will briefly describe one of the proposed DPMs for plant development and evolution—*spatially dependent differentiation (DIF)*—and will discuss a case study in which this plant DPM has been modeled. Later we discuss its potential relationship with a similar DPM in animals and emerging cell aggregates.

The Spatially Dependent Differentiation (DIF) Module and a Modeling Approach to Plant Development

The plant cell wall precludes cell migration and tissue rearrangement, which are important mechanisms for cellular communication and induction in other organisms. However, plant cell walls have channels that allow symplastic movement of molecules between cells, such as transcription factors, peptides, small RNAs, hormones, and other morphogens (Kim and Zambryski, 2005). The presence of theses channels (called plasmodesmata) has actually been identified as a characteristic feature of the multicellular plant body plan (Niklas, 2000).

Table 4.1
Plant dynamic patterning modules (DPMs)

DPM	Characteristic molecules	Physical processes	Evo-devo role
FCW: Future cell wall	Cell wall components, possible mechanosensitive elements	Mechanical stress	Defines the orientation and location of the cell wall
ADH: Cell–cell adhesion	Cell wall components, mainly pectin polysaccharides	Adhesion	Formation of multicellular organisms
DIF: Spatially dependent differentiation	Plasmodesmata components	Diffusion, reaction/ diffusion-like mechanisms	Pattern formation and cell type specification
POL: Polarity and the determination of the apical-basal axis	Auxin, auxin polar transporters, and cell wall components	Mechanical stress	Polarity, axis formation and elongation
BUD: Periodic formation of buds	Auxins, auxin polar transporters, cell wall loosening enzymes, and cell wall components	Lateral inhibition and buckling deformation	Periodic formation of buds and lateral roots
LLS: Formation and shaping of leaf-like structures	Cell wall components	Buckling, compression– expansion	Formation and shaping of leaf-like structures

The flux of molecules across plasmodesmata enables the connection of regulatory networks among cells and the de novo formation of steady and complex morphogen patterns (Kim and Zambryski, 2005; Lucas et al., 2009; Lehesranta et al., 2010). In turn, the patterns generated by the communicating cells result in the specification and consequent differentiation of cell types. The dynamics of spatially dependent differentiation through cell–cell communication constitutes a process in which differentiating cells and tissues thus play an active role in setting up the biochemical pattern rather than just "interpreting" the positional information (Wolpert, 1996; figure 4.1).

The physical process mobilized by symplastic channels of the DIF DPM is passive diffusion which generates concentration gradients. Moreover, it has been proposed that in some plant systems plasmodesmata enable RD and even lateral inhibition mechanisms (e.g., Pesch and Hülskamp, 2004; Jönsson et al., 2005). However, as in the case of animal systems, in plant systems there is also more than a simple pair of reacting and diffusing chemicals (as first proposed by Alan Turing, 1952). Plant regulatory processes underlying pattern formation are dynamically richer, involving sets of genes and gene products, and tissue-scale transport processes conferring dynamical properties that are atypical of simple RD systems (Benítez et al., 2011).

The origin of plasmodesmata in plants is not yet well understood. Current evidence shows that plasmodesmata are present in embryophytes and the charophycean algae (collectively called the streptophytes) and that these intercellular connections are homologous

(Graham et al., 2000). Also, plasmodesmata are found in some members of the Chlorophyta and in the Phaeophyceae, which clearly developed cell walls and multicellularity independently, therefore suggesting that plasmodesmata may have evolved independently in these groups (Raven, 1997). It is clear that these structures are well-conserved within the broad plant lineages and that their role in cell-to-cell communication and pattern formation could thus have been central from the very origin of multicellular plants. Nevertheless, it will be important to study the evolution of not only the molecules that are part of the plasmodesmata structure but also those that are involved in the regulation of molecular trafficking through these channels (Lucas et al., 2009).

Following J. T. Bonner (2001), we consider that mathematical models are great tools to reach for simplicity in the study of development, as well as to identify and postulate potential generic principles in the development of multicellular organisms. In particular, he explored qualitative models that allowed understanding of the overall logic of a developmental process and testing of the relative role of different factors involved in it. We deeply agree with Bonner's appreciation of mathematical modeling in developmental and evolutionary biology and briefly introduce qualitative models that might help illustrate their use in this field. In particular, we comment on a mathematical model that has been useful in the exploration of the DIF DPM dynamics in plants, and in the following section we discuss another model for a similar module but in emerging cellular aggregates.

The first model has been developed for a well-studied system in plant developmental biology, that of cell fate determination and patterning in the epidermis of *Arabidopsis* leaves. On the basis of experimental work performed at several laboratories during the last decades, it has been possible to define and model a gene regulatory network (GRN) containing the key genetic factors involved in the determination of hair (trichome) and nonhair (pavement) cells in the leaf epidermis (see recent reviews in Benítez et al., 2013; Ryu et al., 2013).

An important feature of the GRN characterized for this system is that some of the transcriptional factors that conform it codify for small proteins that can move through plasmodesmata to neighboring cells, thus affecting the overall state of the GRNs in other cells (see figure 4.3). This gives rise to a system of coupled GRNs. Interestingly, the fate that each cell in the epidermis will acquire, either trichome or pavement cell, largely depends on the spatial pattern and concentration of the GRN components. This implies that cells in the epidermis communicate with each other via the passive movement of certain proteins in such a way that they create the spatial pattern that will in turn lead to their differentiation. Moreover, it has been shown that the components of this GRN collectively behave as a RD system although, since there are more than two genes involved, the dynamics of this system are richer than a typical RD system (Pesch and Hülskamp, 2004; Benítez et al., 2011).

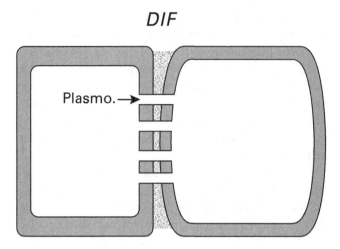

Figure 4.3
Schematic representation of the dynamical patterning module (DPM) for position-dependent cell differentiation (DIF) in plants. Plasmo., plasmodesmata. Figure modified from Hernández-Hernández, et al. (2012).

DPM Modules in the Organization of Multicellular Aggregates: The Case of Spatially Dependent Differentiation

Seeking to explain the formation of the first multicellular organisms and their different patterns of cell differentiation, several authors have pointed out the importance of physical and chemical interactions in this process (e.g., Furusawa and Kaneko, 2002; Newman and Bhat, 2009; Newman et al., 2006). In this section we discuss the potential role of the DIF DPMs in the early organization of multicellular aggregates that can be formed by incomplete division (like plants) or by single cells coming together.

Newman and collaborators (Newman and Bhat, 2009; Newman et al., 2006) explain that, before the origination of the first metazoans, their unicellular ancestors were likely to share a common genetic *toolkit*. A subset of this toolkit was composed of interaction molecules (cadherins, Wnt and their ligands, TGF-beta/BMP, FGF, Hedgehog, and their receptors) which, by mobilizing certain physical processes and effects in the emerging masses of cells, constitute the DPMs. These DPMs, independently or in combination, might then produce different patterns and forms in first multicellular organisms. In addition, Furusawa and Kaneko (2002) have insisted on the role of inter- and intracellular dynamics, which might explain the origin of differentiation and robust developmental patterns in multicellular organisms. In fact, they explain how the coupling of the different internal dynamics and the positional differences of cells in an aggregate are in principle sufficient for cells to reach different steady states or attractors, thought of as cell types (Kauffman, 1969).

Following the ideas of these authors, we have developed a simple model aimed to explore the role of different kinds of communication (i.e., paracrine, juxtacrine, or both) and intensities of cellular adhesion in the formation of patterns underlying cell differentiation in emerging multicellular organisms (Mora Van Cauwelaert et al., 2015). Specifically, we built a scenario with virtual cells with two possible cell states, considered here as different cell types. These states correspond to the steady states of a simple dynamical intracellular network, a two-element network whose elements interact with each other as in an RD system. Cells can communicate with each other via the diffusion of I, either to the medium (indirect communication), directly to and from contacting cells (direct communication), or by a mixed mechanism of direct and indirect communication. When communication is direct, cells form a chess-like pattern with a higher number of blue cells. When it is indirect, a ring of red cells is formed outside the aggregate. Finally, when communication is mixed, the resulting pattern is a chess pattern confined to the center of the aggregate. Interestingly, when virtual cells cannot communicate in any form, all the cells in a simulation reach the same steady state, depending only on their initial intracellular conditions.

These patterns are explained through two main processes. The first one is cell positioning in relation to the medium. Indirect communication (diffusion to medium) is only performed by the external cells of the aggregate. This creates a difference between internal and external cells, ultimately leading to the ring pattern. Indeed, differences in position with respect to the external medium might be fundamental in some developmental processes (Bonner, 2001). Nonetheless, due to the lack of communication the cell types created at the aggregate periphery are not propagated to the rest of the cells. The second important process is cell coupling via the diffusion of a given molecule (I in this case). Direct communication permits the coupling between cells networks, leading to the amplification of local fluctuations and the generation of heterogeneous steady patterns (Turing, 1952; Meinhardt and Gierer, 2000)—such as the chess pattern. When both types of communication are combined in the mixed (direct and indirect communication) scenario, the results are far more interesting; the differences created by the external cells, due to indirect communication, are propagated to the inner part of the aggregate due to direct communication (see figure 4.4). While the modeled aggregates are not assumed to have plasmodesmata-like unions, we postulate that, since diverse steady arrangements of cells with different states are formed by cell communication (through the diffusion of I to other cells or to the medium), the process simulated for virtual cell aggregates is analogous to what has been defined as the DIF DPM in plants.

Some of the processes mentioned above and their resulting patterns resemble those observed in the development of multicellular organisms formed by cellular aggregation, such as *Dictyostelium* species or *Myxoccocus xanthus*. In the case of *Dictyostelium minutum*, at the first stages of their phase of aggregation, primary centers are distributed in a characteristic spotted pattern (Gerisch, 1968). Such pattern is a result both of indirect communication of cAMP and of its reaction and degradation in each of the cells (lateral

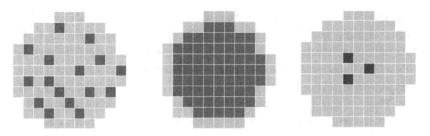

Figure 4.4
Patterns of differentiation in simulated cellular aggregates created using CompuCell3D .

inhibition process) (Gerisch, 1968). Also, during the development of *Dictyostelium dis-coideum*, there is a predetermination in prespores and prestalk cells in the first mass of cells before the formation of the fruiting bodies (Williams et al., 1989). The observed predetermination follows a distinctive centripetal pattern, where prespore cells are confined to the center and the base and the prestalk cells are located in the periphery. The resulting pattern is in turn guided by the concentration of oxygen near to the periphery (that enhances the differentiation in prestalk cells) reinforced by internal ammonia and cAMP (Bonner, 2001).

In the development of *Myxoccocus xanthus*, during colony aggregation—as a response to starvation—cells communicate through the C-signal, a contact-dependent signal (Kaiser, 2001), and cells with a high level of C-signal will turn out into spores. Interestingly, only cells in the inner part of the aggregates will become spore cells; the outer cells will become peripheral cells. This is related to the high concentration of C-signal in this denser part of the colony, propelled by the high cell-contact rate in the center (Julien et al., 2000). Finally, in the inner part of the aggregate, not all the cells become spores. In fact, they form a spotted distribution. This spotted pattern is believed to be partly the result of the alignment between the rod-shaped bacteria (Holmes et al., 2010). Indeed, the alignment is not perfect, which results in different rates of collisions—contact—between cells, and so, in a different concentration of C-signal (leading to the spotted pattern).

Other interaction modules and explanations are required to fully understand the basic principles of cell differentiation in emerging multicellular organisms (e.g., Bahar et al., 2014). Nevertheless, these models prefigure the importance of cell communication, the DIF module, and the DPMs in general when thinking about these principles.

Organization of Cellular Masses via DPMs as a Complementary Approach to the Study of the Transition to Multicellularity

Cooperation among individuals has been considered one of the main catalysts of evolutionary transitions, in particular for the origin of multicellularity (Michod and Roze, 2001).

Actually, our considerations are only valid in this specific transition. Every extrapolation to other levels of organization or social issues will be beyond the scope of our conclusions. However, explaining the appearance and maintenance of cooperation within biological collectives has been one of the major challenges in biology because cooperative communities are easily invaded by the so-called cheaters or defectors. Cheaters use the resources of cooperative individuals without contributing anything in return, and hence increase their fitness in relation to the cooperative units. Multicellular organisms and some eusocial communities are typical examples of cooperative entities organized in larger units. For populations composed of independent organisms, multiple theoretical explanations appear to solve, to some extent, the problem of evolution of cooperation (e.g., kin selection, graph selection, group selection, or more advanced mechanisms, such as control of cheaters or control of the production of the common good; Travisano and Velicer, 2004; Nowak, 2006; Kümmerli and Brown, 2010). Importantly, these scenarios assume that the individuals' behavior (cheater or cooperative) is independent of their interactions with other members of the group during developmental processes, and that it is frequently preestablished.

We argue that not all cells can be accommodated dichotomously as cooperative or defective, as during the development of multicellular organisms in which interactions among cells and intracellular dynamics are highly dynamical and context dependent. For example, a motile cell in *Volvox*, which enables mobility to the whole colony, is often thought of as a cooperator. Nevertheless, if all cells were motile, no cell would reproduce, leading the colony to disappear; the resulting colony would not be an aggregate of cooperators. Moreover, cells in a multicellular organism are not independent agents; cell behaviors are largely dependent on other cells' actions, on the relative position of cells, and on the cell–environment interactions. As illustrated by the models for plant and aggregate development discussed above, the cellular state is often determined by a whole system of coupled internal networks, and the highly nonlinear nature of these systems, what is often identified as an individual behavior, cannot be averaged back to single cells.

Then, while the notions of cooperation and conflict provide a useful framework to think about the maintenance of multicellular organisms at the evolutionary scale (Grosberg and Strathmann, 1998; Michod and Roze, 2001; Folse and Roughgarden, 2010 explanations based only on these concepts are not always necessary (Ispolatov et al., 2012) and are limited by some assumptions (e.g., individual cell behavior as cooperator or defector). Additionally, these explanations do not focus on the developmental mechanisms that accompanied and enabled the emergence of the first cellular patterns in the transition to multicellularity. As illustrated for the cases of epidermal patterning and cellular aggregation described above, considering DPMs in the study of the origin of multicellularity might be a way to study these mechanisms and to understand the strong integration of some multicellular aggregates, even if they are not composed by intrinsically cooperative elements.

One of the central issues in the origin of multicellularity is that of the inheritance and transgenerational stabilization of multicellular patterns and forms. We have suggested that cell differentiation in a developing multicellular organism could be a direct consequence of the interaction between different DPMs (e.g., DIF in plants and newly formed aggregates). The specific patterns produced depend on the different kinds of interaction, involved molecules, and on the size of the aggregates (Bonner, 2001; Furusawa and Kaneko, 2002). These multicellular aggregates could have some evolutionary advantages due to division of labor, and in this case they would be selected as a whole. At the same time, only some of the cells in the aggregate would reproduce (those that, because of coupling, differentiate themselves in germinal cells), but since they would contain the molecules that enable DMPs, they could in principle lead to the recreation of a new aggregate. These mechanisms could be stabilized through genetic assimilation or other similar processes (West-Eberhard, 2005; Newman et al., 2006).

The molecules and properties that conform DPMs are sufficient to render a diversity of patterns and structures observed in multicellular organisms and may have been already present in the unicellular organisms that originated multicellular organisms. We have discussed a dynamical system approach to explore the possibility that DPM were important for the appearance of multicellularity. The results and data reviewed here suggest that this could be the case. Finally, we suggest that the dynamical DPMs framework could challenge or complement other explanations proposed for the evolution of multicellularity.

Acknowledgments

We are grateful to the participants of the Konrad Lorenz Institute workshop on *The Origins and Consequences of Multicellularity,* who provided valuable ideas and criticism regarding this work. This work was written under a national state of concern and distress for the violence prevailing in Mexico; in particular violence against students around the country, who are being killed, disappeared and arbitrarily prosecuted for demonstrating and organizing themselves in the search for a better world.

References

Bahar, F., Pratt-Szeliga, P. C., Angus, S., Guo, J., & Welch, R. D. (2014). Describing *Myxococcus xanthus* aggregation using Ostwald ripening equations for thin liquid films. *Scientific Reports*, 4, 6376.

Barrio, R., Romero-Arias, J. R., Noguez, M. A., Azpeitia, E., Ortiz-Gutiérrez, E., Hernández-Hernández, V., et al. (2013). Cell patterns emerge from coupled chemical and physical fields with cell proliferation dynamics: The *Arabidopsis Thaliana* root as a study system. *PLoS Computational Biology*, 9(5), e1003026.

Benítez, M., Azpeitia, E., & Alvarez-Buylla, E. R. (2013). Dynamic models of epidermal patterning as an approach to plant eco-evo-devo. *Current Opinion in Plant Biology*, 16, 1–8.

Benítez, M., Monk, N., & Alvarez-Buylla, E. R. (2011). Epidermal patterning in *Arabidopsis*: Models make a difference. *Journal of Experimental Zoology. Part B, Molecular and Developmental Evolution*, 316, 241–253.

Bonner, J. T. (2001). *First signals: The evolution of multicellular development*. Princeton, NJ: Princeton University Press.

Caballero, L., Benítez, M., Alvarez-Buylla, E. R., Hernández, S., Arzola, A. V., & Cocho, G. (2012). An epigenetic model for pigment patterning based on mechanical and cellular interactions. *Journal of Experimental Zoology. Part B, Molecular and Developmental Evolution*, 318, 209–223.

Chautá-Mellizo, A., Campbell, S. A., Argenis Bonilla, M., Thaler, J. S., & Poveda, K. (2012). Effects of natural and artificial pollination on fruit and offspring quality. *Basic and Applied Ecology*, 13, 524–532.

Folse, H. J., & Roughgarden, J. (2010). What is an individual organism? A multilevel selection perspective. *Quarterly Review of Biology*, 85, 447–472.

Furusawa, C., & Kaneko, K. (2002). Origin of multicellular organisms as an inevitable consequence of dynamical systems. *Anatomical Record*, 268, 327–342.

Gerisch, G. (1968). Cell aggregation and differentiation in *Dictyostelium*. *Current Topics in Developmental Biology*, 3, 157–197.

Graham, L. E., Cook, M. E., & Busse, J. S. (2000). The origin of plants: Body plan changes contributing to a major evolutionary radiation. *Procedures of the National Academy of Science of the United States of America*, 97, 4535–4540.

Green, P. B. (1962). Mechanism for plant cellular morphogenesis. *Science*, 28, 1404–1405.

Grosberg, R., & Strathmann, R. (1998). One cell, two cell, red cell, blue cell: The persistence of a unicellular stage in multicellular life histories. *Trends in Ecology & Evolution*, 13, 112–116.

Hernández-Hernández, V., Niklas, K. J., Newman, S. A., & Benítez, M. (2012). Dynamical patterning modules in plant development and evolution. *International Journal of Developmental Biology*, 56, 661–674.

Holmes, A., Kalvala, S., & Whitworth, D. (2010). Spatial simulations of myxobacterial development. *PLoS Computational Biology*, 6(2).

Ispolatov, I., Ackermann, M., & Doebeli, M. (2012). Division of labour and the evolution of multicellularity. *Proceedings of the Royal Society*, 279, 1768–1776.

Jablonka, E., & Lamb, M. J. (2005). *Evolution in four dimensions: Genetic, epigenetic, behavioral, and symbolic variation in the history of life*. Cambridge, MA: MIT Press.

Jaeger, J., & Reinitz, J. (2006). On the dynamic nature of positional information. *BioEssays*, 28, 1102–1111.

Jönsson, H., Heisler, M., Reddy, G. V., Agrawal, V., Gor, V., Shapiro, B. E., et al. (2005). Modeling the organization of the WUSCHEL expression domain in the shoot apical meristem. *Bioinformatics*, 21, i232i240.

Julien, B., Kaiser, A. D., & Garza, A. (2000). Spatial control of cell differentiation in *Myxococcus xanthus*. *Proceedings of the National Academy of Sciences of the United States of America*, 97, 9098–9103.

Kaiser, D. (2001). Building a multicellular organism. *Annual Review of Genetics*, 35, 103–123.

Kauffman, S. A. (1969). Metabolic stability and epigenesis in randomly constructed genetic nets. *Journal of Theoretical Biology*, 22, 437–467.

Kim, I., & Zambryski, P. C. (2005). Cell-to-cell communication via plasmodesmata during *Arabidopsis* embryogenesis. *Current Opinion in Plant Biology*, 8, 593–599.

Kondo, S., & Miura, T. (2010). Reaction–diffusion model as a framework for understanding biological pattern formation. *Science*, 329, 1616–1620.

Kümmerli, R., & Brown, S. P. (2010). Molecular and regulatory properties of a public good shape the evolution of cooperation. *Proceedings of the National Academy of Sciences of the United States of America*, 107, 18921–18926.

Lehesranta, S. J., Lichtenberger, R., & Helariutta, Y. (2010). Cell-to-cell communication in vascular morphogenesis. *Current Opinion in Plant Biology*, 13, 59–65.

Lewontin, R. C. (2000). *The triple helix: Gene, organism and environment*. Cambridge, MA: Harvard University Press.

Lucas, W. J., Ham, B. K., & Kim, J. Y. (2009). Plasmodesmata—Bridging the gap between neighboring plant cells. *Trends in Cell Biology*, 19, 495–503.

Lynch, T. M., & Lintilhac, P. M. (1997). Mechanical signals in plant development: A new method for single cell studies. *Developmental Biology*, 181, 246–256.

Mammoto, A., Mammoto, T., & Ingber, D. E. (2012). Mechanosensitive mechanisms in transcriptional regulation. *Journal of Cell Science*, 125, 3061–3073.

Masson, P. H., Tasaka, M., Morita, M. T., Guan, C., Chen, R., Boonsirichai, K. (2002). *Arabidopsis thaliana*: A model for the study of root and shoot gravitropism. *Arabidopsis Book*. 1, e0043.

Meinhardt, H., & Gierer, A. (2000). Pattern formation by local self-activation and lateral inhibition. *BioEssays*, 22, 753–760.

Michod, R. E., & Roze, D. (2001). Cooperation and conflict in the evolution of multicellularity. *Heredity*, 86, 1–7.

Mora Van Cauwelaert, E., Arias del Angel, J. A., Benítez, M., & Azpeitia, E. M. (2015). Development of cell differentiation in the transition to multicellularity: A dynamical modeling approach. *Frontiers in Microbiology*, 6, 603.

Nakayama, N., Smith, R. S., Mandel, T., Robinson, S., Kimura, S., Boudaoud, A., et al. (2012). Mechanical regulation of auxin-mediated growth. *Current Biology*, 22, 1468–1476.

Newman, S. A., & Bhat, R. (2007). Activator–inhibitor dynamics of vertebrate limb pattern formation. *Birth Defects Research. Part C, Embryo Today*, 81, 305–319.

Newman, S. A., & Bhat, R. (2009). Dynamical patterning modules: A "pattern language" for development and evolution of multicellular form. *International Journal of Developmental Biology*, 53, 693–705.

Newman S. A., & Bhat R. (2008). Dynamical patterning modules: Physico-genetic determinants of morphological development and evolution. *Physical Biology*, 5.1, 015008.

Newman, S. A., & Bhat, R. (2009). Dynamical patterning modules: A pattern language for development and evolution of multicellular form. *International Journal of Developmental Biology*, 53, 693. doi:.10.1387/ijdb.072481sn

Newman, S. A., Forgacs, G., & Müller, G. B. (2006). Before programs: The physical origination of multicellular forms. *International Journal of Developmental Biology*, 50, 289–299.

Newman, S. A., & Müller, G. B. (2000). Epigenetic mechanisms of character origination. *Journal of Experimental Zoology*, 288, 304–317.

Niklas, K. J. (2000). The evolution of plant body plans: A biomechanical perspective. *Annals of Botany*, 85, 411–438.

Niklas, K. J. (1992). *Plant biomechanics: An engineering approach to plant form and function*. Chicago: University of Chicago Press.

Nowak, M. A. (2006). Five rules for the evolution of cooperation. *Science*, 314, 1560–1563.

Pesch, M., & Hülskamp, M. (2004). Creating a two-dimensional pattern de novo during *Arabidopsis* trichome and root hair initiation. *Current Opinion in Genetics & Development*, 14, 422–427.

Pigliucci, M., & Muller, G. B. (2010). *Evolution—The extended synthesis*. Cambridge, MA: MIT Press.

Pigliucci, M., Murren, C. J., & Schlichting, C. D. (2006). Phenotypic plasticity and evolution by genetic assimilation. *Journal of Experimental Biology*, 209, 2362–2367.

Poveda, K., Martínez, E., Kersch-Becker, M. F., Bonilla, M. A., & Tscharntke, T. (2012). Landscape simplification and altitude affect biodiversity, herbivory and Andean potato yield. *Journal of Applied Ecology*, 49, 513–522.

Raven, J. A. (1997). Multiple origins of plasmodesmata. *European Journal of Phycology*, 32, 95–101.

Robert, J. S. (2006). *Embryology, epigenesis and evolution: Taking development seriously*. Cambridge, UK: Cambridge University Press.

Ryu, K. H., Zheng, X., Huang, L., & Schiefelbein, J. (2013). Computational modeling of epidermal cell fate determination systems. *Current Opinion in Plant Biology*, 16, 5–10.

Thompson, D. W. (1917). *On growth and form*. Cambridge, UK: Cambridge University Press.

Travisano, M., & Velicer, G. J. (2004). Strategies of microbial cheater control. *Trends in Microbiology*, 12, 72–78.

Tsukaya, H. (2005). Leaf shape: Genetic controls and environmental factors. *International Journal of Developmental Biology*, 49, 547–555.

Tsukaya, H., Shoda, K., Kim, G. T., & Uchimiya, H. (2000). Heteroblasty in *Arabidopsis thaliana* (L.) Heynh. *Planta*, 210, 536–542.

Turing, A. M. (1952). The chemical basis of morphogenesis. *Philosophical Transactions of the Royal Society of London. Series B, Biological Sciences*, 237, 37–72.

Waddington, C. H. (1957). *The strategy of the genes: A discussion of some aspects of theoretical biology*. London: George Allen & Unwin.

West-Eberhard, M. J. (2003). *Developmental plasticity and evolution*. Oxford, UK: Oxford University Press.

West-Eberhard, M. J. (2005). Developmental plasticity and the origin of species differences. *Proceedings of the National Academy of Sciences of the United States of America*, 102, 6543–6549.

Williams, J. G., Duffy, K. T., Lane, D. P., McRobbie, S. J., Harwood, A. J., Traynor, D., et al. (1989). Origins of the prestalk-prespore pattern in *Dictyostelium* development. *Cell*, 59(6), 1157–1163.

Wolpert, L. (1996). One hundred years of positional information. *Trends in Genetics*, 12, 359364.

5 Angiosperm Multicellularity: The Whole, the Parts, and the Sum

Ottoline Leyser

For multicellularity to evolve by natural selection, multicellular organisms must have attributes beneficial to survival and reproduction that are qualitatively or quantitatively different from those of singled-celled organisms. The whole must be more than, or different from, the sum of the parts. Understanding multicellularity therefore requires an understanding of what constitutes the whole and the parts, and perhaps more importantly, what is the sum? How do the parts work together to deliver properties that confer a selective advantage in comparison to those of the parts?

Different chapters in this volume focus on different aspects of these questions. Several are devoted to considerations of the main selective drivers involved in the multicellular habit. What are the most important ways in which the whole differs from the parts? Several interesting ideas are proposed. Knoll and Lahr (this volume) propose that early multicellularity was a defense against eukaryotic phagotrophy. Beyond this, in the Foreword to this volume, Bonner argues that only those iterations of such early multicellularity that solved the problem of achieving further increases in size prospered. Such solutions require specialization of cell types and/or body parts, which simultaneously provides a step change in evolvability by natural selection.

Such cellular specialization or subfunctionalization is an interesting aspect of multicellularity, which begins to highlight questions about the nature of the parts of an organism and their constitution as a whole. A consortium of different specialized cell types has the potential to deliver adaptive functions more effectively than a multifunctional single cell, or to deliver a wider range functions than can be achieved by a single cell, at least not in parallel. A good example is provided by cyanobacteria in the genus *Anabaena*, which grow as chains of cells along which nitrogen fixing cells, termed heterocysts, differentiate with a spacing pattern that has become a classical study system in developmental biology (Golden and Yoon, 2003). The biochemistry for nitrogen fixation is extremely oxygen sensitive and requires anaerobic conditions, which are created in the specialized heterocyst cells. Nitrogen fixation confers an obvious selective advantage that can only be efficiently achieved by cooperation between two cell types, specialized in nitrogen fixation and photosynthesis.

Thus, there are multiple selective drivers for multicellularity with compelling arguments for large size with specialization of parts providing advantages for both acquisition of resources and defense from predation. However, these benefits could in principle be achieved at least in part by a collection of single cells working together in a community. Here slime molds such as *Dictytostelium discoideum* are frequently cited as particularly powerful examples. *Dictytostelium discoideum* has a single-cell free-living amoeboid phase in its life cycle; however, when its bacterial food supply becomes low, the single cells aggregate to produce an impressive multicelluar migratory slug with specialized cell types, which moves to a new location and forms a fruiting body releasing asexual spores (Weijer, 2004). Notably, this means that only a subset of the cells survive the move and colonize the new location. By some definitions, this would make the slug a multicellular organism because it involves cooperation between cells with some being excluded from the reproductive lineage. However, others would argue that the existence of a single-celled free-living stage, and the formation of the migratory slug entirely by emergence involving local interactions with no central coordination rules it out (Arnellos and Moreno, this volume).

This latter argument is particularly interesting because it defines the whole through reference to the nature of the sum of the parts. According to this thesis, true multicellularity requires a particular mode of interaction between the parts. It is not enough that there should be specialization of cell types, or that only a subset of the cells in the collective should contribute to the next generation, but there must be fully integrated control of the whole, mediated by a dedicated regulatory system with centralized control, epitomized by the nervous system and brain of animals (Arnellos and Moreno, this volume).

In all these the narratives, there are distinct features presented by the plant kingdom in comparison to the animal kingdom. These two kingdoms include the most successful iterations of the multicellular habit. Their independent evolution provides opportunities both to draw overarching conclusions about multicellularity, common to all kingdoms, such as the advantages and consequence of increases in size mentioned above, but also to explore differences presumably associated with the different selective pressures at play in plants compared to animals. As animals ourselves, there is a natural tendency to consider the animal kingdom as in some way archetypal, and the plant kingdom as less sophisticated. For example, evolution of the animal lineage can be described as a progression toward increased specialization and centralization of control mechanisms. By comparison, during the evolution of the plant kingdom, while there are specialized regulatory systems integrating activities across the plant body, control is not centralized but distributed. This could be considered as, in some sense, a more primitive state. However, given the extraordinary success of plants in colonizing the planet, arguments for their lack of sophistication require critical examination. I am reminded of a quote from Douglas Adams's (1979) *Hitchhikers Guide to the Galaxy* concerning our superior intelligence to dolphins:

Man had always assumed that he was more intelligent than dolphins because he had achieved so much—the wheel, New York, wars and so on—whilst all the dolphins had ever done was muck about in the water having a good time. But conversely, the dolphins had always believed that they were far more intelligent than man—for precisely the same reasons.

The higher plant manifestation of multicellularity needs to be considered in the context of the selective drivers that shaped it. The profoundly different requirements for success in the two kingdoms, mostly associated with autotrophy versus heterotrophy, have resulted in different configurations of the whole, the parts, and the sum. While in animals, it is relatively easy to define the whole, which consists of distinct and often highly nonredundant organs, largely centrally controlled, plants are much more modular in construction with nonredundant tissues, but largely redundant individual organs, arranged in organ systems and regulated by a distributed control system. These differences can be understood in the context of the selective drivers involved, particularly when considering colonization of the land.

Resource Acquisition and Defense from Predation

Plants are primary producers, generating chemical energy from sunlight and using it to build themselves out of simple chemical constituents collected from the environment, principally water, carbon dioxide, and various minerals. Plants growing in dim light may be energy limited, and, in aqueous environments, remaining sufficiently close to the water surface is important to achieve sufficient energy capture for growth. However, on land many plants have the opposite problem, requiring specific energy-dissipating systems to prevent oxidative damage from excessive light capture (Ort, 2001). In these environments, collecting the carbon dioxide and the mineral nutrients needed to build new cells is more likely to be limiting. This is an important distinction because it means that in many environments, energy efficiency is much less important than effective nutrient capture. Furthermore, on land, it is usually the case that the source of water and mineral nutrients is underground while the source of energy and carbon dioxide is above ground. Land plant evolution is therefore a story of increasing size, including both depth and height accompanied by expansion of below- and above-ground surface area. This is reflected in their two interdependent organ systems: roots and shoots. Roots explore the soil, mining water and nutrients. Shoots explore the air, mining light and carbon dioxide. There are complex relationships between these functions, particularly centered on the biology of water relations. While water is needed as a raw ingredient in photosynthesis, this is a very minor use quantitatively. Much more is needed to maintain turgor in plant cells and to carry nutrients from the roots to the shoots. This latter process is driven by evaporation of water from leaves through open stomatal pores, mostly on their lower surfaces. Meanwhile, these pores are also the entry point for carbon dioxide, resulting in complex trade-offs between water retention and carbon dioxide capture.

Because of these fundamentals of plant biology, size, and particularly large surface areas above and below ground, has been a major driver in land plant evolution. Plasticity in both the above-ground and the below-ground organ systems is also essential for efficient resource capture since, with the exception of carbon dioxide, they are not uniformly distributed in the environment. Because of this, mechanisms that trigger proliferation into pockets of locally high availability, and prevent growth into areas of low availability, confer a considerable selective advantage. This requires environmentally responsive continuous growth and development, achieved by the maintenance of stem cell populations at root and shoot tips, and the establishment of new ones on the growing root and shoot axes, allowing lateral branching and ramification of the root and shoot systems.

This resulting modularity in plant growth makes adaptive plasticity in body plan possible. Simultaneously it provides the additional benefit of protecting the plant from herbivory. The large underground surface areas required for nutrient capture prevent locomotion, making plants sitting ducks. Multiple growth axes, with the ability to generate new ones, make for a high degree of redundancy, rendering loss of any one shoot or root to herbivory survivable. Thus, as in animal systems, increased size for improved nutrient and energy capture, concomitant specialization of organ systems, and the consequences of the evolution of herbivory have made important contributions to the evolution of multicellularity. However, while in animal systems this has led to nonredundant organs with central control by one such organ, in plants it has led to organ systems, each with many redundant parts and a more distributed regulatory system.

The Modularity of Plant Development

While the early stages of plant embryogenesis have many similarities with animal development, involving the establishment of the basic body axes and organs, most plant development is postembryonic. During embryogenesis, small populations of stem cells, the meristems, are established at either end of the main apical–basal axis with the shoot apical meristem at one end and the root apical meristem at the other. After germination of the seed, the root apical meristem gives rise to the entire root system and the shoot apical meristem to the shoot system.

The development of the shoot system provides a particularly good illustration of the modularity of plants (see figure 5.1). Elegant self-organizing patterning systems in the shoot apical meristem drive the formation of leaves and stem while maintaining the meristem as a dome of multipotent cells, which is carried upward as it builds the plant body underneath it (Aichinger et al., 2012). Cells in the center of the meristem, the central zone, divide, replacing themselves and feeding daughter cells into the meristem flanks. This results in a ring of cells surrounding the central zone, termed the peripheral zone. Peripheral-zone cells divide rapidly and can differentiate to form leaves or stem tissue in a spacing

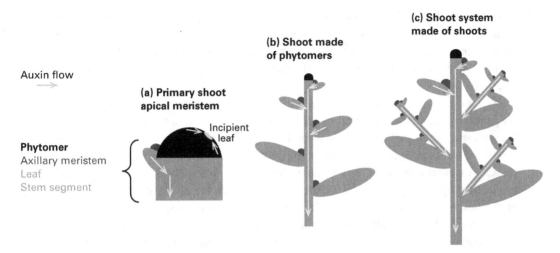

Figure 5.1
Modularity in shoot development. The shoot apical meristem (a) (black) produces phytomers (a, b) consisting a section of stem (gray), a leaf (gray), and a new shoot meristem in the leaf axil (dark gray). Auxin flows (pale gray) pattern the phytomer (a). The primary shoot is made up of a series of phytomers (b) and the activity of its axillary meristems is regulated by auxin flow from the primary meristem (b). Activation of the axillary meristem reiterates the pattern (c).

pattern regulated by the position of the previous leaves. This patterning process involves the dynamic polarization and repolarization of transport of the plant hormone auxin (Reinhardt et al., 2003; Heisler et al., 2005). Leaves initiate at sites of high auxin concentration in the peripheral-zone epidermis formed by the convergence of auxin transport toward those sites, driven by a positive feedback process (Jönsson et al., 2006; Smith et al., 2006). Auxin transport is then internalized, with auxin exporters being expressed in internal tissues and oriented away from the epidermis down into the stem below (Reinhardt et al., 2003; Heisler et al., 2005). This transport becomes organized into narrow files of cells at the site of the future leaf midrib. Altered cell division planes result in the leaf primordium bulging out from the meristem flanks. Transport of auxin toward the incipient leaf primordium depletes auxin from the rest of the peripheral zone, and a boundary is established between the leaf and the surrounding tissues, which become stem. The next leaf initiates at the site farthest from existing leaf primordia, creating a pattern of leaf development, or phyllotaxis, that is dependent on the relative rates of meristem growth and leaf initiation, and the relative sizes of the meristem and the leaf primordia.

Meanwhile, in the newly formed leaf primordia, auxin synthesis is upregulated as the leaf expands and auxin is transported out of the leaf and downward into the stem below (Sawchuk et al., 2013). The stem, and in particular files of cells associated with its vascular bundles (Goldsmith, 1977; Gälweiler et al., 1998), is a high-capacity, highly polar auxin transport path that carries auxin down to the root at average speeds in the order 1 cm/hour (Kramer et al., 2011). This auxin transport route is termed the polar auxin transport stream (PATS), and it is substantially dependent on the high level of expression and strong basal polarization of PIN1 auxin exporters (Gälweiler et al., 1998; Okada et al., 1991; Shinohara et al., 2013).

The transport of auxin from the young expanding leaves into the stem PATS is important for patterning the vascular network of the leaf and connecting it to the vascular system of the stem and, hence, to the rest of the plant. Here again, interesting self-organizing properties of the auxin transport system come into play, again involving positive feedback processes. A central concept is that of auxin transport canalization. This is a hypothesis originally proposed by Tsvi Sachs (1969, 1981) to explain the effects of auxin treatments on the development of vascular strands. In an elegant series of experiments, often using pea shoots, Sachs showed that vascular strand differentiation occurred in a way that connects sources of auxin to sinks. Existing vascular strands are strong sinks since they are the site of the PATS, carrying auxin away, down the stem to the root. Sachs showed that new vascular strands could be induced to form by applying a local auxin source to the side of the stem, and the new strands would always connect the auxin source to the nearest existing stem vascular bundle (Sachs, 1981). Sachs proposed that an initial passive flux of auxin between an auxin source and an auxin sink can be upregulated and polarized in the direction of the flux, resulting in increasingly narrow files of cells connecting the source to the sink. This was proposed to involve the upregulation and polarization of auxin efflux carriers. Some of the resulting high-transport cell files subsequently differentiate into vascular strands, building an organized vascular network connecting auxin sources, principally young expanding leaves in active shoot apices, to auxin sinks, principally established vascular bundles in the stem.

Although the mechanism by which factors proportional to auxin flux regulate auxin transporter levels and polarity is still unknown, the hypothesis has impressive predictive and explanatory power—for example, in generating the patterns of vascular development in leaves (Mitchison et al., 1981; Rolland-Lagan and Prusinkiewicz, 2005). Furthermore, despite being formulated long before any auxin transporters were known, the behavior of PIN1-like auxin efflux carriers in pea follows exactly the behavior proposed by Sachs (Balla et al., 2011). In this way, the auxin transport network plays a central role in building the shoot system. It is involved in patterning leaf formation and in linking each new leaf to the vascular network of the plant, and its export from active apices feeds auxin into the PATS, relaying information about shoot tip activity down the stem and into the root (see figure 5.1).

The pattern of auxin distribution also plays a role in the initiation of new shoot apical meristems, which form in the axil of each leaf, at the site of an auxin minimum at the boundary between the emerging leaf primordium and the meristem central zone (Q. Wang et al., 2014; Y. Wang et al., 2014). These secondary shoot meristems have the same developmental potential as the primary shoot meristem. They can remain active to form an entirely new shoot axis, or they can enter a state of dormancy after the formation of a few leaves as an axillary bud. The leaves produced by these axillary meristems also harbor axillary meristems and so can produce higher order branches. Thus, the shoot system has fractal qualities and, depending on the activity of each axillary meristem, can grow as a single unbranched axis or as a highly ramified bush (see figure 5.1). The shoot can therefore be described in terms of modular units, called phytomers, produced by the shoot apical meristem and consisting of a leaf with its associated axillary bud and stem segment (see figure 5.1). The pattern of phytomer development across the shoot, and in particular the pattern of axillary bud activation, is a major determinant of shoot system architecture.

Because of the modular nature of plant development, the definition of plants as single organisms is problematic, made worse by their extraordinary regenerative capacity. Just as the single cells of slime molds can come together to make a migratory slug, so two plants, even in some cases of different species, can be grafted together to make a single functional unit. Furthermore, shoot cuttings can be taken from a shoot system, which will regenerate roots to produce a new plant. Thus, although the shoot systems of plants are highly dependent on their root systems, they are not exclusively dependent on the root system to which they happen to be currently connected. It is therefore possible to describe plants as colonial organisms of partially independent parts, for example, phytomers (Sachs et al., 1993; Leyser, 2011).

Furthermore, there is good evidence that phytomers compete with one another. This is illustrated by the well-known phenomenon of apical dominance. Apical dominance refers to the ability of an active shoot apex to inhibit the activity of axillary buds in the axils of the leaves below it. It is familiar to gardeners because it is the basis for pruning, in which removing the leading shoot promotes activation of dormant axillary buds. This can be interpreted as competition between shoot apices for nutrients. That there is competition is clear, as demonstrated, for example, by the classical two-branched pea experiments of Snow (1929). Here pea plants are decapitated above their most basal leaf pair, strictly speaking the seed leaves, termed cotyledons. The axillary meristems in the cotyledons activate and can produce a pea plant with two equally vigorous branches. However, it is frequently the case that one comes to dominate the other, halting the latter's growth. Which branch dominates can be influenced, for example, by shading one of the branches, which increases the probability that it will be dominated by the other. Similar experiments in Arabidopsis using two successive nodes of the mature shoot clearly illustrate bud–bud competition (Ongaro et al., 2008). For example, if one bud is removed, the other grows more vigorously. Thus, plants have been described as consisting of redundant competing parts (Sachs et al., 1993).

This, combined with the lack of central control, could call into question the level of functional integration achieved by plants and therefore the extent to which they should be considered as single functional units (Arnellos and Moreno, this volume). However, it is becoming increasingly clear that plants possess sophisticated regulatory systems that allow systemic integration of information across the plant body. These are dedicated regulatory systems that allow similar parts of the plant to respond differently to the same environmental input depending on systemic plant status. Although phytomers in the shoot system compete, there is very little evidence that they compete primarily for nutrients. This is clearly illustrated by considering the life history of a single leaf, which starts as a sink for sucrose, becomes a source of sucrose at maturity, and then at senescence exports all its mobilizable nutrients to the rest of the plant.

Indeed there is very little evidence that plant growth proceeds at its maximal possible rate, but rather growth is regulated and nutrients are allocated to different plant parts as a result of the self-organizing properties of dedicated regulatory systems in which the auxin transport network plays a central part. These principles can be illustrated by consideration of the regulatory systems that control branching in shoots in response to mineral nutrient availability and integrate this with the development of the root system.

The Sum of the Parts

In a plant, the root system gathers mineral nutrients and water while the shoot system gathers light and carbon dioxide. If mineral nutrients are scarce, then growth is focused in the root, with the proportion of biomass in the root increasing relative to the shoot (Scheible et al., 1997). Even within the root system, the pattern of root growth shifts, but this depends on which nutrient is scarce. If phosphate is in short supply, a shallower root system develops. In Arabidopsis this is achieved by suppression of primary root growth and promotion of lateral root growth (Williamson et al., 2001; Linkohr et al., 2002). If nitrate is the limiting resource, a generally longer but more sparse root system develops (Linkohr et al., 2002). These differences can be understood in the context of the general low mobility of phosphate in soil, with high concentrations in top soil, compared to nitrate, which has much higher mobility in soil such that a single root can deplete a large volume of soil (Fitter et al., 2002). If a local patch of high nutrient is provided, lateral roots will proliferate into the patch and their production is suppressed outside the patch (Drew, 1975; Zhang and Forde, 1998; Zhang et al., 1999; Linkohr et al., 2002). However, this does not occur if the supply of nutrients is generally high. This effect is mediated by shoot nutrient status and is most clearly illustrated by experiments involving plants with split root systems (Scheible et al., 1997; Ruffel et al., 2011). In these experiments, the root system of a plant is divided near the shoot and grown into two separate compartments, which can be supplied with different levels of nutrient. If both sides are given low nutrient supply, root growth occurs equally on both sides. This is also the case if both sides receive high levels of nutrient, but here root growth in general is suppressed relative to shoot growth. If one side is supplied with low nutrient levels and the other with high levels, then root growth is promoted on the side with high nutrient availability but suppressed on the side with low levels. Thus, root growth can be promoted or suppressed by both high and low nutrient supply, depending on the nutrient status of the shoot. It is not the case that roots supplied with nutrients grow and those without do not. This is the case even if the root is simultaneously supplied with high levels of sucrose. There is no evidence that roots compete based on their own internal nutrient supply. The response of roots to local fluctuations in nutrient levels is contingent on the status of the shoot. This clearly represents organismal-level functional integration, with communication between shoots and roots responsive to, but distinct from, primary metabolism. This system prioritizes growth by integrating systemic and local information about nutrient availability.

Auxin Transport and Signal Integration

There is a substantial body of evidence that systemic long-range communication between the root and shoot systems is mediated at least in part by plant hormones. In the case of shoot-to-root communication, and communication across the shoot system, auxin and the PATS play a central role with shoot apices competing, not for nutrients, but for access to the PATS.

As described above, growing shoot tips are a major source of auxin, which is transported out of young expanding leaves into the stem below in the PATS in vascular associated cells. Removal of the primary apex removes this auxin source and activates buds (Thimann and Skoog, 1933). Buds can also be activated by blocking auxin transport down the stem by ringing the stem above the bud with pharmacological inhibitors of auxin transport. Furthermore, if auxin is applied to the decapitated stump, buds remain dormant or rapidly reestablish dormancy (Thimann and Skoog, 1933). Interestingly, applying auxin directly to the bud does not inhibit it (Sachs and Thimann, 1967), and indeed auxin can protect actively growing buds from inhibition (Snow, 1937). Furthermore, when radiolabeled auxin is applied to the stump of a decapitated plant, axillary bud activity is inhibited, but radiolabel does not accumulate in the buds to significant levels (Hall and Hillman, 1975; Booker et al., 2003). Indeed basally supplied radiolabeled auxin is delivered to buds more effectively but has no effect on their activity (Booker et al., 2003).

Thus, auxin moving in the PATS in the main stem inhibits the activity of axillary buds but does so indirectly, without entering the bud. In the bud, high auxin is associated with active buds, as would be predicted by the high auxin synthesis observed in young expanding leaves, and beyond this, auxin in buds appears to contribute positively to their activity.

The indirect mode of auxin action in the main stem was for a long time presumed to be due to a second messenger's relaying the auxin signal into the bud (Leyser, 2009). There are two strong candidates for this, namely, the plant hormones cytokinin and strigolactone. Cytokinin generally promotes bud activation, including when it is applied directly to inhibited buds, and auxin can inhibit cytokinin synthesis and the transcription of cyto-kinin biosynthetic genes. Strigolactone generally inhibits bud activation, including when it is applied directly to the bud, and auxin can promote the transcription of strigolactone biosynthetic genes. It has therefore been proposed that bud activity is regulated by the relative levels of cytokinin and strigolactone in buds, both working through their opposing effects on the transcription of a key regulator of bud activity, in Arabiodpsis and pea known as *BRC1* (Dun et al., 2012). However, this attractive idea, while consistent with a substan-tial body of evidence, is unlikely to represent the full story because it cannot explain a range of observations, particularly related to the mode of action of strigolactone.

The second-messenger model requires strigolactone to act as a straightforward inhibitor of bud activity through effects on transcription in the bud. However, in *Arabidopsis*, sup-plying strigolactone to buds does not inhibit their activity unless there is a competing auxin source present, either a second bud or applied apical auxin (Liang et al., 2010; Crawford et al., 2010). In the two-branch system mentioned above, supplying strigolactone results more often in one bud's being inhibited while the other bud remains active. Thus, strigo-lactone does not appear to be a straightforward inhibitor of bud activity, but rather it increases the level of competition between buds, resulting in domination by a smaller number of active buds. Furthermore, mutants in strigolactone synthesis or signaling have intriguing auxin transport–related phenotypes (Bennett et al., 2006; Crawford et al., 2010; Shinohara et al., 2013). They overaccumulate PIN1 in the PATS, and their stems transport larger amounts of applied auxin than wild type. Application of the synthetic strigolactone, GR24, corrects these phenotypes in biosynthetic mutants, but not in signaling mutants, and depletion of PIN1 from the plasma membrane of cells in the PATS is a rapid response to GR24 that occurs in the presence of inhibitors of new protein synthesis, suggesting that it is a primary target for the pathway and not a secondary effect of new gene transcription (Shinohara et al., 2013). A substantial body of evidence supports the central importance of this activity in the increased shoot branching phenotypes of mutants in strigolatone synthesis or signaling. Of particular note, treatment with low levels of pharmacological inhibitors of auxin transport that restore auxin transport levels to wild type also restore wild-type shoot branching levels (Bennett et al., 2006). This result is particularly striking because such treatments promote branching in wild-type plants because they deplete auxin from the stem.

These observations have led to an alternative proposal for the indirect mechanism of action of auxin in the inhibition of bud activity, based on the auxin transport canalization hypothesis described above. In the context of shoot branching, this auxin transport canali-zation is highly relevant because Sachs (1981) showed that the auxin sink strength of existing vascular strands in the stem could be greatly reduced by high levels of auxin moving in the PATS. Specifically, the differentiation of a vascular strand connecting a laterally applied auxin source to the existing stem vasculature could be prevented by feeding auxin into the PATS of the existing vascular bundle. Thus, auxin in the stem can prevent auxin transport canalization from a lateral auxin source. Given that axillary buds are lateral auxin sources, this immediately suggests a mechanism whereby auxin in the main stem can inhibit the activation of lateral buds indirectly, without moving into the bud. If canalization of auxin transport from the bud to the stem is required for the bud to activate, then auxin in the main stem could inhibit bud activity by preventing auxin transport canalization out of the bud (Prusinkiewicz et al., 2009).

This model can explain the strigolactone-related phenomena described above that cannot be explained by a simple second-messenger model alone. Auxin transport canalization is driven by the positive feedback between auxin flux and the upregulation and polarization of auxin transport in the direction of the flux. In terms of PIN1 auxin efflux carriers and buds, this means that an initial low level of auxin transport from a bud to the stem will be upregulated by the accumulation of additional PIN1 proteins into the basal membrane of cells in the PATS of the bud stem. This will in turn increase the flux along this path and drive additional PIN1 accumulation and polarization, supporting bud activation. However, since strigolactones act to remove PIN1 from the plasma membrane, then high strigolactone levels will deplete PIN1, potentially preventing the accumulation of levels sufficient to establish the positive feedback needed for canalization. Here, the system is all about relativity. The initial flux of auxin is dependent on the auxin source strength of the bud and the auxin sink strength of the stem, and strigolactones act against the establishment of canalization between the source and the sink. Very high levels of strigolactone could prevent bud activation under all circumstances, but the effectiveness of strigolactone in bud inhibition will depend on the level of the initial passive flux and on the parameters of the positive feedback between this flux and PIN1 upregulation and polarization. Thus, strigolactone will be much more effective at inhibiting bud activity in the presence of a competing auxin source, as is observed. Our recent results suggest that a similar situation exists for cytokinin, with buds fully able to activate despite simultaneously low levels of cytokinin and high levels of strigolactone, provided that there is no competing auxin source (Müller et al., 2015).

Apart from explaining a wide range of observations not explained by the second-messenger hypothesis, this system confers properties on the network that provide considerable regulatory advantages, precisely because the system involves modulation of competition between shoot apices and is relative, rather than being a straightforward system dealing independently with the absolute concentrations of hormones in each bud. In the context of this model, it is interesting to revisit the observations that auxin regulates the synthesis of both cytokinin and strigolactone. Since active buds export auxin, this would contribute to reducing cytokinin levels and increasing strigolactone levels, thereby making it triply more difficult for further buds to activate. This might be important to prevent runaway bud activation following decapitation. It also provides multiple tuning points for the system such that while the levels of all three hormones can be regulated independently, the result will be a rebalancing of the system to a new stable equilibrium point, supporting a particular number of active buds, and delivering a particular amount of auxin to the roots.

This system provides an excellent illustration of how the hormonal regulatory network allows plastic responses to the environment that only make sense in the context of the whole plant. This can be clearly seen in the context of the modulation of shoot branching by nutrient supply. As described above, low nutrient availability results in the rebalancing of growth between the shoot system and the root system. Root growth is substantially dependent on auxin supply from the shoot. Thus, a vigorous shoot system with many active shoot apices promotes root growth through increased auxin supply to the root. Under low nutrient supply, however, root growth must be protected while shoot growth is restricted. This can readily be achieved via tuning the hormonal network described above. For example, in Arabidopsis, low nitrate supply results in increased export of auxin from each active apex (de Jong et al., 2014). In this way, few axillary buds activate because sufficiently high auxin in the stem to prevent the activation of additional branches is achieved with fewer active branches, but auxin supply to the roots is maintained.

In some species there is evidence that low nitrate also triggers an increase in strigolactone production in roots (Yoneyama et al., 2007). This is certainly more widely the case for low phosphate. Grafting experiments with strigolactone biosynthesis mutants have demonstrated that strigolactone can be transported from the root to the shoot (Leyser, 2009). Here strigolactones enhance competition between buds, thereby limiting the number that can activate. Importantly, this is a systemic effect, indicative of low phosphate/nitrate availability (Umehara et al., 2010; Kohlen et al., 2011). High strigolactone does not dictate which buds should be active. This will be established by local factors in the shoot, giving some buds a competitive advantage in auxin export over others. This system therefore allows the integration of local and systemic information in the decision-making process (Leyser, 2009).

Cytokinins are also affected by nitrate supply, with their levels increasing when nitrate is abundant (Takei et al., 2002). This apparently allows buds to escape the inhibitory effects of an active primary apex, presumably by modulating the competitive ability of axillary buds in some way. It is clear that high cytokinin is not required for bud activation, since our recent results show that mutants deficient in cytokinin synthesis are able to respond normally to decapitation, but they are unable to increase their branching when nitrate is abundant (Müller et al., 2015). Furthermore, cytokinin in the shoot is a key component of the regulatory system modulating root growth in response to nitrate supply described above (Ruffel et al., 2011).

Conclusions

Using this network of interlocked hormone signals moving systemically through the plant, and intimately coupled to the modularity of plant development, plants can adjust their growth in response to the environment and to mitigate the effects of herbivory (decapitation). The essential properties of the network can be best captured by the self-organizing positive feedbacks that model and remodel the auxin transport network, kept in check by negative feedbacks acting through cytokinin and strigolactone. Network modulation by environmental inputs allows dynamic re-equilibration to adapt the balance between root and shoot growth, and growth across the shoot system, to both local and systemic conditions. The system depends on competition between shoot apices for access to auxin transport paths to the root. It is essential that the competition is not directly for nutrients, since the adaptive distribution of resources to the parts of the plant most likely to contribute to the success of the whole requires that the currency of the competition should not be the same as the prize for winning. This competition is therefore not equivalent to the competition between organisms, and it clearly establishes that even though plants can be considered as colonial organisms with considerable resultant difficulty in defining the whole, there is nonetheless a whole, with a deep and sophisticated level of functional integration. Perhaps, in this sense, the colonial nature of plants is more similar to an ant colony than to a slime mold. It is made up of semi-independent multicellular organisms that nonetheless work together through complex self-organizing interactions to forage for resources dispersed unevenly in the environment. However, in plants there is, on the one hand, deeper integration through the physical connectivity across the plant with direct systemic flow of information possible. On the other hand, there is no queen, no central organizing principle. This does not make the system any less integrated. The distributed control system of plants achieves very similar ends to the central control of a queen ant or an animal brain and nervous system. Dynamic plasticity in response to the environment allows local responses to be dictated by systemic information, here relayed by a dedicated self-organizing regulatory network of hormonal signals. It is interesting that the regulatory architecture of this system has striking similarities to brains, where competition and reinforcement dynamically rewire neural connections in response to environmental inputs, resulting in adaptive behaviors.

The nature of the sum, the way in which the parts interact to deliver the properties of the whole, is different in plants and animals, and as a result the nature of the whole is different, being harder to define cleanly in plant systems. This can be readily understood through consideration of the different ways in which the key evolutionary drivers play out in hereotrophs versus autotrophs. One interesting consequence of the large plant solution to the multicellularity problem is the extraordinary flexibility of plants, such that relatively few species can colonize almost every environment on land, whereas most animal species are much more specialized. Thus, it might be constraining to imagine that animals are more fully integrated as multicellular organisms because of their central control. Rather they are just differently integrated, and these very differences provide useful comparative solutions to the problems of multicellularity.

Acknowledgments

OL's research is funded by the Gatsby Charitable Foundation of the UK and the European Research Council. I would like to thank Karl J. Niklas and Stuart A. Newman for organizing such an interesting discussion meeting, and the meeting participants for making it so.

References

Adams, D. (1979). *The hitchhiker's guide to the galaxy*. London: Pan Books.

Aichinger, E., Kornet, N., Friedrich, T., & Laux, T. (2012). Plant stem cell niches. *Annual Review of Plant Biology*, 63, 615–636.

Balla, J., Kalousek, P., Reinöhl, V., Friml, J., & Procházka, S. (2011). Competitive canalization of PIN-dependent auxin flow from axillary buds controls pea bud outgrowth. *Plant Journal*, 65, 571–577.

Bennett, T., Sieberer, T., Willetts, B., Booker, J., Luschnig, C., & Leyser, O. (2006). The *Arabidopsis MAX* pathway controls shoot branching by regulating auxin transport. *Current Biology*, 16, 553–563.

Booker, J. P., Chatfield, S. P., & Leyser, O. (2003). Auxin acts in xylem-associated or medullary cells to mediate apical dominance. *Plant Cell*, 15, 495–507.

Crawford, S., Shinohara, N., Sieberer, T., Williamson, L., George, G., Hepworth, J., et al. (2010). Strigolactones enhance competition between shoot branches by dampening auxin transport. *Development*, 137, 2905–2913.

de Jong, M., George, G., Ongaro, V., Williamson, L., Willetts, B., Ljung, K., et al. (2014). Auxin and strigolactone signaling are required for modulation of Arabidopsis shoot branching by N supply. *Plant Physiology*, 166, 384–395.

Drew, M. C. (1975). Comparison of the effects of a localised supply of phosphate, nitrate, ammonium and potassium on the growth of the seminal root system, and the shoot, in barley. *New Phytologist*, 75, 479–490.

Dun, E. A., de Saint Germain, A., Rameau, C., & Beveridge, C. A. (2012). Antagonistic action of strigolactone and cytokinin in bud outgrowth control. *Plant Physiology*, 158, 487–498.

Fitter, A., Williamson, L., Linkohr, B., & Leyser, O. (2002). Root system architecture determines fitness in an *Arabidopsis* root mutant in competition for immobile phosphate ions but not for nitrate ions. *Proceedings. Biological Sciences*, 269, 2017–2022.

Gälweiler, L., Guan, C., Müller, A., Wisman, E., Mendgen, K., Yephremov, A., et al. (1998). Regulation of polar auxin transport by AtPIN1 in *Arabidopsis* vascular tissue. *Science*, 282, 2226–2230.

Golden, J. W., & Yoon, H.-S. (2003). Heterocyst development in *Anabaena*. *Current Opinion in Microbiology*, 6, 557–563.

Goldsmith, M. (1977). The polar transport of auxin. *Annual Review of Plant Physiology*, 28, 439–478.

Hall, S. M., & Hillman, J. (1975). Correlative inhibition of lateral bud growth in *Phaseolus vulgaris* L. Timing of bud growth following decapitation. *Planta*, 123, 137–143.

Heisler, M. G., Ohno, C., Das, P., Sieber, P., Reddy, G. V., Long, J. A., et al. (2005). Patterns of auxin transport and gene expression during primordium development revealed by live imaging of the *Arabidopsis* inflorescence meristem. *Current Biology*, 15, 1899–1911.

Jönsson, H., Heisler, M. G., Shapiro, B. E., Meyerowitz, E. M., & Mjolsness, E. (2006). An auxin-driven polarized transport model for phyllotaxis. *Proceedings of the National Academy of Sciences of the United States of America*, 103, 1633–1638.

Kohlen, W., Charnikhova, T., Qing, L., Bours, R., Domagalska, M. A., Beguerie, S., et al. (2011). Strigolactones are transported through the xylem and play a key role in shoot architectural response to phosphate deficiency in nonarbuscular mycorrhizal host *Arabidopsis thaliana*. *Plant Physiology*, 155, 679–686.

Kramer, E. M., Rutschow, H. L., & Mabie, S. S. (2011). AuxV: A database of auxin transport velocities. *Trends in Plant Science*, 16, 461–463.

Leyser, O. (2009). The control of shoot branching: An example of plant information processing. *Plant, Cell & Environment*, 32, 694–703.

Leyser, O. (2011). Auxin, self-organisation, and the colonial nature of plants. *Current Biology*, 21, R331–R337.

Liang, J., Zhao, L., Challis, R., & Leyser, O. (2010). Strigolactone regulation of shoot branching in chrysanthemum (*Dendranthema grandiflorum*). *Journal of Experimental Botany*, 61, 3069–3078.

Linkohr, B. I., Williamson, L. C., Fitter, A. H., & Leyser, O. (2002). Nitrate and phosphate availability and distribution have different effects on root system architecture in *Arabidopsis*. *Plant Journal*, 29, 751–760.

Mitchison, G. J., Hanke, D. E., & Sheldrake, A. R. (1981). The polar transport of auxin and vein patterns in plants. *Philosophical Transactions of the Royal Society of London. Series B, Biological Sciences*, 295, 461–471.

Müller, D., Waldie, T., Miyawaki, K., To, J. P., Melnyk, C., Kieber, J. J., et al. (2015). Cytokinin is required for escape but not release from auxin mediated apical dominance. *Plant Journal*, 82, 874–886.

Okada, K., Ueda, J., Komaki, M. K., Bell, C. J., & Shimura, Y. (1991). Requirement of the auxin polar transport system in early stages of Arabidopsis floral bud formation. *Plant Cell*, 3, 677–684.

Ongaro, V., Bainbridge, K., Williamson, L., & Leyser, O. (2008). Interactions between axillary branches of *Arabidopsis*. *Molecular Plant*, 1, 388–400.

Ort, D. R. (2001). When there is too much light. *Plant Physiology*, 125, 29–32.

Prusinkiewicz, P., Crawford, C., Smith, R., Ljung, K., Bennett, T., Ongaro, V., et al. (2009). Control of bud activation by an auxin transport switch. *Proceedings of the National Academy of Sciences of the United States of America*, 106, 17431–17436.

Reinhardt, D., Pesce, E. R., Stieger, P., Mandel, T., Baltensperger, K., Bennett, M., et al. (2003). Regulation of phyllotaxis by polar auxin transport. *Nature*, 426, 255–260.

Rolland-Lagan, A. G., & Prusinkiewicz, P. (2005). Reviewing models of auxin canalization in the context of leaf vein pattern formation in Arabidopsis. *Plant Journal*, 44, 854–865.

Ruffel, S., Krouk, G., Ristova, D., Shasha, D., Birnbaum, K. D., & Coruzzi, G. M. (2011). Nitrogen economics of root foraging: Transitive closure of the nitrate–cytokinin relay and distinct systemic signaling for N supply vs. demand. *Proceedings of the National Academy of Sciences of the United States of America*, 108, 18524–18529.

Sachs, T. (1969). Polarity and the induction of organized vascular tissues. *Annals of Botany*, 33, 263–275.

Sachs, T. (1981). The control of the patterned differentiation of vascular tissues. *Advances in Botanical Research*, 9, 151–162.

Sachs, T., Novoplansky, A., & Cohen, D. (1993). Plants as competing populations of redundant organs. *Plant, Cell & Environment*, 16, 765–770.

Sachs, T., & Thimann, K. (1967). The role of auxins and cytokinins in the release of buds from dominance. *American Journal of Botany*, 54, 136–144.

Sawchuk, M. G., Edgar, A., & Scarpella, E. (2013). Patterning of leaf vein networks by convergent auxin transport pathways. *PLOS Genetics*, 9, e1003294.

Scheible, W.-R., Lauerer, M., Schulze, E.-D., Caboche, M., & Stitt, M. (1997). Accumulation of nitrate in the shoot acts as a signal to regulate shoot–root allocation in tobacco. *Plant Journal*, 11, 671–691.

Shinohara, N., Taylor, C., & Leyser, O. (2013). Strigolactone can promote or inhibit shoot branching by triggering rapid depletion of the auxin efflux protein PIN1 from the plasma membrane. *PLoS Biology*, 11, e1001474.

Smith, R. S., Guyomarc'h, S., Mandel, T., Reinhardt, D., Kuhlemeier, C., & Prusinkiewicz, P. (2006). A plausible model of phyllotaxis. *Proceedings of the National Academy of Sciences of the United States of America*, 103, 1301–1306.

Snow, R. (1929). The transmission of inhibition through dead stretches of stem. *Annals of Botany*, 43, 261–267.

Snow, R. (1937). On the nature of correlative inhibition. *New Phytologist*, 36, 283–300.

Takei, K., Takahashi, T., Sugiyama, T., Yamaya, T., & Sakakibara, H. (2002). Multiple routes communicating nitrogen availability from roots to shoots: A signal transduction pathway mediated by cytokinin. *Journal of Experimental Botany*, 53, 971–977.

Thimann, K. V., & Skoog, F. (1933). Studies on the growth hormone of plants: III. The inhibitory action of the growth substance on bud development. *Proceedings of the National Academy of Sciences of the United States of America*, 19, 714–716.

Umehara, M., Hanada, A., Magome, H., Takeda-Kamiya, N., & Yamaguchi, S. (2010). Contribution of strigolactones to the inhibition of tiller bud outgrowth under phosphate deficiency in rice. *Plant & Cell Physiology*, 51, 1118–1126.

Wang, Q., Kohlen, W., Rossmann, S., Vernoux, T., & Theres, K. (2014). Auxin depletion from the leaf axil conditions competence for axillary meristem formation in *Arabidopsis* and tomato. *Plant Cell*, 26, 2068–2079.

Wang, Y., Wang, J., Shi, B., Yu, T., Qi, J., Meyerowitz, E. M., et al. (2014). The stem cell niche in leaf axils is established by auxin and cytokinin in *Arabidopsis*. *Plant Cell*, 26, 2055–2067.

Weijer, C. J. (2004). Dictyostelium morphogenesis. *Current Opinion in Genetics & Development*, 14, 392–398.

Williamson, L. C., Ribrioux, S. C. P. C., Fitter, A. H., & Leyser, H. M. O. (2001). Phosphate availability regulates root system architecture in Arabidopsis. *Plant Physiology*, 126, 875–882.

Yoneyama, K., Xie, X., Kusumoto, D., Sekimoto, H., Sugimoto, Y., Takeuchi, Y., et al. (2007). Nitrogen deficiency as well as phosphorus deficiency in sorghum promotes the production and exudation of 5-deoxystrigol, the host recognition signal for arbuscular mycorrhizal fungi and root parasites. *Planta*, 227, 125–132.

Zhang, H., & Forde, B. G. (1998). An *Arabidopsis* MADS box gene that controls nutrient-induced changes in root architecture. *Science*, 279, 407–409.

Zhang, H., Jennings, A., Barlow, B. W., & Forde, B. G. (1999). Dual pathways for regulation of root branching by nitrate. *Proceedings of the National Academy of Sciences of the United States of America*, 96, 6529–6534.

AMOEBOZOA, FUNGI, AND RELATED THEORY

6 Cellular Slime Mold Development as a Paradigm for the Transition from Unicellular to Multicellular Life

Vidyanand Nanjundiah

Scope of This Chapter

The cellular slime molds (CSMs) go through a unicellular-to-multicellular transition and back in each life cycle: in them, ontogeny gives every impression of recapturing phylogeny (Haeckel's "biogenetic law"). It seems almost self-evident that a study of CSM development must offer insights into how unicellular ancestors gave rise to multicellular descendants (Bonner, 2000). Oscar Brefeld, who was the first to identify and describe a CSM (*Dictyostelium mucoroides*, in 1869), was a fellow-countryman and contemporary of Haeckel. One cannot help feeling that if only Haeckel had made use of Brefeld's discovery and had known that the development of *D. mucoroides* involved true multicellularity—and not a plasmodial stage—the biogenetic law might have had a stronger impact.

The present essay highlights aspects of CSM development that can be used for hypothesizing a plausible evolutionary route to multicellularity via self-organization. The route will lay stress on the importance of phenotypic plasticity, rather than the recruitment of new genes, for the origin of a trait (the latter viewpoint was explored in Nanjundiah 1985). Here phenotypic plasticity includes the possibility of multiple stable developmental outcomes, namely the possibility of more than one cellular or multicellular phenotype being consistent with a given genotype and given environment (Nanjundiah, 2003). Supporting the notion of a weak correlation between genes and traits in the CSMs, comparative analyses of DNA and protein sequences indicate that morphological traits that were thought to be species-specific in fact exhibit a great deal of variation within a species and have originated independently in different clades (Romeralo et al., 2013). This raises the question whether genetic changes and natural selection could have played an ancillary, not primary, role during the evolution of multicellularity. New genes or new patterns of gene activity may have ensured the reliability of an outcome whose occurrence was made possible by preadaptations that were already present in single cells (Newman and Bhat, 2009). Further, multicellular CSM groups in nature are often genetically heterogeneous (Sathe et al., 2010)—unlike present-day metazoans, which are the clonal products of a fertilized egg. It will be argued that the problem of reproductive asymmetries, that is, of unequal allocations to the primitive "germ line," may not be as serious as one might think (in line with an idea floated recently by Newman, 2011).

What follows begins with a sketch of the standard CSM life cycle and its variants. The sketch will cast doubt on the importance of genetic differences for morphological diversification within the CSMs. Evidence from other microorganisms whose ancestors made the transition from unicellular to multicellular life independently will reinforce the doubt. It will be seen that the range of developmental variations that can be accommodated within a single genotype encompasses much of what have been thought of as defining morphological traits of many CSM species. Next I present observations on CSM development and group behavior under three heads ("A Probabilistic Basis for Cell Differentiation," "The Genetically Heterogeneous Nature of Groups," and "Ways of Resolving Reproductive Conflicts"), in each case ending with an "Interpretation." Taken together, the Interpretations amount to saying that irrespective of whether or not groups consist of clones, phenotypic plasticity at the level of single cells and complex intercellular interactions can sustain multicellular development with reproductive division of labor. The next section, "Generalizing from the CSMs," shows that the Interpretations made in respect of the CSMs either have direct parallels in other organisms or are understandable in terms of fairly general properties of single cells. The pertinent evidence is provided in three corresponding subsections ("Phenotypic Variation of Stochastic Origin," "High Genetic Relatedness is not Essential for Cooperation," and "Group Formation by Self-Organization"). The chapter will conclude by summing up the points made earlier and speculating that the behavioral repertoire of a unicellular ancestor was sufficient for the evolution of multicellularity. The main role of evolution by natural selection may have been to ensure the reliability of multicellular life cycles and make a facultative step constitutive; to improve the signal-to-noise ratio of the unicellular→multicellular transition, so to speak.

Features of Cellular Slime Mold Development

Aggregative Development and Variants

Present-day molecular phylogeny assigns the Dictyostelids or cellular slime molds—the two terms will be used interchangeably here—to the Mycetozoa ("slime molds") along with two sister groups, the Myxogastrids (or Myxomycetes) and Protostelids (Brown and Silberman, 2013; Romeralo and Fiz-Palacios, 2013). It used to be thought that the Dictyostelid lineage diverged from the main animal clade after the higher plants and before the fungi, but molecular phylogenetics makes that appear unlikely. Rather, according to a recently published eukaryotic tree (Burki, 2014), the Mycetozoa fall within the larger eukaryotic "supergroup" Amoebozoa, which constitutes a separate clade from the sister supergroup Ophisthokonta that contains the fungi and animals (see figure 6.1).

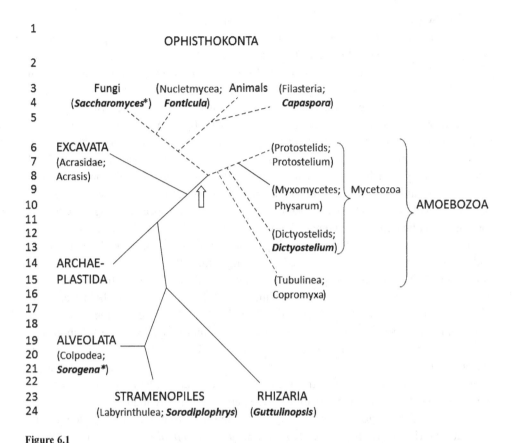

Figure 6.1
Tree of eukaryotes that displays independent origins of aggregative multicellular development; modified from several sources including Shadwick et al. (2009), Brown et al. (2011), Brown and Silberman (2013), Romeralo and Fiz-Palacios (2013), and Burki (2014). Representative generic names are highlighted in italicized bold letters. Not all branches are shown. Names in large capital letters stand for supergroups; the arrow indicates the "most popular hypothesis" for where the tree may be rooted (Burki, 2014; the Stramenopiles, Alveolata, and Rhizaria have recently been clubbed together under a single supergroup with the acronym SAR). Highlighted genera contain species that have a unicellular stage and proceed to build sorocarps after aggregation. (Capaspora shows aggregative multicellularity under laboratory conditions but has not been reported to form a sorocarp.) The asterisk after Sorogena is to indicate that it is a ciliate (Olive and Blanton, 1980); the asterisk after Saccharomyces, a yeast, is meant to indicate that it has been shown to be capable of multicellular development under synthetic laboratory conditions (Ratcliff et al., 2012; Włoch-Salamon, 2014; Conlin and Ratcliff, this volume). A single amoeba of Protostelium can form a sorocarp and Physarum amoebae fuse upon aggregation and form giant multinucleate cells. So far, no case of aggregative multicellularity has come to light in the Archaeplastida.

Our knowledge of the details of CSM development comes largely from the study of *Dictyostelium discoideum*, and most investigations on evolutionary aspects of social behavior have been carried out on *D. discoideum* and *Dictyostelium giganteum*. Unless specified otherwise, the information that follows is based on information gathered from one of these two species. The aspects of cell behavior that are highlighted in this chapter are so general that it seems reasonable to assume that they reflect generic capabilities possessed by the progenitor of the CSMs. The books by Bonner (1967, 2009), Raper (1984), and Kessin (2001) give a comprehensive and, with the exception of information derived from gene, protein, and genome sequences, almost up-to-date account of CSM biology, ecology, and behavior. Bonner (1982, 2009) focuses on evolutionary aspects. Kaushik and Nanjundiah (2003) and Nanjundiah and Sathe (2011, 2013) discuss the evolution of multicellularity in the CSMs from a sociobiological perspective. Original citations for statements that are not referenced below can be found in the above sources.

Dictyostelids are ubiquitous in soils, especially forest soils, and animal dung (Raper, 1984; Sathe et al., 2010). A life cycle comprising a unicellular and multicellular phase along with the formation of a spore-bearing fruiting body is characteristic of CSMs and other sorocarpic amoebae. The sorocarpic amoebae are heterotrophic amoeboid organisms that live freely during a phase of vegetative growth, after which they aggregate and form a structure for dispersal known as the sorocarp or fruiting body. As figure 6.1 indicates, they have been discovered in five supergroups besides the Amoebozoa. On present count they are believed to represent six, not seven, independent origins of sorocarpic aggregative multicellularity (two occur within the Amoebozoa; Brown and Silberman, 2013). Sorocarpic amoebae exhibit diverse morphologies and patterns of differentiation within the same broad life cycle. Myxogastrid amoebae aggregate and fuse to form multinucleate plasmodia that give rise to fruiting bodies (e.g., *Physarum polycephalum*; Guttes et al., 1961). Among Protostelids, a single amoeba can sporulate, releasing an extracellular stalk in the process (Olive and Stoianovitch, 1960). With several caveats, Romeralo and Fiz-Palacios (2013) place the origin of the CSMs to 600–691 MYr ago, that is, well before the Proterozoic–Phanerozoic transition. Kar and Singh (2014) have identified fossil spores dated to ~64–67 MYr ago as Myxogastrids, but to this untrained eye they look as if they belong to a CSM.

Under standard laboratory conditions unicellular and multicellular phases alternate in the CSMs as part of an asexual life cycle (see figure 6.2, left; Bonner, 1967; Raper 1984). Multicellularity is a collective defense against the threat posed by starvation. During the free-living amoeboid stage, cells feed, grow, and divide by mitosis until the food runs out. That triggers the onset of the multicellular or developmental stage. During it, anywhere from a few tens to a million amoebae aggregate, migrate in a group (as "slugs") to the soil surface—probably losing some cells along the way—and transform themselves into a fruiting body consisting of a ball of live spore cells held aloft by an erect stalk of dead cells (or, in some species, a secreted extracellular stalk).

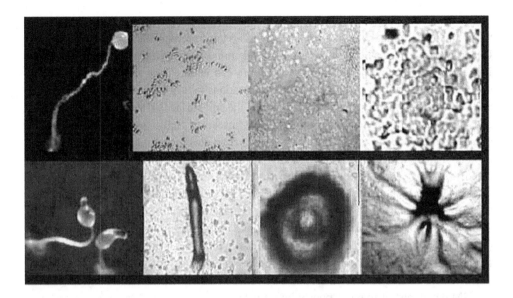

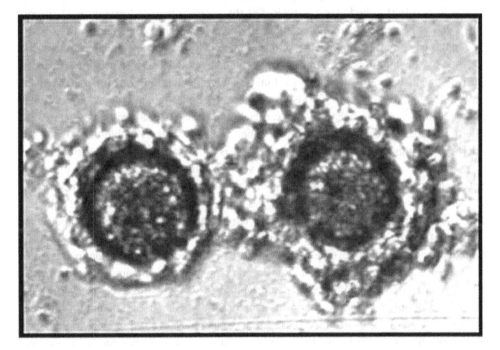

Figure 6.2
Left, asexual life cycle of *D. discoideum*; Right, sexual stage (macrocysts) forming in *D. giganteum*. Going clockwise from the upper left, the life cycle panels show a mature fruiting body with spore mass and stalk, spores, newly germinated amoebae, feeding amoebae, amoebae that have begun aggregation, a finished aggregate, a migrating slug, and a fruiting body being formed. The full sequence typically takes ~24 hr under standard conditions. The fruiting body is ~1 mm tall, as is the slug; amoebae are ~10 μm across and spores ~10 μm long; the macrocysts are ~ 100 μm across. (Photographs by M. Azhar and S. Sathe.)

Spores can be compared to the germ line of animals, and stalk cells to the soma (though the comparison cannot be pushed too far; meiosis is not a prerequisite for spore formation). Under favorable nutrient conditions each spore germinates and releases an amoeba, thereby initiating the next developmental cycle. A spore can endure long periods without food, and the slug and fruiting body are evidently adaptations for effective dispersal. Slugs make use of photo- and thermotaxis in order to reach the soil surface. The spore mass is sticky, and spores can be dispersed passively by passing insects as well as by birds and mammals. The CSM life cycle has all the hallmarks of metazoan development (Bonner, 2000): intercellular signaling, spatially and temporally patterned cell differentiation, polarity, coordinated tissue movements, cell–cell and cell–substrate adhesion and regulated cell death. The role of the biotic environment in the CSM life cycle is largely unexplored. Amoebae move toward food bacteria by sensing the folic acid released by the latter (Pan et al., 1972); *Aerobacter aerogenes* produces a factor that stimulates spore germination in *D. discoideum* (Hashimoto, Tanaka, and Yamada, 1976). Ellison and Buss (1983) discovered that the fungus *Mucor hiemalis* had to be present nearby for multicellular development to occur in a natural isolate of *D. mucoroides*. Such scattered evidence apart, the ecological context of CSM development remains poorly investigated.

The life cycle sketched above is typical of the aggregative development of *D. discoideum* under the usual laboratory conditions. However, variations abound across CSMs. The slug phase may be absent, the stalk may be acellular, and a cellular stalk may be released by the slug during migration. More fundamentally, CSM amoebae can react to starvation in three other ways that do not involve aggregation. In two of them they remain unicellular: unlike metazoans, CSMs do not have to go through an obligate multicellular phase. First, a single amoeba can encyst itself and form a stress-resistant structure, the microcyst (Blaskovics and Raper, 1957). When food becomes available once again, the microcyst germinates and releases the amoeba. It is not known whether microcysts can be dispersed, and if so, how. Second, not all starved amoebae that could have joined an aggregate do so; a small fraction remains single and undifferentiated (Dubravcic et al., 2014). In the long term, an amoeba that stays solitary will die of starvation and dehydration. In the short term, if food reappears soon enough near the same place, an amoeba that opts out of aggregating can begin reproducing before one that has formed a microcyst or joined an aggregate and become a spore. That is because the latter two options entail a substantial waiting time for germination to be completed and the amoeba to reemerge. Third, if an aggregate contains amoebae of complementary mating types, they can fuse and form a diploid "giant cell" that turns cannibalistic, feeds on the remaining amoebae, and ends up encysting itself in the form of a large macrocyst (see figure 6.2, right; Blaskovics and Raper, 1957; Filosa 1979). When food reappears, the macrocyst germinates, undergoing meiosis in the process, and the haploid meiotic products begin feeding and dividing. Microcysts and macrocysts have not been identified in many CSM species and may not occur in them.

Four modes of responding to starvation were listed above. Which ones are chosen must depend on a combination of genetic, environmental, social, and chance factors. Formally, the behavioral alternatives open to a starved CSM amoeba can be described in terms of the conditional probabilities of four mutually exclusive strategies. The amoeba can (1) remain single and undifferentiated, (2) form a microcyst, (3) join an aggregate and become a spore or stalk cell, or (4) join an aggregate and contribute to forming a giant cell and, subsequently, a macrocyst, or be consumed by the giant cell. On this view, the alternatives reflect the operation of an intrinsically stochastic process which is influenced, or biased, by the cell's genotype, its phenotype, its physical and biotic environments, and, for a cell that develops as part of a group, also the social environment provided by the amoebae with which it communicates. In effect, a starved amoeba hedges its bets regarding the best course to adopt. If the probability of a particular strategy is close to one, it means that the corresponding outcome is more or less certain, or quasi-constitutive. By keeping more than one option open, the choice of outcome is made facultative.

Poor Correlation between Genes and Morphology

Even though the evidence is mostly indirect, it appears that in the CSMs, differences in multicellular morphology are poorly correlated with genetic differences. Both between different Dictyostelids and within the same Dictyostelid species, there are variations in multicellular morphology and patterns of differentiation (Bonner, 1967; Olive, 1975; Raper, 1984). Fruiting bodies can consist of an unbranched stalk with a single spore mass at the apex (e.g., *Dictyostelium mucoroides, D. discoideum*) or with spore masses distributed along the length of the stalk (e.g., *Dictyostelium rosarium*). The stalk can be branched, with more than one branch at each horizontal level and a spore mass at the tip of each (e.g., Polysphondylium). The stalk can arise directly from the substrate (*D. giganteum*) or rest on a disc (*D. discoideum*) and can be cellular or, as in Acytostelium, acellular.

In *Fonticula alba*, a sorocarpic amoeba belonging to the Ophisthokonta, a sister group to the Amoebozoa, all amoebae differentiate into spores and form an extracellular stalk (Worley et al., 1979). The Acrasids belong to the Excavata, another sister group of the Amoebozoa. Their fruiting bodies have the appearance of amoebae piled on top of each other, possibly with secondary branches; all amoebae are viable (e.g., *Acrasis rosea*; Olive and Stoianovitch, 1960). Cell death as an obligatory feature of development is absent in the Acrasids but is present in *Guttulinopsis vulgaris* (Olive, 1965), classified under yet another sister group, the Rhizaria.

Evidently the overall strategy of achieving multicellularity and forming a fruiting body can be implemented by using a variety of tactics. It is tempting to categorize the tactics in terms of evolutionarily primitive and derived traits. For example, one might think of the following plausible evolutionary sequence (also see Nanjundiah, 1985): (1) aggregation without differentiation; (2) aggregation followed by a fruiting body that lacks differentiation among cells; (3) a fruiting body that secretes an extracellular stalk; and, last of all, (4) a fruiting body with differentiation into a cellular stalk and spores. Each step involves a modification of the previous step, and changes are cumulative. The Acrasids and Rhizaria seem to have retained ancestral evolutionary features; morphologically distinct spore and stalk cells appear to have arisen in a separate lineage that led to the Dictyostelids. However, molecular studies show that this is the wrong way of looking at these relationships. Recently Romeralo et al. (2013) carried out a comprehensive study of CSM phenotypes in relation to their molecular phylogeny, covering 99 species. The investigation involved amino-acid sequences of 32 orthologous proteins and small subunit ribosomal DNA sequences. The authors concluded that the significant level of plasticity in CSM development is likely to be a consequence of colonial multicellularity. I suggest that on the contrary, plasticity in multicellular outcomes may have been implicit in preexisiting cellular behaviors. Here are some reasons for saying so.

In very small fruiting bodies of *Dictyostelium lacteum* the stalk can be entirely cellular (the usual condition) or partly acellular (Bonner and Dodd, 1962). Multicellular aggregates of Acytostelium erect themselves by secreting an extracellular stalk (Raper and Quinlan, 1958). The extracellular stalk occurs in at least two Acytostelium species that belong to different clades; one of the clades also contains species with cellular stalks (Romeralo et al., 2013). In the course of reporting on compatibilities within and between CSM species Bonner and Adams (1958) remark, "One of the curious facts that comes from our experiments is that specific differences between strains of one species are as great, or even greater, than those between different species" (p. 352). *Protostelium* fruiting bodies are mostly unicellular but sometimes consist of two to four cells (Raper, 1984); sparse cultures of *Acytostelium* occasionally contain single-celled fruiting bodies (observation by B. M. Shaffer, cited in Bonner, 1967). Romeralo et al. (2013) say, on the basis of a comparison of traits across 99 CSM species referred to earlier, that there is "a fairly scattered distribution of character states … with many states reappearing multiple times in different clades" (referring mainly, though not exclusively, to the size and shape of multicellular structures; Discussion, p.7 in Romeralo et al., 2013). They go on to observe that the morphology and distribution of fruiting bodies (unbranched/branched, solitary/clustered) is strongly dependent on initial cell density. If this is what a study of development under a tightly controlled laboratory setting shows, it is not unreasonable to think that when monitored across a range of ecological conditions, the correlation between the details of development among the CSMs (and likely other protists that show aggregative sorocarpic development), in particular morphological outcomes, and genetic differentiation, is weak at best.

Thus, with regard to the "standard" multicellular morphologies in the CSMs, interspecies differences are blurred by the variations in developmental pattern that can be elicited from the same species. Genetic differences between species are at best weakly correlated with the morphologies that are considered typical of them. When considered within a narrow range of physical, social, and biotic conditions, the same genotype can give rise to phenotypes that are sufficiently distinct for them to have been thought of as species-specific traits. The conventional attitude is to regard them as evolutionary consequences of natural selection having acted on phenotypes that came about because of genetic variations that happened to occur independently in different lineages. I argue that there is also a case for ascribing their origin to the dynamics that operates among cells that possess certain properties to begin with, properties which can lead to diverse stable multicellular states. As we have seen, the range of morphologies displayed by fruiting bodies in other sorocarpic amoebae overlaps with what is seen in the CSMs. It is apparent that the same "characteristic" of multicellular life has evolved more than once and in different clades. At the same time, other aspects of morphology, for example those related to cytological differences between (dead) stalk and (live) spore cells, as well as differences in biochemistry and physiology, would seem to demand the incorporation of new patterns of gene activity or of new genes altogether, via conventional adaptive evolution.

Next I turn to findings from *D. discoideum* and *D. giganteum* that direct attention to cellular properties that potentiate successful multicellularity.

A Probabilistic Basis for Cell Differentiation

On the face of it, differentiation into stalk and spore is contingent on the normal life cycle being followed. It would appear that a cell has to pass through the multicellular phase in order for its choice of differentiation pathway to become evident. Nevertheless, three lines of evidence indicate that there is an underlying probabilistic basis for differentiation that is autonomous to the cell and can be influenced by environmental signals but does not require multicellular development.

First, meaningful differences in cell phenotype exist in pre-aggregation amoebae too (reviewed in Nanjundiah and Saran, 1992). The differences are evocative of the classical embryological concept of a determined state preceding terminal differentiation, except that here determination appears to reflect a cell-autonomous decision. Experiments dealing with pre-aggregation nutritional status (Leach et al., 1973), cell cycle phase at starvation (McDonald and Durston, 1984), and cellular calcium (Azhar et al., 1996) indicate as much. When amoebae grown on a high concentration of sugar are mixed in equal proportions with amoebae grown without sugar, the "high-sugar" cells are in a majority in the eventual spore population (Leach et al., 1973). Cells that are starved at different phases of the cell cycle show differences in their spore- and stalk-forming tendencies after they are mixed (McDonald and Durston, 1984). When a clone of cells is grown in a liquid medium with shaking, so that all of them are exposed to the same environment, their calcium levels follow a bimodal distribution. "Low-calcium" cells tend to form spores whereas "high-calcium" cells show a tendency to die and contribute to the stalk; cell cycle phase and calcium show the expected correlation (Saran, 1999; Azhar et al., 2001). In all these instances, it is not the internal state of a cell per se that determines developmental fate but its state relative to that of the other cells. If all cells are grown on high (or low) glucose or are harvested at approximately the same cell cycle phase or are from the high-calcium (or low-calcium) class, multicellular development proceeds more or less normally (except that the proportions of differentiated cells can be slightly altered; Leach et al., 1973; Baskar et al., 2000). Once again, one can make a comparison with a concept from embryology: the determined state is reversible, and determined, but not terminally differentiated; prestalk and prespore cells can interconvert (see figure 6.3; Raper, 1940). Predispositions that are present before the onset of multicellularity have a significant, but not irreversible, impact on the course of differentiation after a cell becomes part of a collective. The predispositions can be detected among cells of the same genotype that are in the same environment. They reflect phenotypic differences that arise spontaneously within a clonal population that is maintained in a spatially homogeneous external environment. As with differences among embryonic cells that adopt different developmental pathways, the phenotypic differences are not based on genetic differences. However, if genetic differences exist among pre-aggregation cells, there is no reason why they should not have a bearing on post-aggregation differences in phenotype. Over and above the differences that arise autonomously within a genetically uniform population, genetic variation can act as an additional source of bias on developmental cell fate (Bonner and Adams, 1958; Sathe et al., 2014).

Second, with a certain probability, _D. discoideum_ amoebae can differentiate after being spread on a surface at such low density that for all practical purposes they are isolated from one another. A chemical stimulus is provided via the liquid medium that surrounds the cells. The probability with which the stimulus induces a cell to differentiate is influenced by factors that influence the cell's physiology—for instance, its nutritional status (Thompson and Kay, 2000). The nature of the stimulus decides whether amoebae differentiate into stalk cells (in which case DIF-1 or differentiation-inducing factor is a possible stimulus; Town et al., 1976) or spore cells (in which case the stimulus includes cyclic AMP; Kay, 1982). The proportion of amoebae that differentiate increases with the strength of the stimulus. Differentiation is all or none; cells that do not differentiate presumably die. The most straightforward interpretation is that the stimulus substitutes for signals that normally derive from other cells and that differentiation itself is a probabilistic event that takes place once a stimulus threshold has been crossed.

The third piece of evidence in support of differentiation as a probabilistic process comes from precise counts of stalk and spore cell numbers carried out on tiny fruiting bodies consisting of ~100 or fewer cells (Nanjundiah and Bhogle, 1995). The small numbers make it possible to estimate the mean proportion of cells that differentiate into spores (or stalk) and, crucially, the extent to which the proportions fluctuate about the mean—the signal and the noise, as it were. The results can be compared with theoretical models for the proportioning of cell types. The simplest possible model, namely, that the differentiation of a single cell is an autonomous event that takes place with a fixed probability, already

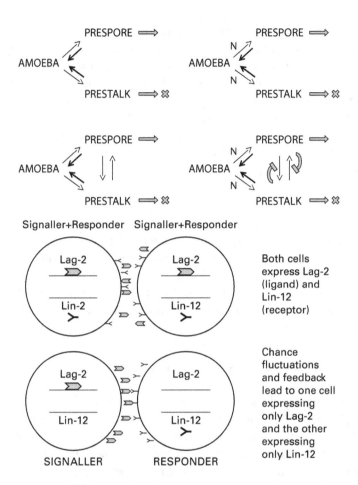

Figure 6.3
Top panel, progressively complex formal models for the regulation of cell type proportions in *D. discoideum*; bottom, the anchor cell/ventral uterine precursor decision in *Caenorhabditis elegans*. Top: "prespore" and "prestalk" stand for a presumptive spore and stalk cell that is undifferentiated but on the way to becoming part of the spore mass or stalk, respectively. A third cell type, the basal disc, consists of dead cells and its cell number appears to be regulated coordinately with that of stalk cells (Nanjundiah and Bhogle, 1995). Thin lines ending in arrows indicate differentiation steps (including interconversions between prespore and prestalk cells) that occur on a probabilistic basis; thick arrows stand for terminal differentiation (presumptive spore to spore and presumptive stalk cell to dead stalk cell, indicated by cross). Arrows directed to "AMOEBA" indicate that when presented with food, a prespore or prestalk cell can revert to become a feeding amoeba. Thick lines ending in short bars show negative feedbacks. "N" conveys that the probability of the relevant decision is dependent on the total number of cells in the aggregate. The spatial pattern that develops is assumed to follow from a sorting out of similar cell types and does not form part of the models (adapted from Nanjundiah and Bhogle, 1995). The bottom panel is based on Wilkinson et al. (1994) and represents the formal basis behind a cell differentiation event in the *C. elegans* hermaphrodite gonad. Spontaneous fluctuations (autonomous to each cell) followed by feedbacks (due to the other cell) lead to the differentiation of complementary cell types. See the text for details.

yields a rough fit to experimental measurements—to the signal. But the noise, the extent to which proportions vary about the mean, is much lower than what the simplest model predicts. The fit between theory and observation improves considerably after the model is made more realistic by adding negative feedbacks between differentiated cells and a mechanism for sensing the overall size of the group, the sorts of inputs one might anticipate from intercellular communication. It appears that the decision to become a stalk or spore cell is made probabilistically. To a first approximation, it is also taken autonomously. To a further approximation, the probability is modified as a result of signals from other amoebae (see figure 6.3, top panel).

Interpretation

An isolated amoeba can differentiate into one of two complementary cell types, either a spore or stalk cell, autonomously and with probabilities that depend only on its physiology and the external environment. The presence of other amoebae modulates these probabilities in a manner that depends on their own phenotypes. The influences are reciprocal, so cellular behavior feeds back on itself. The differentiated phenotype of a cell is ultimately dependent on the phenotypes of all cells with which it interacts (Nanjundiah and Sathe, 2011, 2013).

The Genetically Heterogeneous Nature of Groups

Multicellular CSM groups in nature are often genetically heterogeneous (i.e., chimaeras). Despite complexities in the natural multipredator–multiprey system (Ketcham et al., 1988) in which CSMs participate, it seems that many species occupy the same ecological niche. Filosa (1962) discovered heterocaryosis, the coexistence of more than one genotype within the same spore mass, in single spore masses of *D. mucoroides*. Eisenberg (1976) showed that a 6-mm soil core could yield up to 4 CSM species and 100 CSM propagules; the numbers varied substantially between adjacent samples. The species did not discriminate between different bacterial prey (when tested in the laboratory). Both interspecific and intraspecific variation occurred over small length scales: Ketcham and Eisenberg (1989) obtained clonally diverse isolates of *Polysphondylium pallidum* from 1.1-cm soil cores. Fortunato et al. (2003) made a similar finding with *D. discoideum* over a distance of 6 mm. Sathe et al. (2010) found that a speck (~1 mm) of forest soil or a fragment of animal dung could contain many genetically distinct CSM strains belonging to the same species as well as more than one CSM species. On the other hand, Gilbert et al. (2009) discovered a single clone of *D. discoideum* distributed across 12 m of a cattle pasture. To put the distances in perspective, a typical CSM aggregate is ~1 mm across, and migrating slugs can cover several centimeters. These observations show that multiclonal and clonal CSM groups occur in nature, the former perhaps more frequently. By using random primers to amplify polymorphic DNA sequences (RAPD), Kaushik and Nanjundiah (2003) discovered at least 9 *D. giganteum* clones in one fruiting body. When collections of undisturbed forest soil (250 μm–1 mm) or dung (2 cm) from a variety of wild mammals were transported to the laboratory without disturbing them, CSM fruiting bodies belonging to one or more of 12 species appeared on them in a matter of days, and more often than not, they were chimaeras (Sathe et al., 2010). Among those that were examined in detail, all 6 *Dictyostelium purpureum* fruiting bodies and 9 out of 11 of *D. giganteum* contained 3 to 7 clones. The remaining 2 fruiting bodies of *D. giganteum* yielded a single clone. The limitations of sampling mean that the measured number of clones per fruiting body must be taken to be lower limits (Sathe et al., 2010).

Interpretation

In nature, the multicellular stage of CSMs often involves several genotypes.

Ways of Resolving Reproductive Conflicts

Does reproductive asymmetry among the members of a chimaeric group in respect of the efficiency of spore formation rule out long-term coexistence? Given that spores and stalk cells are analogous to the animal germ line and soma, respectively, the occurrence of multiclonal groups raises the question of differential reproductive success over life cycles. Sathe et al. (2014) followed the course of development after mixing equal numbers of amoebae of *D. giganteum* (from 5 genetically distinct natural isolates) and *D. purpureum* (from 3 genetically distinct natural isolates) in all 28 pair-wise strain combinations. Not surprisingly, members of the two species were more or less incompatible. Interspecies mixes sorted out almost entirely into separate cell masses, even when they had formed a single composite aggregate. The development of a chimaera composed of wild-type strains of the same species was superficially indistinguishable from the development of a clone of any strain by itself. Intraspecies mixes showed two patterns of behavior. Some pairs were highly compatible and formed chimaeric fruiting bodies most or all of the time. The mutual compatibility of the remaining pairs of strains fell midway between the above two extremes. A subset of intraspecies mixes formed chimaeric fruiting bodies whereas in others the spores came predominantly from one or the other strain. Irrespective of the degree of compatibility between two strains of a species, most of the time there was a reproductive skew. The component genotypes were disproportionately represented in the spore population; the more compatible a pair of strains, the lower was the skew. In short, the genetic differences that exist between individuals that form chimaeric multicellular groups are not trivial but have a bearing on their (relative) fitness.

That raises the obvious question, can different genotypes of the same species coexist stably, and if so how? A trivial answer is that the observed genetic polymorphisms are misleading because they are transient. The one piece of direct evidence we have indicates that the trivial answer may be wrong. Francis and Eisenberg (1993) found that allowing for DNA sequence changes on account of the movement of transposable elements, the same 39 distinct genotypes (out of 54 isolates) of *D. discoideum* were present in repeated samples from the same location. However, it might be argued that what Francis and Eisenberg found was exceptional, and that in the absence of countervailing factors, selection ought to lead to a single genotype dominating in each species (allowing for new variations introduced by rare mutations). It might equally well be argued that intraspecies genetic polymorphisms are neutral and keep shifting as the genetic makeup of the population changes on account of mutation, recombination (allowing for sexual reproduction), and drift. From what we know about CSMs, the first possibility appears implausible. The second is unlikely, too, because mixing experiments in the laboratory show that wild isolates differ in fitness-related traits. Selective differentials on account of reproductive skew are quite large, at least in pair-wise mixes, and sexual reproduction appears to be rare (Fortunato et al., 2003; Kaushik et al., 2006; Sathe et al., 2014). Were that all, natural selection should rapidly lead to the strain that is most successful at forming spores to eliminate the others. Let us consider the alternative, which is that the observed polymorphisms are stable outcomes of balancing selection. Stability can be accounted for in more than one way:

1. Stability may result from complex interactions in multigenotype assemblages. Kaushik et al. (2006) showed what could happen: among three strains of *D. giganteum*, each 1:1 pair-wise mix displayed a significant reproductive skew, but the skew was reduced significantly in a 1:1:1 mix of all three strains.

2. Mixes in which the components are present in equal numbers at the start constitute a special case and are almost certainly restricted to experiments conducted in the laboratory. In nature the components would be expected to occur in all possible proportions. Under those circumstances, when considered over different strain combinations, the dynamics of intercellular interactions may stabilize the population-wide proportions of genotypes. For instance, on mixing amoebae of a pair of *D. mucoroides* isolates in different proportions, Buss (1982) discovered that one of them formed fewer spores than expected when the proportion of its amoebae was high and more spores when the proportion was low. Predictably, there was a stable intermediate composition at which the proportions did not change. We have found a similar frequency-dependent efficiency of spore formation in *D. giganteum* isolates (Sathe et al., unpublished).

3. The relative degree of success in differentiating into a spore is only one of the many factors that go into deciding the lifetime fitness of an amoeba. The rate of growth, pace of development, and ability to reach the soil surface also impinge on fitness. The existence of trade-offs between fitness-dependent traits would imply that spore-forming efficiency by itself is an incomplete measure of fitness. Trade-offs of different kinds have been put

forward as possible explanations for the stability of genetically diverse multicellular groups in *D. discoideum* (Tarnita et al., 2015; Wolf et al., 2015).

4. Potential fitness imbalances between strains could be offset if the same genotype is sometimes in a clonal aggregate and sometimes in a multiclonal aggregate. Genotypes whose members suffer a loss in fitness when they develop in multiclonal groups may recoup their numbers during life cycles in which they develop clonally. This could be another factor contributing to long-term coexistence (Kawli and Kaushik, 2001; Sathe et al., 2014), similar to what Chuang et al. (2009) have demonstrated with bacteria.

5. We have only a superficial understanding of the ecological context within which CSMs live. Factors ignored in our analysis, such as the role of the physical environment, interactions with predator and prey species, and modes of dispersal may permit the long-term persistence of genotypes.

6. Finally there is the intriguing possibility raised by Bonner (1967), based on Filosa's (1962) finding, that heterocaryosis may confer the benefit of functional complementation and indeed act as a partial substitute for sexuality.

Interpretation

Individuals of one genotype may do better than another when tested for spore formation in 1:1 mixes. This does not rule out the stable long-term coexistence of many genotypes. The multiclonal aggregate is a viable entity in the CSMs.

Generalizing from the CSMs

It has been estimated that the unicellular-to-multicellular transition occurred independently in evolution on at least 25 occasions (Grosberg and Strathman, 2007). At least two of them took place in the amoebozoan lineage, one each in the Tubulinids and in the ancestors of the Mycetozoa (see figure 6.1). Bonner (see the foreword to this volume) states that the ancestor of today's metazoans is unlikely to have had a CSM-like life cycle. Fisher et al. (2013) reach a similar conclusion on the basis of a comparative analysis. They claim that obligatory multicellular life with many cell types, including sterile (somatic) cells, was more likely to have originated in clonal groups than in nonclonal groups like the CSMs. The common inference is that colonial multicellularity, not aggregative multicellularity, was the precursor to the metazoan form of life. Newman (this volume) has advanced an interesting line of argument to support the hypothesis that the egg, traditionally placed at the beginning of metazoan development, could have evolved from a nonclonal assemblage of cells. The hypothesis appears to fit well with observations that support the so-called hourglass model of development (see Kalinka et al., 2010; Piasecka et al., 2013, offer a mixed assessment). Irrespective of how the story of metazoan origin(s) unfolds, an organism that switches routinely between unicellular and multicellular phases should offer useful insights into single-cell properties that were responsible for mediating the transition from a unicellular ancestor to a metazoan progenitor. In this chapter I have tried to put together information on CSM development in order to deduce cellular properties that could have fostered the origin of multicellularity in these organisms, and by extension, in other protists. One conclusion is that the probabilistic (stochastic) generation of phenotypes played an important role; another is that the relevant probabilities were influenced (biased) by environmental factors and feedbacks based on cell–cell interactions; a third is that as long as cellular phenotypes were compatible, shared genetic interests among cells may not have played a determining role in the origin of life cycles with a multicellular component. I proceed to cite examples of cell behavior in other systems that support these conclusions before outlining plausible mechanistic models that support the overall line of reasoning.

Phenotypic Variation of Stochastic Origin

On top of the commonly invoked deterministic (albeit context-dependent) influence of the cell's own genotype, the phenotype of a cell depends on at least three additional inputs: one is entirely probabilistic or stochastic, another is the external physical environment, and the third is the social environment comprising other cells of the same type. Stochastically generated phenotypic variation used to be disregarded as developmental "noise," a word that conveys a hint of something that is a nuisance and a distraction from what is significant—namely, the underlying "signal." This attitude is no longer tenable. Stochastic phenotypic variation is a pointer to the existence of phenotypic plasticity, which in turn can be significant for evolutionary change via at least two routes (Nanjundiah, 2003). One is genetic assimilation, which can operate if the initial population is genetically heterogeneous and reproduces sexually. The other is the Baldwin effect, which does not require genetic variation or sexual reproduction. Stochastic developmental variation is a ubiquitous and phylogenetically ancient phenomenon (Vogt, 2015). The stochastic or "noisy" component of a cell's phenotype is ultimately based on the inherently probabilistic nature of the chemical processes that go on within the cell, including gene expression, interactions between genes and gene products, and metabolic pathways. Thanks to interactions and feedbacks, which are common in cellular biochemistry, a process that is basically stochastic is capable of yielding a rich array of outcomes, most importantly multistability. This means that there can be more than one outcome, each of them locally stable. At the same time, either on account of spontaneous fluctuations or an external trigger, one outcome can switch to another (reviewed in Qian, 2012).

Two examples from bacteria are striking for the mechanistic insights they provide. Because of the switch-like functioning of the lac operon of *Escherichia coli*, a genetically uniform population of cells displays two bistable phenotypes that originate stochastically (Novick and Wiener, 1957). The operon can be in one of two states, either uninduced ("off") or fully induced ("on"). The states vary randomly from cell to cell with the relative proportions of the two cell types depending sensitively on the extracellular concentration of the inducer. The states are propagated faithfully from mother to daughter, but occasionally, on account of positive feedback, a cell switches spontaneously from one state to the other (Ozbudak et al., 2004). A finding made by Pradhan and Chatterjee (2014) is especially relevant for our purpose. They discovered that reversible phenotypic switching in *Pseudomonas syringae* and *Xanthomonas campestris* bacteria occurred during the cell density–dependent phenomenon known as quorum sensing and led to the multicellular cooperative group known as a biofilm. Cells transited stochastically between states that were either responsive to quorum sensing signals or not, which means that a bacterium switched between phases that were predisposed to exhibit social or asocial behavior. Switching can also come about as a consequence of intercellular interactions. It can result from negative cross-feedbacks or a combination of positive and negative feedbacks between a pair of cells. Vulval development in *Caenorhabdits elegans* (Wilkinson et al., 1994) illustrates a parallel to the formal model proposed for *D. discoideum* (see figure 6.3, top panel). The choice of becoming an anchor cell (AC) or ventral uterine precursor (VU) is open to both of two undifferentiated neighbors in the hermaphrodite gonad; mutual interactions result in the two making distinct choices (Wilkinson et al., 1994; figure 6.3, bottom panel). The critical players are a cell surface receptor, LIN-12, and its cognate ligand, LAG-2. In a cell in which lin-12 activity has slightly increased on account of a spontaneous fluctuation, transcription from the lin-12 gene is further stimulated (positive feedback) and transcription from the lag-2 gene of the same cell is inhibited. This cell becomes a presumptive VU cell and does not make LAG-2. The LIN-12 receptor on its neighbor remains unbound and the lin-12 gene of the neighbor remains unactivated. The neighbor makes only LAG-12 and adopts the AC pathway. Thus exactly one cell becomes AC and the other becomes VU (Wilkinson et al., 1994). Once differentiation has occurred, the phenotypes of the two cells complement one another. Another example, remarkably reminiscent of the CSMs, comes from studies on a pancreatic cell line. When undifferentiated human PANC-1 cells (derived from an adenocarcinoma) are starved by depriving them of serum, they differentiate, exhibit the capacity to communicate via diffusible signals, and display glucose-dependent calcium oscillations (Hiram-Bab et al., 2012). The course of differentiation is not uniform but leads to a heterogeneous cell population with—once again—complementary cell types. A few cells make fibroblast growth factor (FGF2, which is released by the cell) but not its receptor. Other cells make the receptor (FGFR, which gets incorporated on the cell surface) but not the growth factor. The binding of extracellular FGF2 to receptor induces chemotactic movement toward the source of FGF2, which results in cell aggregation followed by the formation of a hormone-secreting pancreatic tissue-like precursor (Hardikar et al., 2003).

High Genetic Relatedness Is not Essential for Cooperation

Multicellular organisms are believed to represent the acme of social behavior. Not surprisingly, as with many contemporary discussions of the evolution of social behavior, a common genetic interest between the unicellular constituents is believed to be an essential requirement for the long-term stability of a multicellular life cycle. The analogy between the reproductive and nonreproductive individuals in a social group, on the one hand, and the germ line and soma of a multicellular animal, on the other, makes this clear. However, when the analogy is used to infer that the efficiency of group life (multicellular or social) must necessarily improve with an increase in the proportion of shared genes among its constituents, one is making a mistake. Equally, it is a mistake to infer from the analogy that just because it is genetically heterogeneous, a group is vulnerable to successful exploitation by a genetic variant that derives the benefit of group living without paying the cost. Here is a brief explanation of why, using examples drawn from aggregative multicellularity and social groups in multicellular organisms.

A possible reason was given by Darwin (1871/1981) and was put in the language of population genetics by Haldane (1932) among others: a trait that lowers the fitness of an individual within a group can spread in a population if groups containing such individuals are sufficiently more successful than groups in which the individuals are represented in lower proportions. Chuang et al. (2009) provided an elegant demonstration of this in a synthetic bacterial system. Another reason is the (at first) counterintuitive notion that natural selection can foster stability as often as it can lead to evolution. The supposed genetic advantage conferred by haplodiploid inheritance, which resulted in an average relatedness of 75% between sterile workers who were full sisters, was initially seized upon as the key element behind the evolution of eusociality in the Hymenoptera (Wilson, 1975). Today one knows that the workers in a single nest need not all be the offspring of a single mother and single father; therefore, the average relatedness in a hive can be quite low, even lower than 50% (see below). This realization has contributed to renewed interest in thinking about group-level advantages of social behavior as against thinking of groups only as collectives of individuals with potentially conflicting interests (Wilson and Wilson, 2007). Loope et al. (2014) found that effective paternity levels in colonies of the wasp *Vespula* varied between 1.5 and 2.1 for small colonies (*V. rufa* species group) and ~3.1 for large colonies (*V. flavopilosa*), and the mean intracolony relatedness was significantly less than 0.5 (the cutoff value below which helping the mother to raise sisters gives a worker a lower genetic reward than mating and reproducing herself) for the larger colonies. The authors point to enhanced genetic diversity as a possible advantage of multiple paternity. Even individuals of different species are reported to cooperate in forming bacterial communities (Sachs and Hollowell, 2012) and social insect nests (Hunt, 2009).

In the CSMs, some studies on *D. discoideum* have been interpreted as showing that high relatedness guards against "cheating" (Gilbert et al., 2007) whereas other studies are interpreted as showing poor correlation of relatedness and the propensity to "cheat" (Saxer et al., 2010). Observations on *D. giganteum* show that natural multicellular groups tend to be polyclonal (Kaushik and Nanjundiah, 2003; Sathe et al., 2010), are composed of genotypes whose individuals differ in fitness-related traits, and, as shown in pair-wise as well as three-way mixing experiments, form chimaeric groups in which rather than kinship per se, complex nonlinear interactions are involved in determining the allocation to the reproductive and nonreproductive pathways (Kaushik et al., 2006; Sathe et al., 2014). As mentioned before, it could be that genetic diversity in a multicellular group is an advantage, not a disadvantage. It has been conjectured that a trade-off between traits may be involved (Sathe et al., 2014). Perhaps the different genotypes complement each other functionally, as Haldane (1955) suggested in the case of heterokaryosis in fungi—incidentally, an example of aggregation leading to genetic diversity within a single multinucleate cell, as also happens in in the Myxomycetes. The word "cheater" conveys a misleading notion of what is going on and is best avoided. If genetic heterogeneity in the multicellular aggregate is maintained by natural selection, a "cheater" will have a lower lifetime reproductive fitness than the wild type and is better termed a "loser" (Kaushik and Nanjundiah, 2003).

Similar to social insects of different genotypes that cooperate within the same nest, cells belonging to different genotypes can function in concert and develop into a multicellular organism in spite of their genetic diversity. Chimaeric aggregates made using embryonic cells of goat and sheep (Fehilly et al., 1984), or of tissue primordia of chick and quail embryos, embryonic cells, and quail–chick chimaeras (Teillet et al., 1999) develop, to all purposes normally, and give rise to chimaeric adults even though the lineages of the two "parents" separated several million years ago. Ascidian chimaeras, which form readily by embryo fusion, often contain more than two members, and multicomponent chimaeras may indeed do better in some respects than two-component chimaeras (Rinkevich and Shapira, 1999). These cases show that given a certain level of conservation or compatibility of intercellular signals and adhesion mechanisms, genetic differences need not be a barrier to cells or incipient individuals living together as a multicellular entity.

Group Formation by Self-Organization

Mathematical models show that heterogeneous units can continue to coexist through cycles of individual and collective phases in spite of reproductive differentials during the individual phase of the life cycle. Kaneko's models (Kaneko, 2003, this volume) represent a cell as an oscillatory dynamical state (limit cycles) in the phase space defined by internal biochemistry. Cells are mutually coupled via diffusible intermediates. Isolated cells exhibit multistability: the same chemical constituents can exist in more than one stable configuration ("cell type"). When a cell divides into two, sister cells are clustered on the limit cycle, that is, their oscillations are quasi-synchronous. As numbers increase, different subsets of cells begin to form distinct clusters on the same cycle and, eventually, split up into different limit cycles ("differentiation"). Cell–cell adhesion allows spatial structure to emerge, and differential strengths of adhesion lead to the appearance of spatial patterns. The transition from one cell to an organized group of many cells is simply a consequence of the preexisting intracellular dynamics and elementary interactions between cells. The hidden agency behind differentiation is increase in size, or more accurately cell number, which harks back to a theme long advocated by Bonner (1998, 2003, 2006). Garcia and DeMonte (2013) consider "fission–fusion" events in which the collective or multicellular stage comes about by the random coalescence of single cells. An aggregate breaks up later to give rise to individual cells that constitute the next generation. Differences in adhesive strength ("stickiness") between cells make them more or less likely to join a collective. A sticky cell improves the cohesion of an aggregate (and so makes it last longer) but pays the cost of making the adhesion factor(s). According to the model, single cells and collectives can persist over generations, which is reminiscent of what is found in at least one CSM (Dubravcic et al., 2014; Tarnita et al., 2015). As De Monte and Rainey (2014) emphasize, the collectives of successive cycles are linked by genealogy (an overlap of individuals between collectives) rather than inheritance. In other words, for the long-term persistence of the system it is sufficient if each individual in a generation is descended from some individual in the previous generation, which makes genetic relatedness and the heritability of fitness differences among collectives a nonissue. These models demonstrate that a transition from one cell to many is sustainable on the basis of certain assumptions pertaining to single-cell properties and system dynamics alone. Similar to the CSM situation, aggregation and clumping (Włoch-Salamon, 2014), and multicellularity with spatial differentiation (Conlin and Ratcliff, this volume), occur in yeast, and multicellularity driven by phenotypic heterogeneity is found in bacteria (e.g., Pradhan and Chatterjee, 2014).

Summing Up

Natural selection is not evolution. (Opening sentence of The Genetical Theory of Natural Selection by R. A. Fisher; 2nd edition; Dover, New York, 1958)

Natural selection is very much behind a part of the picture sketched here, but it is not responsible for the whole picture. Since the goal is to explain a transition from solitary to group living in entities that almost certainly differed in fitness-related traits, natural selection is likely to have been present right from the start. Balancing natural selection has been invoked to account for the long-term persistence of multicellular aggregates. But at the onset of multicellularity, the outcomes of some developmental decisions are hypothesized to have been relatively neutral when averaged over many life cycles. Examples would be the choice between remaining single or joining an aggregate, joining one incipient aggregate or another, or developing a fruiting body with one or the other morphology. In fact, Bonner (2013) has proposed that morphological variation is on the whole neutral in microorganisms but not in macroorganisms. It is generally assumed that only that part of phenotypic variation that is underpinned by genetic variation is of interest for evolution by natural selection. However, today we know that variations in the phenotype that are nongenetic in origin can be inherited, at least for some generations, and when not, can be evoked de novo generation after generation (Waddington, 1953; Jablonka and Raz, 2009). Phenotypic variation makes it possible for natural selection to act, and the persistence of a phenotype, even in the short term, opens the field for genetic changes to stabilize it subsequently and so delink it from whatever caused it to appear initially.

In this chapter I have drawn attention to one way in which phenotypic variation can arise without genetic change, namely, via the spontaneous generation of alternative discrete phenotypes in the same genetic background and the same environmental conditions. Even when they occur in the same environment, different phenotypes need not mean different fitnesses. They may represent different equilibria (or steady states) of a multistable system, random phenotypes in the sense of Bonner (2013). As he has emphasized, they are the appropriate null models one should use when trying to explain the origin of phenotypic differences. The ultimate basis of spontaneous variation of this sort is the switch-like nature of genetic circuits, which in turn rests on the facts that (1) gene expression is an inherently stochastic process and (2) interactions between genes, proteins, and cells commonly involve feedbacks. In the course of time, if natural selection strongly favors one phenotypic alternative over another, a genetic change can make the favored alternative develop constitutively—that is, make its appearance autonomous and reliable. But the favored phenotype would have originated on account of generic properties of the system (Newman & Comper, 1990; Müller, 2007).

Acknowledgments

It is obvious how much the thoughts shared in this essay owe to the contributions of John Tyler Bonner, and impossible to convey the extent of my gratitude to him for many years of patient advice, instruction, and criticism. I thank Gerd B. Müller, Werner Callebaut, and colleagues at the KLI Institute for the convivial atmosphere which made it possible to develop these ideas; Stuart A. Newman and Karl J. Niklas for allowing me to present them; and Stuart A. Newman, H. Sharat Chandra, Laasya Samhita, Santosh Sathe, and an anonymous referee for useful comments on an earlier draft. This text is based on a talk given at the meeting The Origins and Consequences of Multicellularity held at the KLI Institute, Klosterneuburg, Austria, September 25–28, 2014.

References

Azhar, M., Kennady, P. K., Pande, G., Espiritu, M., Holloman, W., Brazill, D., et al. (2001). Cell cycle phase, cellular Ca2+ and development in *Dictyostelium discoideum. International Journal of Developmental Biology*, 45, 405–414.

Azhar, M., Kennady, P. K., Pande, G., & Nanjundiah, V. (1996). A Ca2+-dependent early functional heterogeneity in amoebae of *Dictyostelium discoideum*, revealed by flow cytometry. *Experimental Cell Research*, 227, 344–351.

Baskar, R., Chhabra, P., Mascarenhas, P., & Nanjundiah, V. (2000). A cell type–specific effect of calcium on pattern formation and differentiation in *Dictyostelium discoideum. International Journal of Developmental Biology*, 44, 491–498.

Blaskovics, J. C., & Raper, K. B. (1957). Encystment stages of *Dictyostelium. Biological Bulletin*, 113(1), 58–88.

Bonner, J. T. (1967). *The cellular slime molds* (2nd ed.). Princeton, NJ: Princeton University Press.

Bonner, J. T. (1982). Evolutionary strategies and developmental constraints in the cellular slime molds. *American Naturalist*, 119, 530–552.

Bonner, J. T. (1998). The origins of multicellularity. *Integrative Biology*, 1, 27–36.

Bonner, J. Y. (2000). *First signals: The evolution of multicellular development*. Princeton, NJ: Princeton University Press.

Bonner, J. T. (2003). On the origin of differentiation. *Journal of Biosciences*, 28, 523–528.

Bonner, J. T. (2006). *Why size matters*. Princeton, NJ: Princeton University Press.

Bonner, J. T. (2009). *The social amoebae*. Princeton, NJ: Princeton University Press.

Bonner, J. T. (2013). *Randomness in evolution*. Princeton, NJ: Princeton University Press.

Bonner, J. T., & Adams, M. S. (1958). Cell mixtures of different species and strains of cellular slime moulds. *Journal of Embryology and Experimental Morphology*, 6, 346–356.

Bonner, J. T., & Dodd, M. R. (1962). Aggregation territories in the cellular slime molds. *Biological Bulletin*, 122, 13–24.

Brown, M. W., & Silberman, J. D. (2013). The non-dictyostelid sorocarpic amoebae. In M. Romeralo, S. Baldauf, & R. Escalante (Eds.), *Dictyostelids: Evolution, genomics and cell biology* (pp. 219–242). Heidelberg, Germany: Springer.

Brown, M. W., Silberman, J. D., & Spiegel, F. W. (2011). "Slime molds" among the Tubulinea (Amoebozoa): Molecular systematics and taxonomy of Copromyxa. *Protist, 162*, 277–287.

Burki, F. (2014). The eukaryotic tree of life from a global phylogenomic perspective. *Cold Spring Harbor Perspectives in Biology*, 6, a016147. http://cshperspectives.cshlp.org/.

Buss, L. W. (1982). Somatic cell parasitism and the evolution of somatic tissue compatibility. *Proceedings of the National Academy of Sciences of the United States of America*, 79, 5337–5341.

Chuang, J. S., Rivoire, O., & Leibler, S. (2009). Simpson's paradox in a synthetic microbial system. *Science*, 323, 272–275.

Darwin, C. (1981). *The descent of man, and selection in relation to sex*. Princeton, NJ: Princeton University Press. (Original work published 1871).

De Monte, S., & Rainey, P. B. (2014). Nascent multicellular life and the emergence of individuality. *Journal of Biosciences*, 39, 237–248.

Dubravcic, D., van Baalen, M., & Nizak, C. (2014). An evolutionarily significant unicellular strategy in response to starvation in *Dictyostelium* social amoebae. http://f1000research.com/articles/3-133/v2

Eisenberg, R. M. (1976). Two-dimensional microdistribution of cellular slime molds in forest soil. *Ecology*, 57, 380–384.

Ellison, A. M., & Buss, L. W. (1983). A naturally occurring developmental synergism between the cellular slime mold, *Dictyostelium mucoroides* and the fungus, *Mucor hiemalis*. *American Journal of Botany*, 70, 298–302.

Fehilly, C. B., Willadsen, S. M., & Tucker, E. M. (1984). Interspecific chimaerism between sheep and goat. *Nature*, 307, 634–636.

Filosa, M. F. (1962). Hetercytosis in cellular slime molds. *American Naturalist*, 96, 79–91.

Filosa, M. F. (1979). Macrocyst formation in the cellular slime mold *Dictyostelium mucoroides*: Involvement of light and volatile morphogenetic substance(s). *Journal of Experimental Zoology*, 207, 491–495.

Fisher, R. M., Cornwallis, C. K., & West, S. A. (2013). Group formation, relatedness and the evolution of multicellularity. *Current Biology*, 23, 1120–1125.

Fortunato, A., Strassmann, J. E., Santorelli, L., & Queller, D. C. (2003). Co-occurrence in nature of different clones of the social amoeba, *Dictyostelium discoideum*. *Molecular Ecology*, 12, 1031–1038.

Francis, D., & Eisenberg, R. (1993). Genetic structure of a natural population of *Dictyostelium discoideum*, a cellular slime mould. *Molecular Ecology*, 2, 385–391.

Garcia, T., & De Monte, S. (2013). Group formation and the evolution of sociality. *Evolution*, 67, 131–141.

Gilbert, O. M., Queller, D. C., & Strassmann, J. E. (2009). Discovery of a large clonal patch of a social amoeba: Implications for social evolution. *Molecular Ecology*, 18, 1273–1281.

Gilbert, O. M., Foster, K. R., Mehdiabadi, N. J., Strassmann, J. E., & Queller, D. C. (2007). High relatedness maintains multicellular cooperation in a social amoeba by controlling cheater mutants. *Proceedings of the National Academy of Sciences of the United States of America*, 104, 8913–8917.

Grosberg, R. K., & Strathman, R. R. (2007). The evolution of multicellularity: A minor major transition? *Annual Review of Ecology Evolution and Systematics*, 38, 621–654.

Guttes, E., Guttes, S., & Rusch, H. P. (1961). Morphological observations on growth and differentiation of *Physarum polycephalum* grown in pure culture. *Developmental Biology*, 3, 588–614.

Haldane, J. B. S. (1932). *The causes of evolution*. London: Longman, Green.

Haldane, J. B. S. (1955). Some alternatives to sex. *New Biologist*, 19, 7–26.

Hardikar, A. A., Marcus-Samuels, B., Geras-Raaka, E., Raaka, B. M., & Gershengorn, M. C. (2003). Human pancreatic precursor cells secrete FGF2 to stimulate clustering into hormone-expressing islet-like cell aggregates. *Proceedings of the National Academy of Sciences of the United States of America*, 100, 7117–7122.

Hashimoto, Y., Tanaka, Y., & Yamada, T. (1976). Spore germination promoter of *Dictyostelium discoideum* excreted by *Aerobacter aerogenes*. *Journal of Cell Science*, 21, 261–271.

Hiram-Bab, S., Shapira, Y., Gershengorn, M. C., & Oron, Y. (2012). Serum deprivation induces glucose response and intercellular coupling in human pancreatic adenocarcinoma PANC-1 cells. *Pancreas*, 41, 238–244.

Hunt, J. H. (2009). Interspecific adoption of orphaned nests by *Polistes* paper wasps (Hymenoptera: Vespidae). *Journal of Hymenoptera Research*, 18, 136–139.

Jablonka, E., & Raz, G. (2009). Transgenerational epigenetic inheritance: Prevalence, mechanisms, and implications for the study of heredity and evolution. *Quarterly Review of Biology*, 84, 131–176.

Kalinka, A. T., Varga, K. M., Gerrard, D. T., Preibisch, S. C., David, L., Jarrells, J., et al. (2010). Gene expression divergence recapitulates the developmental hourglass model. *Nature*, 468, 811–814.

Kaneko, K. (2003). Organization through intra–inter dynamics. In G. Müller & S. A. Newman (Eds.), *Origination of organismal form: Beyond the gene in developmental and evolutionary biology* (pp. 195–220). Cambridge, MA: MIT Press.

Kar, R., & Singh, R. S. (2014). Earliest record of slime moulds (Myxomycetes) from the Deccan Intertrappean beds (Maastrichtian), Padwar, India. *Current Science*, 107, 1237–1239.

Kaushik, S., Katoch, B., & Nanjundiah, V. (2006). Social behaviour in genetically heterogeneous groups of *Dictyostelium giganteum*. *Behavioral Ecology and Sociobiology*, 59, 521–530.

Kaushik, S., & Nanjundiah, V. (2003). Evolutionary questions raised by cellular slime mould development. *Proceedings of the Indian National Sciences Academy*, B69, 825–852.

Kawli, T., & Kaushik, S. (2001). Cell fate choice and social evolution in *Dictyostelium discoideum*: Interplay of morphogens and heterogeneities. *Journal of Biosciences*, 26, 130–133.

Kay, R. R. (1982). cAMP and spore differentiation in *Dictyostelium discoideum*. *Proceedings of the National Academy of Sciences of the United States of America*, 79, 3228–3231.

Kessin, R. H. (2001). *Dictyostelium: Evolution, cell biology, and the development of multicellularity*. Cambridge, UK: Cambridge University Press.

Ketcham, R. B., & Eisenberg, R. M. (1989). Clonal diversity in populations of *Polysphondylium pallidum*, a cellular slime mold. *Ecology*, 70, 1425–1433. doi:.10.2307/1938201

Ketcham, R. B., Levitan, D. R., Shenk, M. A., & Eisenberg, R. M. (1988). Do interactions of cellular slime mold species regulate their densities in soil? *Ecology*, 69, 193–199.

Leach, C. K., Ashworth, J. M., & Garrod, D. R. (1973). Cell sorting out during the differentiation of mixtures of metabolically distinct populations of *Dictyostelium discoideum*. *Journal of Embryology and Experimental Morphology*, 29, 647–661.

Loope, K. J., Chien, C., & Juh, M. (2014). Colony size is linked to paternity frequency and paternity skew in yellowjacket wasps and hornets. *BMC Evolutionary Biology*, 14, 277. doi:.10.1186/s12862-014-0277-x

McDonald, S. A., & Durston, A. J. (1984). The cell cycle and sorting behaviour in *Dictyostelium discoideum*. *Journal of Cell Science*, 66, 195–204.

Müller, G. B. (2007). EvoDevo: Extending the evolutionary synthesis. *Nature Reviews. Genetics*, 8, 943–950.

Nanjundiah, V. (1985). The evolution of communication and social behaviour in *Dictyostelium discoideum*. *Proceedings of the Indian Academy of Sciences. Animal Science (Penicuik, Scotland)*, 94, 639–653.

Nanjundiah, V. (2003). Phenotypic plasticity and evolution by genetic assimilation. In G. Müller & S. A. Newman (Eds.), *Origination of organismal form: Beyond the gene in developmental and evolutionary biology* (pp. 244–263). Cambridge, MA: MIT Press.

Nanjundiah, V., & Bhogle, A. S. (1995). The precision of regulation in *Dictyostelium discoideum*: Implications for cell-type proportioning in the absence of spatial pattern. *Indian Journal of Biochemistry & Biophysics*, 32, 404–416.

Nanjundiah, V., & Saran, S. (1992). The determination of spatial pattern in *Dictyostelium discoideum*. *Journal of Biosciences*, 17, 353–394.

Nanjundiah, V., & Sathe, S. (2011). Social selection and the evolution of cooperative groups: The example of the cellular slime moulds. *Integrative Biology*, 3, 329–342. doi:.10.1039/C0IB00115E

Nanjundiah, V., & Sathe, S. (2013). Social selection in the cellular slime moulds. In M. Romeralo, S. Baldauf, & R. Escalante (Eds.), *Dictyostelids: Evolution, genomics and cell biology* (pp. 193–217). Heidelberg, Germany: Springer.

Newman, S. A. (2011). Animal egg as evolutionary innovation: A solution to the "embryonic hourglass" puzzle. *Journal of Experimental Zoology. Part B, Molecular and Developmental Evolution*, 316, 467–483.

Newman, S. A., & Comper, W. D. (1990). "Generic" physical mechanisms of morphogenesis and pattern formation. *Development*, 110, 1–18.

Newman, S. A., & Bhat, R. (2009). Dynamical patterning modules: A "pattern language" for development and evolution of multicellular form. *International Journal of Developmental Biology*, 53, 693–705.

Novick, A., & Wiener, M. (1957). Enzyme induction as an all-or-none phenomenon. *Proceedings of the National Academy of Sciences of the United States of America*, 43, 553–566.

Olive, L. S. (1965). A developmental study of *Guttulinopsis vulgaris* (Acrasiales). *American Journal of Botany*, 52, 513–519.

Olive, L. S. (1975). *The Mycetozoans*. New York: Academic Press.

Olive, L. S., & Blanton, R. L. (1980). Aerial sorocarp development by the aggregative ciliate, *Sorogena stoianovitchae*. *Journal of Protozoology*, 27, 293–299.

Olive, L. S., & Stoianovitch, C. (1960). Two new members of the Acrasiales. *Bulletin of the Torrey Botanical Club*, 87, 1–20.

Ozbudak, E. M., Thattai, M., Lim, H. N., Shraiman, B. I., & Van Oudenaarden, A. (2004). Multistability in the lactose utilization network of *Escherichia coli*. *Nature*, 427, 737–740.

Pan, P., Hall, E. M., & Bonner, J. T. (1972). Folic acid as a second chemotactic substance in the cellular slime moulds. *Nature: New Biology*, 237, 181–182.

Piasecka, B., Lichocki, P., Moretti, S., Bergmann, S., & Robinson-Rechavi, M. (2013). The hourglass and the early conservation models—Co-existing patterns of developmental constraints in vertebrates. *PLOS Genetics*. doi:.10.1371/journal.pgen.1003476

Pradhan, B. B., & Chatterjee, S. (2014). Reversible non-genetic phenotypic heterogeneity in bacterial quorum sensing. *Molecular Microbiology*, 92, 557–569.

Qian, H. (2012). Cooperativity in cellular biochemical processes: Noise-enhanced sensitivity, fluctuating enzyme, bistability with nonlinear feedback, and other mechanisms for sigmoidal responses. *Annual Review of Biophysics*, 41, 179–204.

Raper, K. B. (1940). Pseudoplasmodium formation and organization in *Dictyostelium discoideum*. *Journal of the Elisha Mitchell Scientific Society*, 56, 241–282.

Raper, K. B. (1984). *The Dictyostelids*. Princeton, NJ: Princeton University Press.

Raper, K. B., & Quinlan, M. S. (1958). *Acytostelium leptosomum*: A unique cellular slime mould with an acellular stalk. *Journal of General Microbiology*, 18, 16–32.

Ratcliff, W. C., Denison, R. F., Borrello, M., & Travisano, M. (2012). Experimental evolution of multicellularity. *Proceedings of the National Academy of Sciences of the United States of America*, 109, 1595–1600.

Rinkevich, B., & Shapira, M. (1999). Multi-partner urochordate chimeras outperform two-partner chimerical entities. *Oikos*, 87, 315–320.

Romeralo, M., & Fiz-Palacios, O. (2013). Evolution of *Dictyostelid* social amoebas inferred from the use of molecular tools. In M. Romeralo, S. Baldauf, & R. Escalante (Eds.), *Dictyostelids: Evolution, genomics and cell biology* (pp. 167–182). Heidelberg, Germany: Springer.

Romeralo, M., Skiba, A., Gonzalez-Voyer, A., Schilde, C., Lawal, H., Kedziora, S., et al. (2013). Analysis of phenotypic evolution in Dictyostelia highlights developmental plasticity as a likely consequence of colonial multicellularity. *Proceedings. Biological Sciences*, 280, 20130976.

Sachs, J. L., & Hollowell, A. C. (2012). The origins of cooperative bacterial communities. *mBio*, 3(3), 00099–12. doi:.10.1128/mBio

Saran, S. (1999). Calcium levels during cell cycle correlate with cell fate of *Dictyostelium discoideum*. *Cell Biology International*, 23, 399–405.

Sathe, S., Kaushik, S., Lalremruata, A., Aggarwal, R. K., Cavender, J. C., & Nanjundiah, V. (2010). Genetic heterogeneity in wild isolates of cellular slime mold social groups. *Microbial Ecology*, 60, 137–148. doi:.10.1007/s00248-010-9635-4

Sathe, S., Khetan, N., & Nanjundiah, V. (2014). Interspecies and intraspecies interactions in social amoebae. *Journal of Evolutionary Biology*, 27, 349–362. doi:.10.1111/jeb.12298

Sathe, S., Altenberg, L., & Nanjundiah, V. (unpublished). Complex interactions and trade-offs in heterogeneous social groups of *Dictyostelium giganteum*.

Saxer, G., Brock, D. A., Strassmann, J. E., & Queller, D. C. (2010). Cheating does not explain selective differences at high and low relatedness in a social amoeba. *BMC Evolutionary Biology*, 10, 76. doi:.10.1186/1471-2148-10-76

Shadwick, L. L., Spiegel, F. W., Shadwick, J. D. L., Matthew, W., Brown, M. W., & Silberman, J. D. (2009). Eumycetozoa=Amoebozoa?: SSUrDNA phylogeny of protosteloid slime molds and its significance for the amoebozoan supergroup. *PLoS One*, 4(8), e6754.

Tarnita, C. E., Washburne, A., Martinez-Garcia, R., Sgro, A. E., & Levin, S. A. (2015). Fitness tradeoffs between spores and nonaggregating cells can explain the coexistence of diverse genotypes in cellular slime molds. *Proceedings of the National Academy of Sciences of the United States of America*, 112, 2776–2781.

Teillet, M. A., Ziller, C., & Le Douarin, N. M. (1999). Quail-chick chimeras. *Methods in Molecular Biology (Clifton, N.J.)*, 97, 305–318.

Thompson, C. R. L., & Kay, R. R. (2000). Cell-fate choice in *Dictyostelium*: Intrinsic biases modulate sensitivity to DIF signalling. *Developmental Biology*, 227, 56–64.

Town, C. D., Gross, J. D., & Kay, R. R. (1976). Cell differentiation without morphogenesis in *Dictyostelium discoideum*. *Nature*, 262, 717–719.

Vogt, G. (2015). Stochastic developmental variation, an epigenetic source of phenotypic diversity with far-reaching biological consequences. *Journal of Biosciences*, 40, 159–204.

Waddington, C. H. (1953). Genetic assimilation of an acquired character. *Evolution; International Journal of Organic Evolution*, 7, 118–126.

Wilkinson, H. A., Fitzgerald, K., & Greenwald, I. (1994). Reciprocal changes in expression of lin-12 (receptor) and lag-2 (ligand) prior to commitment in a *C. elegans* cell fate decision. *Cell*, 79, 1187–1198.

Wilson, E. O. (1975). *Sociobiology: The new synthesis*. Cambridge, MA: Harvard University Press.

Wilson, D. S., & Wilson, E. O. (2007). Rethinking the theoretical foundation of sociobiology. *Quarterly Review of Biology*, 82, 327–348.

Włoch-Salamon, D. (2014). Sociobiology of the budding yeast. *Journal of Biosciences*, 39, 225–236.

Wolf, J. B., Howie, J. A., Parkinson, K., Gruenheit, N., Melo, D., Rozen, D., et al. (2015). Fitness trade-offs result in the illusion of social success. *Current Biology*, 25, 1086–1090.

Worley, A. C., Raper, K. B., & Hohl, M. (1979). *Fonticula alba*: A new cellular slime mold (Acrasiomycetes). *Mycologia*, 71, 746.

7   Trade-offs Drive the Evolution of Increased Complexity in Nascent Multicellular Digital Organisms

Peter L. Conlin and William C. Ratcliff

The transition to multicellularity was a major step in the evolution of large, complex life on Earth (Maynard Smith and Szathmáry, 1995). Unlike other major evolutionary transitions, which have occurred only once (e.g., prokaryotes to eukaryotes), multicellularity has evolved multiple times in diverse lineages including archaea (Jahn et al., 2008), bacteria (Velicer and Vos, 2009; Overmann, 2010; Schirrmeister et al., 2011), and eukaryotes (Bonner, 1998; King, 2004; Grosberg and Strathmann, 2007; Herron et al., 2009; Herron et al., 2013). Prior work suggests that the formation of simple clusters of cells, the first step in the transition to multicellularity, may be adaptive under a number of distinct ecological scenarios. For example, clusters may provide protection from predation (Kessin et al., 1996; Boraas et al., 1998), protection from environmental stress (Smukalla et al., 2008), or improved utilization of diffusible nutrients (Pfeiffer and Bonhoeffer, 2003, Koschwanez et al., 2011; Koschwanez et al., 2013). Nevertheless, how and why nascent multicellular lineages evolve increased complexity remains a fundamental question in evolutionary biology. Progress has been impeded by a lack of experimental systems due to the fact that most nascent multicellular lineages have been lost to extinction.

To sidestep this historical limitation, we (and colleagues) have been using experimental evolution to re-create this major transition under controlled laboratory conditions (reviewed in Ratcliff and Travisano, 2014). Starting with outbred diploid unicellular yeast, we selected for cluster formation by favoring yeast that settle rapidly through liquid medium. In all 10 replicate populations, cluster-forming "snowflake" yeast readily evolved and displaced their unicellular ancestors. Snowflake yeast consists of daughter cells that remain attached after mitotic division, forming spherical branched structures of genetically identical cells. Over the next several hundred generations, several traits of interest evolved as snowflake yeast further adapted to this selection regime.

In response to selection for rapid settling, snowflake yeast first evolved to form clusters that contain more cells. Later, snowflake yeast evolved a 2.1-fold increase in the volume of individual cells, further increasing cluster biomass and thus settling speeds (Ratcliff, Pentz, and Travisano, 2013). Large-bodied yeast also evolved higher rates of programmed cell death, hereafter referred to as apoptosis. Prior experiments suggest that these dead cells act as "weak links" in the chains of cells that make up the cluster, resulting in greater reproductive asymmetry (i.e., smaller propagules relative to cluster size). This conclusion is based on comparisons between high- and low-apoptosis strains, direct experiments modifying the frequency of apoptosis chemically, and the observation that dead cells are found at the site of propagule scission ~12 times more frequently than is expected by chance (Ratcliff et al., 2012).

Fitness trade-offs, while central to all of life history theory (Roff, 2002), are thought to take on a particularly important role during major evolutionary transitions such as the evolution of multicellularity (Michod et al., 2006). Specifically, trade-offs between survival and reproduction may drive increases in complexity and cellular differentiation (Michod et al., 2006). Perhaps the best-known example comes from the evolution of multicellularity in the volvocine algae: individual cells cannot reproduce and phototax simultaneously (Koufopanou, 1994), favoring the evolution of divided labor through germ–soma differentiation (Koufopanou, 1994; Solari et al., 2006). More generally, simple clusters of cells may benefit from increased size (e.g., reduced consumption by predators [Boraas et al., 1998; Becks et al., 2012]), but cellular clusters face greater diffusional limitation than single cells, impeding resource uptake from their environment (Lavrentovich et al., 2013). Our experimental results suggest that trade-offs also play a role in the evolution of multicellularity in snowflake yeast. Evolving larger clusters increases settling speed but decreases growth rates, likely because cells in the interior of large clusters become resource limited as a result of greater diffusional impedance (Ratcliff et al., 2012; Lavrentovich et al., 2013). Similarly, increasing cell size may decrease the rate at which individual cells are produced, again because larger cells have a proportionally greater surface area to volume ratio (but see Jorgensen et al., 2002). The effects of apoptosis are a bit more complicated. A small fraction of the cells in the cluster (~1.5%–2.5%) die, a direct viability cost. However, by producing proportionally smaller propagules, large clusters produce offspring that are less diffusionally limited. Thus, apoptosis increases growth rates but decreases survival during settling selection and will only be adaptive when the sum of these effects is positive.

Here we investigate the role of simple trade-offs during the evolution of increased multicellular complexity in snowflake yeast by modeling the evolution of simple multicellular digital organisms. We find that apoptosis, which results in the production of smaller propagules at the expense of the acting cell's life, is adaptive under a broad suite of conditions. This is because it can increase growth rates enough to compensate for the loss of apoptotic cells and reduced survival during settling selection. In our models, competition for faster settling results in an evolutionary arms race that drives a modest (maximum of 150 cells) increase in cluster size and apoptosis. Much larger clusters only evolved if the size required for surviving settling selection was increased through time. Using a two-player tournament-style evolutionary algorithm, we find that snowflake yeast that are initially mismatched in size will niche partition, with the smaller strain evolving into a growth specialist and the larger strain a settling specialist. Finally, we find that increasing the dimensionality of the multicellular trait space from two (cells per cluster and apoptosis) to three (adding cell size) increases the degree to which competing strains in a single population will diverge. This work demonstrates that multicellular complexity readily arises when trade-offs between group size and growth rates are ameliorated by the evolution of novel multicellular traits.

Model Description

Cluster Growth and Reproduction

The model we develop here considers competition occurring between two genetically distinct snowflake yeast strains within a single population. As in our laboratory experiments, the transfer cycle involves two discrete phases: growth and settling selection (summarized in figure 7.1). For each time step, clusters grow by adding cells in proportion to their initial number following the equation

$$n' = n(2 - nd) - na,$$ (7.1)

where n is cell number per cluster, d is the diffusional limitation cost (ranging from 0.001 to 0.002), and a is the rate of apoptosis (see table 7.1 for a summary of model parameter). In comparison to single cells, clusters are diffusionally limited and thus grow less rapidly (Ratcliff et al., 2012). For simplicity, we model the cost of diffusional limitation as a linear trade-off between cluster size (number of cells) and growth rate, such that a single cell doubles during each time step, and larger clusters grow to size $n(2 - nd)$ cells. Cluster growth is offset by apoptosis where the number of cells that undergo cell death at each time step is calculated as

$$n(a) = \frac{n(\alpha - 0.5)}{50},$$ (7.2)

Table 7.1
Summary of model parameters

Parameter	Description	Base value
a	Rate of apoptosis	0.001
α	Reproductive asymmetry	0.55
d	Cost of diffusional limitation	0.001
n_{min}	Minimum number of cells within a cluster	1
n_{max}	Maximum number of cells within a cluster	2,000
N	Population size	$8 \cdot 10^6$
r	Size at reproduction	150
s	Size threshold for settling selection	140

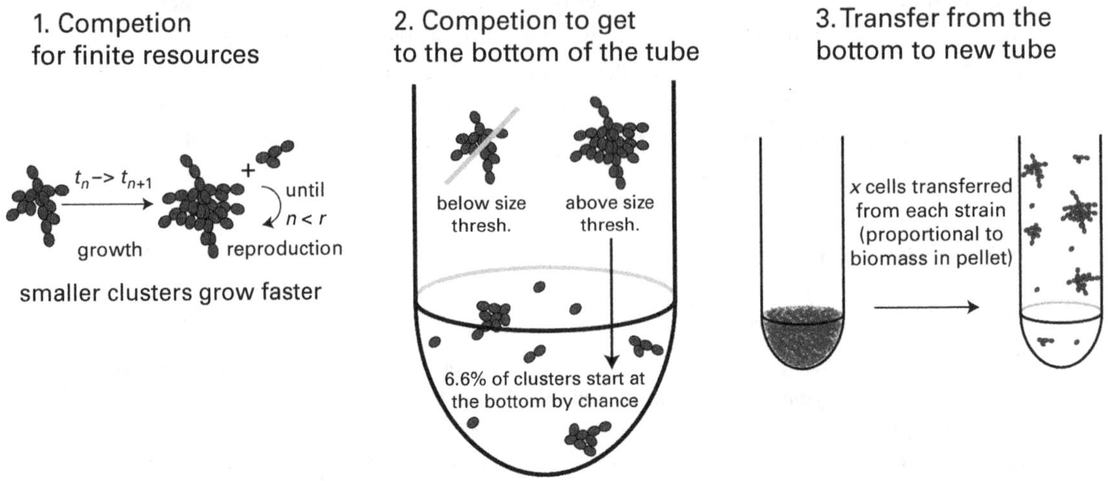

1. Competion for finite resources

$t_n \rightarrow t_{n+1}$

growth

+ until $n < r$

reproduction

smaller clusters grow faster

2. Competion to get to the bottom of the tube

below size thresh.

above size thresh.

6.6% of clusters start at the bottom by chance

3. Transfer from the bottom to new tube

x cells transferred from each strain (proportional to biomass in pellet)

Figure 7.1
Model schematic. The model is separated into two distinct phases. First, clusters grow, competing for finite resources. Larger clusters face greater diffusional limitation and thus gain proportionally fewer cells during each time step. If a cluster's growth causes it to exceed its reproductive size r, then propagules are produced sequentially until cell number $n < r$. Settling selection is applied once resources are exhausted. All clusters above size threshold (thresh.) s settle to the bottom of the tube. Not all cells at the bottom are large, however: 6.6% of the clusters in the population simply start out there by chance. Finally, clusters are transferred to fresh medium. To allow for sufficient growth between rounds of settling selection, we transfer 1/20 of the stationary phase biomass to fresh medium. Clusters are transferred in proportion to the biomass of each strain in the pellet.

where α is reproductive asymmetry, a parameter that specifies the propagule size when a cluster undergoes reproduction (discussed below). For most of our simulations, the rate of apoptosis is equal to ~0.001 unless otherwise noted. For a 148-cell cluster growing up from a 10-cell propagule during a single culture cycle, this corresponds to a cumulative death rate of 1.93%, which is similar to what we observe in our experiments (Ratcliff et al., 2012). Importantly, this step occurs before new cells are added, so only cells that are one generation old or older are capable of dying from apoptosis.

When a cluster grows larger than r cells, the cluster splits, producing daughter propagules sequentially until they are smaller than the reproductive threshold. Offspring size depends on the cluster's reproductive asymmetry, α. Specifically, mean propagule size is $n(1-\alpha)$. Because snowflake yeast produce offspring that vary in size, we implemented a stochastic smoothing function into our model, such that asymmetry at each reproductive event is drawn from a uniform probability distribution bounded by $(\alpha - 0.05, \alpha + 0.05)$. Similarly, a cluster's size at reproduction is drawn from a uniform probability distribution bounded by $(r - 5, r + 5)$. This has the effect of preventing the accumulation of many clusters of exactly the same size, which can result in simulation artifacts (e.g., abrupt changes in fitness with small changes in traits). For all simulations, asymmetry was bounded between 0.5 and 0.9.

Population Growth and Settling Selection

We model our experimental regime as two distinct phases, a growth phase and a settling phase. During the growth phase, resources are consumed by both strains of yeast until they are exhausted (in most simulation runs, we allow for 8×10^6 cellular reproductions) with each yeast strain growing and producing propagules according to the equations given above. After resource exhaustion, settling selection is applied.

In our laboratory experiments, we transfer the cells found in the lower 100 µl of a microcentrifuge tube after settling selection to 10 ml of fresh medium. There are two ways that clusters can get to the bottom of the test tube: first, a small fraction of clusters (~6.6%) simply start out there after the tube is mixed. These clusters need not be large—they are simply lucky. Second, clusters can settle rapidly enough to sink to the bottom of the tube. Here, we impose a simple threshold, such that clusters containing more than s cells make it to the bottom of the test tube, and those smaller do not. In our laboratory experiments, settling selection results in a 20- to 25-fold dilution per day for relatively fast-settling snowflake yeast. We model this by selecting clusters from each strain (in proportion to their biomass in the pellet) until 1/20 of the carrying capacity of the population is met. These clusters are then transferred to fresh medium and the cycle repeated.

In reality, settling selection is less precise: small clusters starting out just above the lower 100 µl may still join the pellet while larger cluster starting at the very top of the tube may fail to make it to the bottom. Still, this simplifying assumption does not change the basic dynamics of size selection favoring larger sized clusters. Selection can be made more or less stringent by increasing or decreasing the threshold cluster size, s.

Results

Local Fitness Landscapes Reveal the Conditions Favoring Elevated Apoptosis

We first examined the snowflake yeast fitness landscape under different nutrient diffusion regimes, d, as a function of both the rate of apoptosis and cluster size at reproduction. In each case, we competed a single strain (demarcated by the black circle in figure 7.2A–D) against 1,554 different competitor strains that varied in these traits. We measure relative fitness as the change in frequency of strain 1 cells relative to strain 2 cells between stationary phase in transfer 2 and stationary phase in transfer 3. The threshold for surviving settling selection s was 140 cells, which is similar to what we have observed in early snowflake yeast (1–3 weeks of evolution). Competition between smaller clusters with little diffusional limitation ($d = 0.001$) favors larger clusters with negligible apoptosis (see

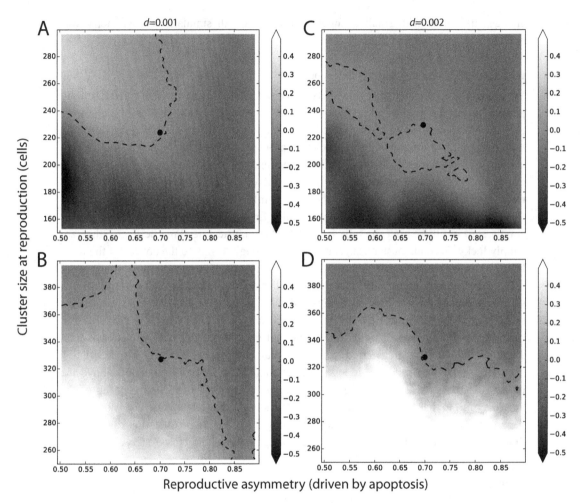

Figure 7.2
Fitness landscapes vary depending on the extent of resource diffusion and on cluster size. In each fitness landscape plotted above, a single strain (filled circle) competes against 1,554 competitor strains varying in cluster size at reproduction and apoptosis. Large size and low apoptosis are favored in small clusters with little diffusion limitation (A), while increasing the growth cost of diffusion favors smaller clusters with higher rates of apoptosis (C). Increasing cluster size at reproduction by 100 favors smaller cluster size (B, D). Apoptosis provides more of a benefit to larger clusters (lower region of B, D). Here $s = 140$, $d = 0.001$, and the growth phase contains sufficient resources for the production of 8×10^6 cells. The dashed line demarcates a relative fitness of 0.

figure 7.2A). Increasing the severity of diffusional limitation ($d = 0.002$) favors elevated apoptosis (see figure 7.2C). We note, however, that the genotype with the highest fitness under these conditions still has the lowest rate of apoptosis. Among clusters that are much larger than the size necessary to survive settling selection (see figure 7.2B and D; which start out at size 250, but need only be 140 cells in size to survive selection), smaller size is beneficial. Importantly, higher rates of apoptosis can be selectively advantageous among these larger clusters, even when diffusional limitation is mild ($d = 0.001$).

Snowflake yeast compete in two key arenas: for resources during the 24 h of batch culture, and for a spot at the bottom of the tube during settling selection. It is in these two arenas where fitness trade-offs are realized. For example, large clusters settle quickly but grow slowly. By increasing reproductive asymmetry (reducing propagule size), increased rates of apoptosis should increase growth rates at the expense of survival during settling selection. Whether or not these traits are adaptive for a given environmental context depends on the benefit of faster growth relative to the cost of reduced survival during settling. To directly compare the magnitude of this fitness trade-off, we calculated the selection rate constants (following Travisano and Lenski, 1996) for growth and settling for the landscape in figure 7.2C. This approach allows us to compare each phase of competition using fitness as a common currency. As expected, elevated apoptosis increased growth rates, but reduced survival for settling selection, and larger cluster size was uniformly favored during settling (see figure 7.3). The benefits of faster growth outweighed the cost of slower settling for part of the trait space (asymmetry between 0.62 and 0.75 and cluster size between 200 and 220 cells; see figure 7.2C). This result also highlights the fine line being walked during the evolution of elevated apoptosis: mutations of large effect may produce strains with too much apoptosis, such that the costs of slower settling exceed the costs of faster growth.

Competition Drives an Evolutionary Arms Race for Increased Cluster Size

Static fitness landscapes (e.g., see figures 7.2 and 7.3) are useful for examining the interaction between traits and fitness over only a limited range of conditions because relative fitness is contextual and changes over evolutionary time along with the traits of the two competitors. To examine how cluster size and apoptosis rates might coevolve as the competitor also changes, we implemented a two-player evolutionary algorithm. For each time step, 10 derivatives of each snowflake yeast genotype were generated, each with a 90% chance of mutation in cluster size at maturity or reproductive asymmetry. Each mutation was drawn from a normal distribution, with the mean being the former trait value with standard deviation of 1 (for cluster size at reproduction) or 0.003 (for reproductive asymmetry). All 10 variants of strain 1 were competed against last-round's strain 2 winner, and then all 10 variants of strain 2 are competed against the best strain 1 variant. The strain with the highest relative fitness after three transfers was selected as the parent strain for

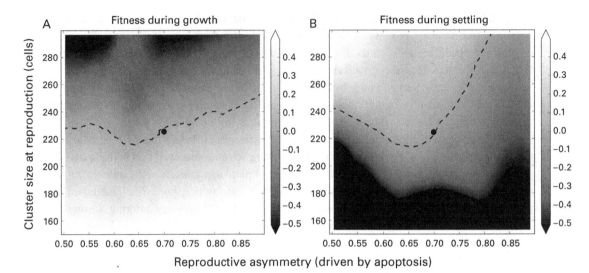

Figure 7.3
Disentangling fitness contributions from growth (A) and settling (B). We calculated the relative fitness consequences (as selection rate constants) during growth and settling for the landscape shown in figure 7.2C. Smaller cluster size at maturity and apoptosis increases fitness during growth at a cost to settling. Here $s = 140$, $d = 0.002$, and the growth phase contains sufficient resources for the production of 8×10^6 cells. The dashed line demarcates a selection rate constant of 0.

the next round. For all simulations, the size threshold for surviving settling selection (s) was 140.

When competition occurs between two similarly sized strains, both readily evolve larger cluster size and higher rates of apoptosis (see figure 7.4A and B). In contrast, when we compete two strains that vary substantially in size (150 vs. 200 cells at reproduction), snowflake yeast rapidly partition their niches (see figure 7.4C). Specifically, the 150-celled strain evolved to form clusters that were ~15 cells smaller than its ancestor over 150 rounds of competition against a 200-celled strain, while the same starting genotype evolved to form clusters that were an average of 67 cells larger when competed against another 150-celled strain (see figure 7.4A vs. 7.4C; $t_{157} = 114$, $p < 0.0001$, Bonferroni-corrected two-way t test). This effect appears to be due to competitive exclusion during settling selection, driving the smaller strain to evolve smaller size and increased competitiveness during the growth phase of competition. We examined the size difference required for niche partitioning to occur, varying strain 1's starting size from 150 to 160 cells while leaving strain 2's starting size at 200 cells. For each competition, we ran 100 simulations for 150 transfers. The percentage of runs in which strain 1 evolved to be a growth specialist declined linearly as their size increased ($y = 1503.9 - 9.3x$, $r^2 = 0.97$, $F_{1,10} = 254.9$, $p < 0.001$; see figure 7.4C, insert). Interestingly, we also found that the 200-celled strain evolved to form clusters that were ~30 cells larger when competing against another 200-

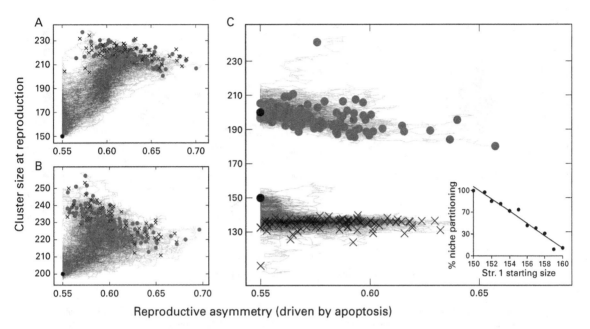

Figure 7.4
Arms races and niche partitioning. Larger clusters with higher rates of apoptosis evolve when both starting strains are similarly sized (A, B). If the initial size difference is substantial, arms-race dynamics are prevented, and instead the smaller strain evolves smaller size, becoming a growth specialist (C). The frequency of niche partitioning declines linearly as strain (Str.) 1's starting size increases from 150 to 160 (c, insert). Plotted are 100 simulations for each strain (strains 1 and 2 are demarcated by dark X's and light circles, respectively) over 150 transfers. Here $s = 140$, $d = 0.0015$, and the growth phase contains sufficient resources for the production of 2×10^6 cells.

celled strain, but not against a 150-celled strain (see figure 7.4B vs. 7.4C; $t_{196} = 30$, $p <$ 0.0001, Bonferroni-corrected two-way t test), further illustrating the importance of coevolution in our model. One caveat of this simulation is that it ensures coexistence of the two competing strains. It is possible that extinction in real populations would limit the ability for the evolution of niche partitioning.

The results of the two-player games (see figure 7.4) illustrate the importance of arms-race dynamics among similarly sized competitors in the evolution of increased cluster size. The extent of directional change is limited, however, as the trade-off between settling and growth components of fitness result in stable coexistence at modest (250–340 cell) cluster size and low apoptosis (reproductive asymmetry ≈ 0.57; see figure 7.5A). To examine how size and apoptosis coevolve in response to different size-selection thresholds, we simulated competition in environments where the size threshold for surviving settling selection, s, varied from 50 to 750. We initialized each competition with two identical strains whose maximum size was just 10 cells larger than that required for settling and then allowed

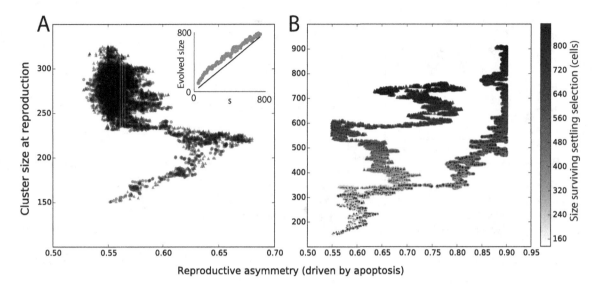

Figure 7.5
Pushing the envelope—the evolution of large clusters. (A) When the environment is constant ($s = 140$ for all 7,600 transfers), equilibrium dynamics rapidly establish themselves with the evolution of modest cluster size and low apoptosis. For a range of s from 50–750, cluster size at equilibrium (1,000 transfers, purple circles in 7.5A inset) is only modestly larger than required for surviving settling selection (black line, inset). (B) Slowly ratcheting up the threshold for settling to the bottom of the tube (s starts at 140 and increases by 1 every 10 transfers) results in the evolution of very large clusters with high rates of apoptosis. Filled circles and triangles refer to the two strains in competition. In these simulations the growth phase contains sufficient resources for the production of 2×10^6 cells and $d = 0.001$.

them to come to equilibrium (1,000 transfers per competition). While larger clusters readily evolved, they rarely got more than 150 cells larger than the threshold for surviving settling selection (see figure 7.5A, insert). Larger size clearly evolves in response to selection for faster settling, but imposing severe selection for rapid settling on small clusters can be counterproductive: selection cannot favor faster settling clusters if none survive. How then do large clusters evolve? In our experiments, we periodically increased the strength of settling selection (Ratcliff, Pentz, and Travisano, 2013), favoring the progressive evolution of faster settling. We thus reran the above simulation, starting with a small snowflake yeast ($r = 140$), and increased the size of settling selection by 1 cell every 10 time steps. Here, much larger (>900-celled) clusters evolved, along with maximal rates of apoptosis (reproductive asymmetry ≈ 0.9 in strain 1; see figure 7.5B).

Modeling the Evolution of Increased Cell Size

In our experiments, in addition to evolving an increased number of cells per cluster, snowflake yeast also evolved a 2.1-fold increase in cell size within 2 months (~400 generations;

Ratcliff, Pentz, and Travisano, 2013), increasing the settling speed of clusters. We thus modified the model to allow for the evolution of increased cell size, changing the settling selection step to count cluster biomass equivalents in addition to cell number. Further, we imposed a linear growth penalty for larger cells, assuming that cells with twice the volume grow at 98% the speed of a wild-type cell. We repeated the two-player tournament simulations, allowing for mutations that increase cell size to occur with 90% probability (the distribution of mutational effect sizes was identical to that of reproductive asymmetry). Increasing the strength of settling selection was again essential for the evolution of both large and many-celled snowflake yeast (see figure 7.6).

Increasing the dimensionality of the multicellular trait space makes it possible for genetically and phenotypically distinct strains to arrive at the same multicellular solution (i.e., strains differing in cell size and cell number can nonetheless evolve the same overall cluster biomass). To examine how allowing a third multicellular trait to evolve affected within-population divergence, we calculated the Euclidian distance of normalized trait values between competitors in each microcosm, either with or without cell size mutations, after equilibrium was reached (1,000 time steps, $s = 140$). Increasing the dimensionality of the multicellular trait space increased the opportunity for within-population diversification: competitors were an average of 32% more phenotypically divergent ($t_{196.7} = 3.07$, $p = 0.0024$, two-sided t test) when cell size was allowed to evolve (see figure 7.7). Because there was only a 1% difference in cluster biomass (mean of 272 and 279 wild-type cell equivalents for 2-D and 3-D simulations, respectively) and no difference in apoptosis (mean of 55.9% for both 2-D and 3-D simulations), it appears that yeast capable of evolving both greater cell number per cluster and larger cell size took different, though ecologically equivalent, paths to increased cluster size. This demonstrates that the evolution of additional multicellular traits may, as a side effect, increase the population's capacity to support the coexistence of ecologically equivalent (though genetically and phenotypically distinct) isolates.

Discussion

One of the most surprising results from our yeast experiments is the rapidity of adaptation after simple multicellularity evolves. After just ~400 generations, snowflake yeast evolve to form clusters containing twice as many cells, higher rates of apoptosis, and cells that are more than twice as large as the ancestor. These clusters settled 28% faster, on average (Ratcliff, Pentz, and Travisano, 2013). Further, we have found that after a similar length of time (but in a different experiment), 9/10 replicate populations contained at least two strains that varied in size (Rebolleda-Gomez et al., 2012). The modeling results presented above help to explain both of these observations.

In both this model and in evolving populations of yeast (Ratcliff et al., 2012; Ratcliff, Pentz, and Travisano, 2013), initially small clusters of cells are under strong selection for

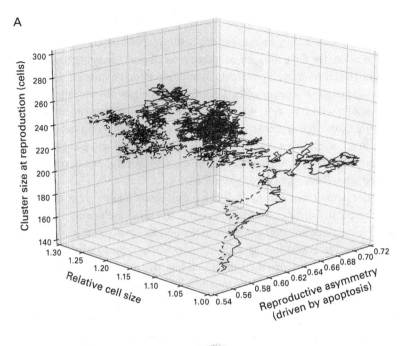

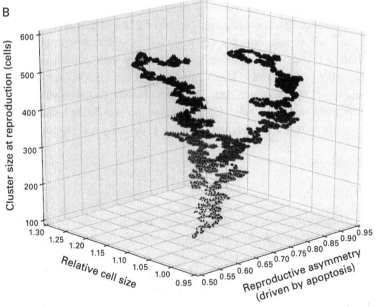

Figure 7.6
Evolution of larger cell size. Coevolution in a static ($s = 140$; A) or gradually more stringent size-selective environment (s starts at 140 and increases by 1 every 10 transfers; B). Increasing the strength of settling selection favors the evolution of all three key multicellular traits: large cluster size at reproductive maturity, high rates of apoptosis, and large individual cells. Here the growth phase contains sufficient resources for the production of 2×10^6 cells and $d = 0.001$. Dashed and solid lines represent the two competitors in A, while circles and triangles are used in B. As in figure 7.5B, marker shading in 7.6B indicates cluster size surviving settling selection.

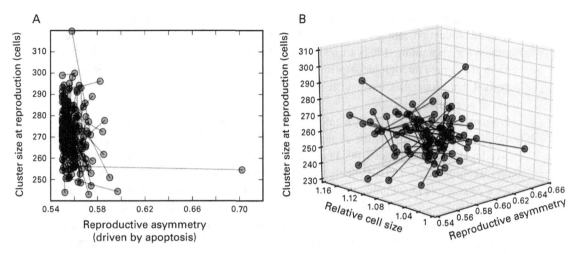

Figure 7.7
Divergence in 2-D versus 3-D games. Competing pairs of snowflake yeast evolved more divergent multicellular traits, measured as the Euclidian distance of each competitor after 1,000 generations. Plotted are the results of 2-D simulations where cell size was held constant (A) and 3-D simulations where cell size was allowed to evolve (B).

increased size. The model described here shows why: even small increases in settling speed dramatically improve relative fitness, and at this small size, diffusional limitation is not yet very restrictive. As a result, competition among coevolving yeast results in a short-term arms race for increased size. As larger cluster size evolves, apoptosis becomes increasingly beneficial, allowing large-bodied yeast to produce proportionally smaller offspring that are temporarily freed from strong diffusional limitation. Competition in a static environment (no change in s) thus results in a modest increase in both size and apoptosis. However, we find that for very large, high apoptosis clusters to evolve, the strength of settling selection must increase through time. Long-term experiments running in our lab suggest that this increase in the strength of settling selection is also required for the in vitro evolution of large snowflake yeast clusters (Ratcliff, unpublished).

Apoptosis readily evolves in this simple simulation model, ameliorating the growth rate cost incurred by large body size. This effect appears to be due to the fact that apoptosis produces proportionally smaller offspring that have a growth rate advantage when resource diffusion is more limiting. Specifically, increased rates of apoptosis are favored under conditions of low resource penetration into the cluster (high d), or for low values of d, as a consequence of selection favoring the evolution of progressively larger slower growing clusters (but see Duran-Nebreda and Solé, 2015, for a nonadaptive explanation for elevated rates of apoptosis). Apoptosis may therefore be a general solution to the biophysical constraint of diffusionally limited growth so long as (1) selection favors larger clusters of cells, (2) larger clusters grow less rapidly than smaller clusters, (3) cell death results in propagule production, and (4) propagules that are produced by apoptosis are related (high Hamilton's r; Hamilton, 1964) to apoptotic cells. Both (3) and (4) likely require that clusters are formed through incomplete mother–daughter cell separation. Finally, it is also important that propagules have sufficient time to grow large enough to survive the next round of size-based selection.

Programmed cell death (PCD; a category of cell death mechanisms that includes apoptosis) plays a critical role in the evolution of multicellular complexity. In independently evolved multicellular lineages, PCD is used to modify multicellular form during development (Jacobson et al., 1997; Pennell and Lamb, 1997; Umar and Van Griensven, 1997), plays a central role maintaining the multicellular body (e.g., removal of damaged [Jacobson et al., 1997], infected [Lam, 2004], or cancerous cells [Lee and Bernstein, 1995]), and can be useful in coordinated multicellular behaviors (e.g., leaf abscission [Bleecker and Patterson, 1997]). Indeed, PCD plays such an important role in multicellular organization that it is difficult to imagine complex multicellular life in its absence. PCD, however, is not a multicellular invention: many diverse unicellular organisms possess active, genetically regulated cell death mechanisms (Nedelcu et al., 2011). Comparative work demonstrates that some PCD pathways in multicellular organisms arose in their unicellular ancestors (Nedelcu, 2009), suggesting they were co-opted for novel use after the transition to multicellularity. The work presented here provides a simple, general explanation for how unicellular PCD can be co-opted for a novel multicellular purpose. See Duran-Nebreda et al. (this volume) for a discussion of how multicellular complexity can emerge as a consequence of interactions among the component cells.

Within-population diversity is present in both our lab experiments (Rebolleda-Gomez et al., 2012) and this simulation model. The model predicts that some of this diversity may simply be the result of different lineages' taking divergent trajectories during adaptation (see figures 7.5 and 7.6). This may be especially important when the multicellular trait space is high dimensional because, all else equal, these contain a greater number of ecologically equivalent trajectories. There are a number of potential multicellular traits that we did not include in the model but which may be relevant and would further increase the dimensionality of the multicellular trait space. For example, the architecture of snowflake yeast clusters (Libby et al., 2014) might change, affecting both d and how size relates to surviving during settling selection. This can be due to simple alterations, like the shape of individual cells, or more complex modifications. Indeed, after 227 days of selection, we see the evolution of more spherical, hydrodynamic clusters that settle 35% faster per increase in unit mass (Ratcliff, Pentz, and Travisano, 2013). We do not yet know the mechanistic basis of this trait, but it is potentially due to a modified branching pattern, or changes in the location and timing of cellular apoptosis.

How multicellular complexity arises in evolution is a fundamental question in biology. In combination with our experimental work (Ratcliff et al., 2012; Rebolleda-Gomez et al.,

2012; Ratcliff, Pentz, and Travisano, 2013; Ratcliff et al., 2015), the results described here demonstrate that multicellular complexity readily arises as a solution to a trade-off between selection for fast growth and large size. While selection for rapid sedimentation is likely not a good proxy for natural systems, selection for large size is certainly common in microbial populations (e.g., cluster formation increases survival in the face of predators [Kessin et al., 1996; Boraas et al., 1998; Becks et al., 2012], but cluster formation slows growth [Yokota and Sterner, 2011; Becks et al., 2012]). This work, along with other pioneering work experimentally evolving novel multicellularity (Boraas et al., 1998; Koschwanez et al., 2011; Koschwanez et al., 2013; Ratcliff, Herron, et al., 2013; Hammerschmidt et al., 2014), has shown that the first steps in this transition readily occur. The challenge before us is to determine how more complex, functionally integrated multicellular individuals (e.g., organisms that consists of multiple cell types whose multicellular life cycle is developmentally regulated) evolve from simple multicellular ancestors.

Acknowledgments

We would like to thank Ben Kerr, Vidyanand Nanjundiah, Kayla Peck, and Jennifer Pentz for thoughtful comments on this manuscript and the Konrad Lorenz Institute for funding/hosting the meeting where this work was first presented. This work was supported by NASA Exobiology grant #NNX15AR33G.

References

Becks, L., Ellner, S. P., Jones, L. E., & Hairston, N. G. (2012). The functional genomics of an eco-evolutionary feedback loop: Linking gene expression, trait evolution, and community dynamics. *Ecology Letters*, 15, 492–501.

Bleecker, A. B., & Patterson, S. E. (1997). Last exit: Senescence, abscission, and meristem arrest in *Arabidopsis*. *Plant Cell*, 9, 1169–1179.

Bonner, J. T. (1998). The origins of multicellularity. *Integrative Biology Issues News and Reviews*, 1, 27–36.

Boraas, M. E., Seale, D. B., & Boxhorn, J. E. (1998). Phagotrophy by a flagellate selects for colonial prey: A possible origin of multicellularity. *Evolutionary Ecology*, 12, 153–164.

Duran-Nebreda, S., & Solé, R. (2015). Emergence of multicellularity in a model of cell growth, death and aggregation under size-dependent selection. *Journal of the Royal Society, Interface*, 12, 20140982.

Grosberg, R. K., & Strathmann, R. R. (2007). The evolution of multicellularity: A minor major transition? *Annual Review of Ecology Evolution and Systematics*, 38, 621–654.

Hamilton, W. D. (1964). The genetical evolution of social behavior. I. *Journal of Theoretical Biology*, 7, 1–16.

Hammerschmidt, K., Rose, C. J., Kerr, B., & Rainey, P. B. (2014). Life cycles, fitness decoupling and the evolution of multicellularity. *Nature*, 515, 75–79.

Herron, M. D., Hackett, J. D., Aylward, F. O., & Michod, R. E. (2009). Triassic origin and early radiation of multicellular volvocine algae. *Proceedings of the National Academy of Sciences of the United States of America*, 106, 3254–3258.

Herron, M. D., Rashidi, A., Shelton, D. E., & Driscoll, W. W. (2013). Cellular differentiation and individuality in the "minor" multicellular taxa. *Biological Reviews of the Cambridge Philosophical Society*, 88, 844–861.

Jacobson, M. D., Weil, M., & Raff, M. C. (1997). Programmed cell death in animal development. *Cell*, 88, 347–354.

Jahn, U., Gallenberger, M., Paper, W., Junglas, B., Eisenreich, W., Stetter, K. O., et al. (2008). *Nanoarchaeum equitans* and *Ignicoccus hospitalis*: New insights into a unique, intimate association of two archaea. *Journal of Bacteriology*, 190, 1743–1750.

Jorgensen, P., Nishikawa, J. L., Breitkreutz, B. J., & Tyers, M. (2002). Systematic identification of pathways that couple cell growth and division in yeast. *Science*, 297, 395–400.

Kessin, R. H., Gundersen, G. G., Zaydfudim, V., & Grimson, M. (1996). How cellular slime molds evade nematodes. *Proceedings of the National Academy of Sciences of the United States of America*, 93, 4857–4861.

King, N. (2004). The unicellular ancestry of animal development. *Developmental Cell*, 7, 313–325.

Koschwanez, J. H., Foster, K. R., & Murray, A. W. (2011). Sucrose utilization in budding yeast as a model for the origin of undifferentiated multicellularity. *PLoS Biology*, 9, e1001122.

Koschwanez, J. H., Foster, K. R., & Murray, A. W. (2013). Improved use of a public good selects for the evolution of undifferentiated multicellularity. *eLife*, 2, e00367.

Koufopanou, V. (1994). The evolution of soma in the Volvocales. *American Naturalist*, 143, 907–931.

Lam, E. (2004). Controlled cell death, plant survival and development. *Nature Reviews. Molecular Cell Biology*, 5, 305–315.

Lavrentovich, M. O., Koschwanez, J. H., & Nelson, D. R. (2013). Nutrient shielding in clusters of cells. *Phyical Review E*, 87, 062703.

Lee, J. M., & Bernstein, A. (1995). Apoptosis, cancer and the p53 tumour suppressor gene. *Cancer and Metastasis Reviews*, 14, 149–161.

Libby, E., Ratcliff, W., Travisano, M., & Kerr, B. (2014). Geometry shapes evolution of early multicellularity. *PLoS Computational Biology*, 10, e1003803.

Maynard Smith, J., & Szathmáry, E. (1995). *The major transitions in evolution*. Oxford, UK: Oxford University Press.

Michod, R. E., Viossat, Y., Solari, C. A., Hurand, M., & Nedelcu, A. M. (2006). Life-history evolution and the origin of multicellularity. *Journal of Theoretical Biology*, 239, 257–272.

Nedelcu, A. (2009). Comparative genomics of phylogenetically diverse unicellular eukaryotes provide new insights into the genetic basis for the evolution of the programmed cell death machinery. *Journal of Molecular Evolution*, 68, 256–268.

Nedelcu, A. M., Driscoll, W. W., Durand, P. M., Herron, M. D., & Rashidi, A. (2011). On the paradigm of altruistic suicide in the unicellular world. *Evolution*, 65, 3–20.

Overmann, J. (2010). The phototrophic consortium "*Chlorochromatium aggregatum*"—A model for bacterial heterologous multicellularity. In P. C. Hallenbeck (Ed.), *Recent advances in phototrophic prokaryotes* (pp. 15–29). New York: Springer.

Pennell, R. I., & Lamb, C. (1997). Programmed cell death in plants. *Plant Cell*, 9, 1157–1168.

Pfeiffer, T., & Bonhoeffer, S. (2003). An evolutionary scenario for the transition to undifferentiated multicellularity. *Proceedings of the National Academy of Sciences of the United States of America*, 100, 1095–1098.

Ratcliff, W. C., Denison, R. F., Borrello, M., & Travisano, M. (2012). Experimental evolution of multicellularity. *Proceedings of the National Academy of Sciences of the United States of America*, 109, 1595–1600.

Ratcliff, W. C., Fankhauser, J. D., Rogers, D. W., Greig, D., & Travisano, M. (2015). Origins of multicellular evolvability in snowflake yeast. *Nature Communications*, 6, 6102.

Ratcliff, W. C., Herron, M. D., Howell, K., Pentz, J. T., Rosenzweig, F., & Travisano, M. (2013). Experimental evolution of an alternating uni-and multicellular life cycle in *Chlamydomonas reinhardtii*. *Nature Communications*, 4, 2742.

Ratcliff, W. C., Pentz, J. T., & Travisano, M. (2013). Tempo and mode of multicellular adaptation in experimentally-evolved *Saccharomyces cerevisiae*. *Evolution*, 67, 1573–1581.

Ratcliff, W. C., & Travisano, M. (2014). Experimental evolution of multicellular complexity in *Saccharomyces cerevisiae*. *Bioscience*, 64, 383–393.

Ratcliff, W. C. (2015). [Long-term experiments on strength of settling selection]. Unpublished raw data.

Rebolleda-Gomez, M., Ratcliff, W., & Travisano, M. (2012). Adaptation and divergence during experimental evolution of multicellular *Saccharomyces cerevisiae*. *Artificial Life*, 13, 99–104.

Roff, D. A. (2002). *Life history evolution*. Sunderland, MA: Sinauer Associates.

Schirrmeister, B. E., Antonelli, A., & Bagheri, H. C. (2011). The origin of multicellularity in cyanobacteria. *BMC Evolutionary Biology*, 11, 45.

Smukalla, S., Caldara, M., Pochet, N., Beauvais, A., Guadagnini, S., Yan, C., et al. (2008). *FLO1* is a variable green beard gene that drives biofilm-like cooperation in budding yeast. *Cell*, 135, 726–737.

Solari, C. A., Kessler, J. O., & Michod, R. E. (2006). A hydrodynamics approach to the evolution of multicellularity: Flagellar motility and germ–soma differentiation in Volvocalean green algae. *American Naturalist*, 167, 537–554.

Travisano, M., & Lenski, R. E. (1996). Long-term experimental evolution in *Escherichia coli*: IV. Targets of selection and the specificity of adaptation. *Genetics*, 143, 15–26.

Umar, M. H., & Van Griensven, L. (1997). Morphogenetic cell death in developing primordia of *Agaricus bisporus*. *Mycologia*, 89, 274–277.

Velicer, G. J., & Vos, M. (2009). Sociobiology of the myxobacteria. *Annual Review of Microbiology*, 63, 599–623.

Yokota, K., & Sterner, R. W. (2011). Trade-offs limiting the evolution of coloniality: Ecological displacement rates used to measure small costs. *Proceedings. Biological Sciences*, 278, 458–463.

8 The Paths to Artificial Multicellularity: From Physics to Evolution

Salva Duran-Nebreda, Raúl Montañez,
Adriano Bonforti, and Ricard Solé

Summary

Multicellularity is a collective phenomenon involving the generation of organized cell diversity under an embodied, spatial structuring. The transition from totipotent free-living unicellular organisms to cell-specialized consortia involves novel difficulties for the organism in the form of cheating strategies and increased costs for coordination. In spite of that, multicellularity is known to be a recurrent event in the history of life, and it is also facultatively exhibited by some microbial species. Although most efforts have been focused on unraveling the genetic control of developmental programs, many relevant aspects of development are grounded in purely physicochemical processes. Here we present a simple model approach to the problem of emergence of complexity in multicellular aggregates, where cooperating lower level entities (cells) develop higher order, collective properties. We show how a minimal set of generative rules allowing for differential adhesion, phenotypic plasticity, ecological interactions, and evolutionary dynamics is capable of displaying a meaningful repertoire of spatial patterns, which are selected and shaped by evolutive processes.

Introduction

The emergence of multicellular organisms is a delightful puzzle in its elegance yet maddening by its accessibility. The transition to multicellularity is tied to the emergence of stable interactions among previously isolated cells that provide novel advantages to the collective. Among the major transitions in evolution, it is one of the most frequent innovations that has taken place in multiple independent ways. It fairly well illustrates the idea of emergent complexity as defined by Phil Anderson in his classic paper "More Is Different" (Anderson, 1972; Schuster, 1996). A multicellular system defines a new level of organization whose global properties (grounded in cell–cell interactions) cannot be reduced to the properties of the individual units (the cells). The evolutionary paths toward multicellularity have been traced by means of a diverse array of approximations that include molecular and cell biology comparative studies as well as phylogenetic analyses. Moreover, artificially evolved multicellular aggregates (Ratcliff et al., 2012), synthetic multicellular analogs (Maharbiz, 2012), and their model counterparts (Libby and Rainey, 2013; Duran-Nebreda and Solé, 2015; Libby and Ratcliff, 2014) have shown the potential for novel explorations of an old issue.

While considering the requirements for multicellularity to emerge, an often forgotten component is the requirement of a physical embodiment of cellular and multicellular structures. However, any relevant model should actually take into account the role of generic physical mechanisms associated with cell–cell interactions and their consequences. By generic we mean mechanisms and processes not associated to complex regulatory processes—that is, physical constraints involving gravity, adhesion, or diffusion (Newman and Comper, 1990; Goodwin, 1994; Forgacs and Newman, 2005; Solé et al., 2007; Solé, 2009). Such fundamental constraints pervade both early stages of multicellular life forms as well as current morphogenetic processes (Alberch, 1980; Bonner, 2001; Forgacs and Newman, 2005; Newman and Bhat, 2008). Even under very simplistic assumptions, complex structures are often obtained as soon as physical models of interactions are put in place (Niklas, 1997; Eggenberger, 1997; Coen et al., 2004; Cummings, 2006; Doursat, 2008; Kaandorp et al., 2008).

What are the minimal requirements to define a proper model for the emergence of multicellular systems in the world before developmental programs? What is the contribution of each component to the richness of possible structures and their evolvability? What is the impact of external constraints in canalizing such richness? These are the main questions that we address here. Along with bare physical interactions (associated to cell–cell adhesion and communication) other basic components of cellular behavior can be taken into account. One includes phenotypic plasticity: cell states do not need to remain in a fixed state over time but are known to swap—aided by environmental cues or not—into other available phenotypes in an effort to increase their survivability.

Here we consider a scenario (grounded in existing experiments involving prokaryotic phase variation) where two cell types exist, one of them unable to metabolize waste while the second can. Processing waste to effectively remove it from the system implies a limited growth of those cells performing the waste removal. Waste processing can then be interpreted as a public good, and, accordingly, the risk for a tragedy of the commons exists (Hardin, 1968; Queller, 2000; Hudson et al., 2002). However, since less waste enhances growth of cells not capable of dealing with toxic metabolites, we should ask what type of multicellular solutions to this problem emerge. As shown below, most solutions to the problem result in complex spatial patterns, suggesting that as soon as a small number of basic requirements are in place, consortia and structure may be present and selected long before developmental programs take control. In the next sections, we consider increasingly complex scenarios of multicellular organization emerging from the introduction of physical interactions, phenotypic transitions driven by stochastic switches, ecological dynamics associated with limited resources and waste, and finally the role played by evolutionary dynamics.

The Physical Setting: Energy Functions

As a first step in our analysis of pattern formation associated with multicellular structures, we will consider the role played by preferential adhesion, using a cellular Potts model similar to the one used by Steinberg (1975) (figure 8.1). We consider a population, made of a limited number of cell types, occupying a given two-dimensional grid. These cells interact through adhesion molecules, and their movements are driven by a tendency to minimize their interaction energy. If M cell types are considered, to be indicated as $\sigma_n \in \Sigma = \{0,1,2,...,M\}$, we can define the strength of the interactions among different types by means of an adhesion matrix:

$$J = \begin{pmatrix} J_{00} & \cdots & J_{0M} \\ \vdots & \ddots & \vdots \\ J_{M0} & \cdots & J_{MM} \end{pmatrix}.$$

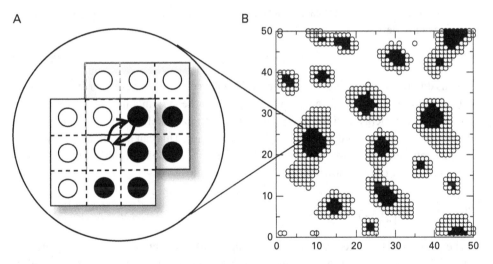

Figure 8.1
Pattern formation through differential adhesion. Here the system is composed by two types of cells (open and filled circles) which are initially scattered at random on a two-dimensional lattice. Cells move and can swap their location with a neighboring cell (A) provided that the final energy is reduced. Eventually, a stable arrangement of ordered cell assemblies (B) is obtained, in a configuration depending on the values of the adhesion matrix. In this example, the adhesion parameters satisfy $J_{12} > (J_{11} + J_{22}) = 2$ and $J_{11} > J_{22}$.

Here J_{ab} is the interaction strength associated between two neighboring cells belonging to types a and b, respectively. For our current system we will consider three possible kinds of cells: σ_0 (gray, medium), σ_1 (white, fast growing), and σ_2 (black, waste processing). The underlying idea here is that cells will tend to attach with their neighbors with a higher probability whenever this allows the system to reach a lower energy per unit area (Steinberg, 1975). It can be shown that an energy function in a given position i, j can be defined as follows:

$$\mathcal{H}_{i,j} = \sum_{S_{k,l} \in \Gamma_{i,j}} J_{S_{kl}, S_{ij}} \,, \tag{8.1}$$

where $\Gamma_{i,j}$, in a square lattice, is the set defined by the eight nearest neighbors of a given cell in position i, j (Moore's neighborhood), each of them occupying a position k, l, and having a defined type $S_{k,l}$. By evaluating this energy function, we can calculate the probability that the cell in i, j will move to a given neighboring location i', j', thus swapping states with that particular neighbor. In order to do this we first calculate the energy function in the case in which no swap occurs; here the energy term, named for convenience $\mathcal{H}_{ORI}$, consists of a part involving the cell in its original position and its neighborhood set $\Gamma_{i,j}$, and of a second part involving the cells' neighbor, located in i', j', with its own neighborhood $\Gamma_{i,j}$. Then we virtually swap the positions of the cell with its neighbor in i', j', and perform again the energy calculation in case of swap, $\mathcal{H}_{SWAP}$. The energy difference then, is readily defined as

$$\Delta \mathcal{H} = \mathcal{H}_{SWAP} - \mathcal{H}_{ORI} \,. \tag{8.2}$$

If this difference is positive, the probability that the cell actually performs the movement will be small, whereas if it is negative, it means that the system's new state would have an overall lower energy, and the swap will be thus be more likely to occur. The higher the energy difference, the more accentuated the probabilities. This can be formalized by using a Boltzmann function. If we indicate as $P(S_{ij} \rightarrow S_{i'j'})$ the probability that our cell moves from i, j to i', j', it can be shown that

$$P\left(S_{ij} \rightarrow S_{i'j'}\right) = \frac{1}{1 + e^{\Delta \mathcal{H}/T}} \,. \tag{8.3}$$

The parameter T is fixed and acts as a "temperature," essentially tuning the degree of determinism of our system. As we can see, if the energy change is positive, the so-called Boltzmann factor $e^{\Delta \mathcal{H}/T}$ will be high, and the denominator of the previous equation will be large. As a consequence, the transition probability will be small. If the difference is negative (the new state implies a reduction of energy), the Boltzmann factor will be small and the probability high: the cell displacement is likely to happen. If no difference is present, $\Delta \mathcal{H} = 0$ and the probability is 1/2 (as in a coin toss).

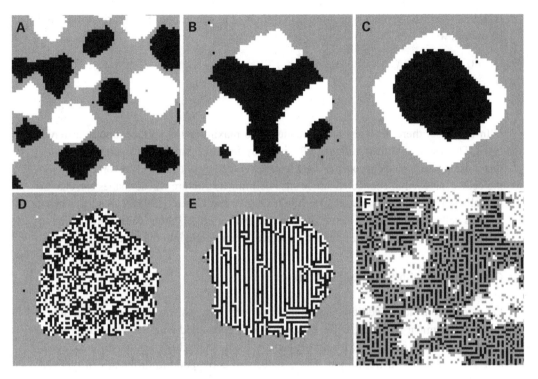

Figure 8.2
Different patterns that can be observed in a cell sorting system with two cell types: separate blobs (A), contiguous domains (B), "onion" layers (C), random mass (D), labyrinth (E), and holed structures (F). Such patterns can be determinable from the adhesion matrix, and although it may take some time to reach a stable configuration, these tend to form a single solid mass of cells (except for F-like patterns).

In figure 8.2b we display the result of a cell sorting process using T = 10 and a matrix favoring the attachment of cells of the same class. Specifically, we apply the following adhesion matrix:

$$J = \begin{pmatrix} 0 & 0 & 0 \\ 0 & -20 & -10 \\ 0 & -10 & -20 \end{pmatrix}.$$

The result of this simulation is a well-defined pattern of spatial segregation resulting from the differential adhesion. The initial condition is a well-mixed system with equal amounts of white and black cells (types 1 and 2 respectively), and half of the sites of the lattice unoccupied (type 0). At this time there are no growth and death processes; accordingly, the number of cells of each type remains unchanged throughout the simulation. For comparison, let us consider an opposite scenario, where cells of one type try to attach to cells belonging to the other type more than to cells of its same state. The matrix now would be as follows:

$$J = \begin{pmatrix} 0 & 0 & 0 \\ 0 & -10 & -20 \\ 0 & -20 & -10 \end{pmatrix}.$$

In this case, there is an inevitable tendency to maximize the surface among different types. The result, as shown in figure 8.2e, is a labyrinth-like pattern. It is worth mentioning that more sophisticated approaches were later developed by Glazier and coworkers, involving cells composed of connected lattice sites defining a structure whose shapes can change in realistic ways (Graner and Glazier, 1992; Glazier and Graner, 1993). Having a surface enables cells to squeeze and bend; they can be stiff or engulf other cells, allowing us to incorporate a whole range of relevant biological phenomena into the model.

Cell Adhesion with Switching Cell States

Let us now consider an additional element in our previous scheme. This component introduces a dynamical switching between states that is uncoupled with the adhesion process. Specifically, let us consider that cells are capable of switching between two different available states in a stochastic fashion. These states will correspond to different cell types of the adhesion matrix J.

This choice is inspired by the dynamics exhibited by the so called phase variation in prokaryotes, which has been identified in several experiments involving aggregation, antigenicity, and pathogenicity among others (Henderson et al., 1999; Darmon and Leach, 2014). Specifically, these cells are known to transit between different phenotypes in a way that can be history independent—requires no memory—or needs no external chemical cues. As a result of this process, a fixed fraction of a population will consistently display a particular phenotype, even after the removal of that particular subpopulation. This phenomenon introduces phenotypic plasticity in the system: even if some phenotypes are selected against and disappear, the complementary phenotype will regenerate the original population with both states present, increasing the robustness of the whole in a strategy that has been related to bet hedging (Veening et al., 2008).

Since these transitions between states are not dependent on the previous sequence of states of the cells, the whole mechanism can be aptly modeled by a very simple Markovian process (see figure 8.3). As new parameters of the system we now include the transition probabilities between states p and q and a scaling factor κ. This last parameter is introduced to regulate the relative speeds at which the two presently used processes (physics, i.e., cell sorting by adhesion, and phenotypic switching) do happen. Equations 8.4 and 8.5 describe the fundamental dynamics of the two populations in the form of ordinary differential

A

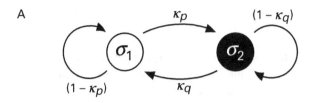

B

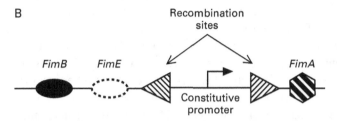

Figure 8.3
Description of the very simple genetic processes that cells undergo in our model (A) with fixed probabilities of transition between states σ_1 and σ_2 that do not depend on their history and are not regulated by signaling mechanisms. Similar phenomenon (B) implemented by DNA reorganization in type I fimbriae in *E. coli*. Once the DNA is inverted (mediated by the recombinases: FimB or FimE), the promoter changes its orientation, disabling the transcription of gene FimA (Gally et al., 1993).

equations, including cells' growth and death, although it must be noted that these two processes are still not considered at this stage:

$$\frac{\partial P_{\sigma_1}}{\partial t} = \mu_{\sigma_1}(P)P_{\sigma_1} + \kappa(qP_{\sigma_2} - pP_{\sigma_1}) - \eta_{\sigma_1}P_{\sigma_1},$$ (8.4)

$$\frac{\partial P_{\sigma_2}}{\partial t} = \mu_{\sigma_2}(P)P_{\sigma_2} + \kappa(pP_{\sigma_1} - qP_{\sigma_2}) - \eta_{\sigma_2}P_{\sigma_2}.$$ (8.5)

The terms $\mu_{\sigma_n}(P)P_{\sigma_n}$ represent the growth functions of the separated populations of the two types of cells, namely, P_{σ_1} and P_{σ_2}, and depend on the global population value P. The terms η_{σ_1} and η_{σ_2} indicate the specific death rate of each of the two populations. As already mentioned, the term in the middle describes the phenotypic switching between the two states, depending on the transition probabilities p and q and the scaling factor κ. At the steady state, given that $\mu_{\sigma_n}(P)P_{\sigma_n} - \eta_{\sigma_1}P_{\sigma_n} = 0$, the ratio between the two populations is given by

$$P_{\sigma_1}/P_{\sigma_2} = q/p.$$ (8.6)

Introducing a conservation equation of the form $P_{\sigma_1} + P_{\sigma_2} = 1$—valid once there is no growth and we normalize P_{σ_n}—gives the values for each subtype:

$$P_{\sigma_1} = \frac{q}{p+q} P_{\sigma_2} = \frac{p}{p+q} .$$

(8.7)

Remarkably, once phenotypic switching is enabled, new kinds of patterns become possible within the system. A hallmark of self-organization in embryology and pattern formation, Turing patterns (Turing, 1952) were first described as a possible mechanistic explanation for the skin pigmentation in mammals. Later refined by others (Meinhardt and Gierer, 2000; Gray and Scott, 1990; Murray, 1981), Turing patterns rely on very simple chemical processes, namely, the asymmetrical roles of two chemical species dubbed morphogens and their diffusive properties. In essence, one of the two morphogens (referred to as the activator) enhances its own synthesis as well as that of the other morphogen (the so-called inhibitor). The second morphogen, contrariwise, impedes the synthesis of the activator and diffuses more rapidly. This very simple motif has been shown to operate in various systems and has been used to model disparate processes at very different scales: from vegetation patterns (Lejeune and Tlidi, 1999) to digit formation (Sheth et al., 2012).

Our model would seem quite different from the classical reaction–diffusion scheme just presented, as it does not include any diffusion asymmetry nor specific roles for cellular states. Other interesting mechanisms for the generation of periodic structure can be found in the literature. For instance, it has been shown that contact-mediated communication between cells is formally equivalent to a reaction–diffusion process (Babloyantz, 1977). Nevertheless, differently from what was proposed by Babloyantz, in our model the molecules on the surface of cells—essentially attachment molecules—do not affect rates of reactions in their neighbors; phenotypic switching occurs in a truly Markovian way (independently from the past and from the neighbors' states). Yet, as shown in figure 8.4, cell sorting with switching states consistently leads to spatial patterns with dominant wavelength, highly dependent on the transition probabilities p, q and the scaling factor κ.

Specifically, close to $p = q$ the balanced dynamic equilibrium establishes a maze-like pattern (see figure 8.4b, c). Increasing disparity between the transitions leads to more striped patterns and then, on the extremes, to spotted arrangements (see figure 8.4e, f). Additionally, the scaling factor κ plays a central role in defining the particular wavelength of the pattern. For high scaling between genetic processes and physical ones (i.e., approaching the original situation where only physical processes are relevant) the system behaves like immiscible liquids separating in distinct phases. On the other side, when the stochastic switching is fast enough, the pattern dissolves into the randomness of switching (see figure 8.4d). However, for intermediate values of κ, a stable wavelength exists which decreases with the scaling factor (see figures 8.4d–f).

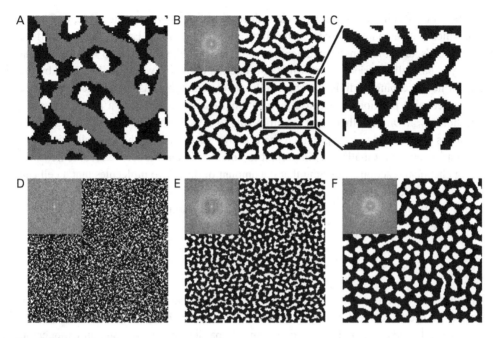

Figure 8.4
Constant wavelength patterns resulting from preferential attachment rules using an "onion" adhesion matrix and half empty spaces (A) or a diagonal matrix of couplings and no empty positions (B–F). A zoomed region is offered in (C) to show the rounded ends of the distinct phases. The patterns emerge from a population of only σ_2 cells (black) in a 200×200 toroidal lattice. The transition parameters are $p = 0.5$ and $q = 0.25$ (A, D–F) or $p = q = 0.5$ (B, C). The scaling factor is $\kappa = 1$ (d), $\kappa = 0.01$ (e), or $\kappa = 0.0001$ (A–C, F). We also show, as insets for some configurations, the existence of a dominant frequency in Fourier space. The insets were created applying the Fiji fast Fourier transform algorithm on a radial gradient masked version of the pictures.

Adhesion, Switching and Resources: Ecology and Competition

Let us include a new level of interaction between cells, namely, the existence of chemical substances that can lead to cell growth or death. These simulated molecules will be incorporated in the form of a discretized field, in such a way that they exist both inside and outside cells. Diffusive processes, consumption, and synthesis will be the main drivers of concentration change of these chemicals. Specifically, we will include two such molecules: a resource needed for survival and division, or Nutrient (N), and a toxic by-product of metabolism, or Waste, which can cause death to both types of cells when its concentration reaches a given threshold in a given point. We will also introduce phenotypic variation in the way the two different cell types interact with these molecules. One of them (σ_1) will be able to consume nutrient, thus generating waste, but will not be able to degrade waste. The other (σ_2) will display a linear trade-off between the ability to absorb nutrient and the ability to degrade waste. The variable e regulates the trade-off between these two abilities for σ_2 cells. We also need to introduce an *internal currency* molecule, or Energy (E), which is created internally to each cell by absorbing nutrient. Energy does not diffuse outside cells, and it is transformed again to nutrient when the cell dies, so we can say that energy is basically nutrient transformed into internal currency, which will be used by the cell for housekeeping or division. Thus, cells can die either because of an excess of waste in the occupied position or because their internal energy falls below a given starvation threshold θ_{DE}. On the other hand, a cell is able to divide when its internal energy level surpasses a given value θ_E, provided there is free space around the cell, allowing it to generate a new cell. Finally, it is important to note that in our model we assume nonsaturating chemical reactions, meaning that the amount of these two molecules that a cell can take in a time step is always proportional to the concentration of molecules in the occupied lattice position. The change in a finite element of the variable N_{ij} for a given position i, j will be

$$\frac{\partial N_{ij}}{\partial t} = D_N \nabla^2 N_{ij} + \mu \delta_{(\sigma_0, S_{ij})} - N_{ij}(\eta_N + \rho \delta_{(\sigma_1, S_{ij})} + \varepsilon \rho \delta_{(\sigma_2, S_{ij})}), \tag{8.8}$$

meaning that there is a fixed addition of nutrient m if that particular position is empty (hence the use of Kronecker's delta function $\delta_{(\sigma_0, S_{ij})}$, which equals 1 in the case where $S_{ij} = \sigma_0$ and 0 otherwise). There is diffusion at a particular rate D_N and an exponential decay η_N that ensures reaching a fixed value of $N = \mu / \eta_N$ in the absence of cells. Finally, there are the two possible absorption rates for the two types of cells: ρ (σ_1) and, altered by a linear trade-off, $\varepsilon \rho$ (σ_2). Likewise, the change in a finite element of the variable E_{ij} for a given position i, j will be

$$\frac{\partial E_{ij}}{\partial t} = N_{ij}\left(\rho \delta_{(\sigma_1, S_{ij})} + \varepsilon \rho \delta_{(\sigma_2, S_{ij})}\right) - \eta_E E_{ij}. \tag{8.9}$$

As mentioned, we consider that energy does not diffuse and can only exist inside cells. It is created at exactly the same rate by which N_{ij} is absorbed, and it decays exponentially at rate η_E. If any cell were to increase its energy value beyond a certain fixed threshold θ_E and an empty position exists in its neighborhood, a new cell would be created with exactly the same properties, and the energy would be split between the two. Conversely, should any cell have less than θ_{DE} energy, it would die because of starvation. Finally, the change in a finite element of the variable W_{ij} for a given position i,j will be

$$\frac{\partial W_{ij}}{\partial t} = D_W \nabla^2 W_{ij} + N_{ij}\left(\rho \delta_{(\sigma_1, S_{ij})} + \varepsilon \rho \delta_{(\sigma_2, S_{ij})}\right) - W_{ij}(\eta_W + (1-\varepsilon)\rho \delta_{(\sigma_2, S_{ij})}). \tag{8.10}$$

Waste has its own diffusion coefficient D_W, decays proportionally to η_W, and is also created at the same rate that nutrient is consumed. Still, σ_2 cells can degrade waste at a rate dictated by a linear trade-off $(1-\varepsilon)$ between growth and degradation. If a cell were to experience a waste greater than a preestablished threshold (θ_{DW}), it would die, and its contents (i.e., its energy, transformed again into nutrient) would be returned to the surrounding media.

Besides the examined methods of death, we assume in our model that cells have a constant, age-independent, probability of dying because of external factors $\xi = 5 \cdot 10^{-5}$. Finally, the diffusion operator $D_x \nabla^2 X_{ij}$ is computed using the standard discretization:

$$D_x \nabla^2 X_{ij} = D_x \sum [X_{kl} - X_{ij}]. \tag{8.11}$$

The presence or absence of these generic resources and toxic molecules that intervene in the birth and death dynamics of cells adds a coupling and a communication of sorts that clearly goes beyond the local action of the adhesion processes described before. Diffusion of these molecules introduces a higher range of interaction between cells, potentially leading to novel types of ordering. Furthermore, cells can now compete or coordinate in order to deal with problems arising from lack of nutrient or excess of waste, which, again, can push the system toward the kinds of spatial and temporal structuring that have been described before in ecological systems, such as waves and spirals (Solé and Bascompte, 2006). Figure 8.5 shows an instance in which the system behaves as excitable media.

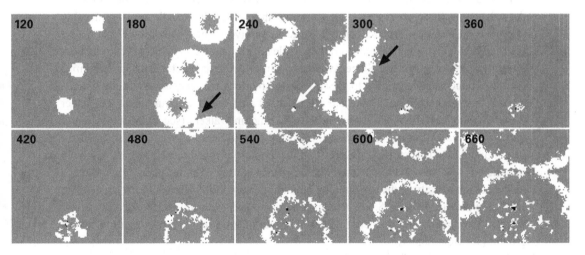

Figure 8.5
Waves of cell growth propagating through the simulation space. The time of each snapshot, in algorithm iterations, is displayed at the top left of each image. Due to the toroidal conditions, the wave annihilates itself when the fronts collide (black arrows). Yet black cells—here created at a very low rate—survive the pulse of waste and serve as generators of the following waves (white arrow). At the end of the sequence, several nucleating clusters of black cells compete, each generating new white cells which fail to propagate.

Evolving Multicellularity in a Limiting Environment

As the last step, we incorporate all the previously discussed mechanisms into a genetic algorithm (GA), so that some of the parameters are mutable and inheritable by cells. Since our interest lies in understanding the emergence of multicellularity, the initial conditions of these simulations are designed to remain as neutral in that regard as possible. That is, cells will begin as blank slates for the terms of the adhesion matrix and for the phenotypic switching probabilities, meaning $p = q = 0$ and $J_{ab} = 0$ at the beginning of the simulation. The transitions for phenotypic switching are individual properties; however, the adhesion process DH now needs to take into account properties from pairs of individuals that might have different associated values for adhesion. Let us then define a new energy function that averages the corresponding adhesion values:

$$
\mathcal{H}_{ij} = \sum_{S_{kl} \in \Gamma_{ij}} \begin{cases} 0, \text{ if } S_{ij} = S_{kl} = 0 \\ \dfrac{\left[J^{ij}_{S_{ij}, S_{kl}} + J^{kl}_{S_{ij}, S_{kl}} \right]}{2 - \delta_{(\sigma_0, S_{ij})} - \delta_{(\sigma_0, S_{kl})}} \text{ otherwise} \end{cases}.
\tag{8.12}
$$

Here, the superscripts in J are used to indicate positions in the lattice (so the adhesion matrix for a particular finite element).

For the first experiment with GA, the trade-off between nutrient consumption and waste degradation is fixed at $\varepsilon = 0.4$. This ensures that although σ_2 cells grow noticeably slower, it makes them process enough waste to survive higher densities. In that sense, black cells would display a fixed κ strategy while their counterparts invest in reproduction, thus falling into an r strategy. Since e will not be an evolving parameter, cells will optimize the other dimensions available to them, finding solutions that accommodate the "preassigned" roles to each type. Figure 8.6 recapitulates a representative run of the GA under these conditions. Cells begin as essentially σ_1 random walkers (see figure 8.6b), but a variant that maximizes its contact surface to the empty medium is quickly selected for (see figure 8.6c). This strategy ensures that the amount of medium will diffuse passively into the cell and waste will be excreted at the maximum rate. Yet this phenotype becomes unfit once the toxicity of the surrounding medium becomes threatening, and σ_2 free-living start to appear (see figure 8.6c). Among those, mutants with high $\sigma_2 - \sigma_2$ cohesion are selected for, maximizing the benefit from the investment in reducing waste at the cost of growth (see figure 8.6D).

We could say that, from a game theoretic perspective, cells can become intelligent players and choose to play against other strategies that benefit them the most, reducing exploitation from cheaters. Then a "parasite" appears (see figure 8.6E) with new adhesion properties: σ_1 cells attaching to σ_2 (that degrade waste for them and accordingly increase their survivability). In this final stage stable consortiums develop between σ_1 and σ_2 that

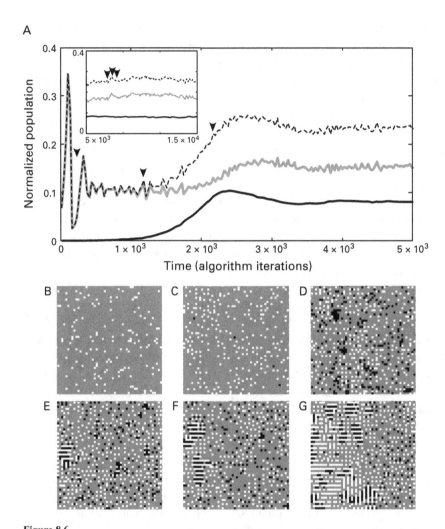

Figure 8.6
Evolution of virtual organisms in a genetic algorithm with mutating adhesion parameters (J) and transition probabilities (p, q). In this simulation the trade-off between growth and waste degradation for σ_2 cells (ε) was fixed to 0.4. The top plot (A) shows the normalized population of σ_1 cells (gray line), σ_2 (black line), and the total (dashed black line). The system converges to a stable configuration with a dynamical equilibrium made of different species that coexist. The snapshots shown—also marked by arrows in the temporal occupation plot—are representative subregions of the 250×250 lattice at 250 (B), 1,250 (C), 2,500 (D), 7,500 (E), 8,000 (F), and 8,500 (G) algorithm iterations. The initial conditions are a random seed of σ_1 cells (amounting to 5% of the grid positions) with $p = q = 0$ and $J_{ij} = 0$ (all entries are null).

share the same genotype, the first ones optimized for resource absorption/reproduction (germ line) and the latter for maintenance of a proper environment free from waste (soma, albeit they can still divide at a significantly slower rate). The sequence in figure 8.6E–G shows the growth of such a consortia followed by a breakage of the aggregate, suggesting the first stages of a common life cycle and reproduction at the group level. At the same time, other "unicellular" species inhabit this virtual world, making use of the public good performed by σ_2 cells.

Although a fixed ε would seem to force preassigned roles to each cellular state, stochastic phenotypic switching introduces a relevant twist to the concept of division of labor in these proto-organisms. This is also the case for reproductive fitness, since cells that are performing soma functions may switch to the reciprocal state and make more copies of themselves for a time. Indeed, we observe high values of both p and q, suggesting that successful cells make use of a mixed, bet-hedging strategy—meaning that they can aggregate if enough cells with the same adhesion properties are around but, at the same time, do have mild success in a free-living environment.

In the second experiment (see figure 8.7), ε becomes an evolvable parameter alongside the previously used one, so on this occasion cells are able to evolve their strategy. Starting from the same initial conditions as in the first experiment, the σ_1 are rapidly selected for maximum surface in contact with the medium (see figure 8.7C). Then, as with the previous setup, σ_2 cells start to appear (see figure 8.7C). Aggregates of waste-degrading cells are also selected for, since they can extend the metabolic protection to their progeny (see figure 8.7D), yet at this stage the history of these two systems diverges. What ensues is a two-species system, one with high p, low q, and ε that does not reach the fixed value of the previous simulation but converges at approximately 0.82, while the second species is high q, low p, and has wide ranges of e. The aggregates of σ_2 displayed in figure 8.7D correspond to the first entity while the sparse structures composed of σ_1 make up the second one, whose strategy is to capture as much substrate as possible.

Collectively, the two-species system displays oscillatory dynamics (see figure 8.7A; a full oscillation is shown in snapshots in figure 8.7E–G) due to the ecological interactions between the components. The slow-growing cells create a protective environment in which the second species can thrive, displacing the former in the competition for nutrient. Yet, as more σ_2 cells die, waste accumulates and the white cell population collapses in a sudden fashion, creating the space and resources for slow growers to take over again. This forced life cycle due to cheaters that do not contribute to the public good has been shown in operation before (Hammerschmidt et al., 2014), proposed as a universal mechanism to ensure periodic genetic bottlenecks in proto-organisms and unavoidable division of labor in aggregates of cells.

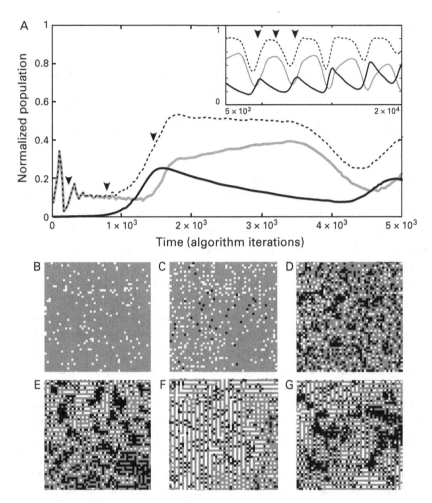

Figure 8.7
Execution of the genetic algorithm with free J, p, q, and ε. The initial conditions are the same as those used in figure 8.6. The starting value of ε was set to 1, so that initially σ_1 and σ_2 cells are phenotypically identical. The top plot (A) shows normalized populations of σ_1 (gray), σ_2 (black), and total (dashed black line) cell types. Again snapshots are representative subregions of the whole simulation matrix at times 250 (B), 750 (C), 1,500 (D), 8,500 (E), 12,000 (F), and 15,250 (G) iterations.

Discussion and Prospects

Artificial settings for the study of evolutionary processes—artificial life (Lenski et al., 1999; Adami et al., 2000), artificial evolution (Ratcliff et al., 2012; Lenski and Travisano, 1994), and synthetic biology (Chuang, 2012; Maharbiz, 2012) among others—offer a unique opportunity to witness innovation in biology unfold. The case of the major transitions in evolution seems especially appealing for such treatment, since the sudden changes in the organizing principles of organisms and ecologies left little trace and still lack proper understanding.

In this chapter we have presented one possible way to build from the ground up a plausible scenario for the emergence of multicellularity. By adding one by one the different layers of complexity that make this model, we are able to better understand their role and how they interact together. One particular feature that strikes us as rather untapped is stochastic phenotype switching. A very relevant trait in modern bacteria for pathogenic activity and mixed strategies in the form of phase variation, one can suspect that ancestral multicellular precursors may have used it as a form of temporal division of labor, later refined into a spatial division of labor through "irreversible" cell differentiation.

The integration of the layers provided new ways for ordering: from spatial to temporal regularities, a hallmark of self-organized biological processes. Yet for each component, a set of parameters could be devised so that a particular process remained completely inconsequential, making the different layers orthogonal. This is to say, each process can only increase the potential number of patterns created by the model (as illustrated qualitatively in figure 8.8), though in the final evolutive step very few were selected for. In other words, the physicochemical mechanisms included in the present work act as enablers of patterning (Newman and Comper, 1990; Newman et al., 2006) while evolutive forces discard many unfit possibilities.

Particularly, in the spatial patterns section we showed that periodic structures can be created by very simple processes (essentially a free energy minimization problem pushed constantly outside equilibrium by the phenotypic switching). We concede that cells forming the pigmentation of animals do not change their adhesion properties stochastically, and thus, the mechanism presently discussed most likely will not be responsible for the skin patterns of mammals. Yet the bigger picture presented here puts together elementary mechanisms—displayed by current and past unicellular organisms—in order to obtain meaningful structures observed in our biosphere.

Lastly, this work can be expanded in several interesting ways. Introducing genetic regulation in the form of complex transcription factor networks might offer a glimpse of the opportunities and risks associated with building genetic programs at the expense of phenotypic plasticity. Moreover, cell–cell explicit communication was not included at the present time and, through its interaction with genetic and metabolic processes, might come to play a central role in organizing the protobody plans of artificial creatures. Again, the

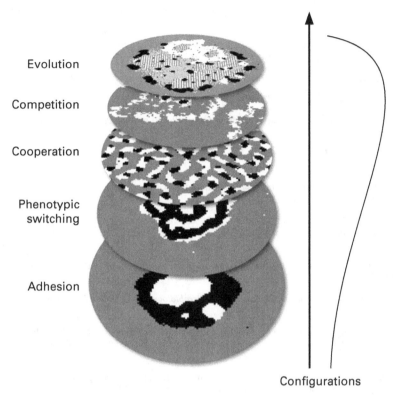

Evolution

Competition

Cooperation

Phenotypic
switching

Adhesion

Configurations

Figure 8.8
As we increase the layers of complexity in our virtual world, the number of different patterns displayed by the system increases: first, "simple" arrangements due to adhesion; then, tunable wavelength Turing patterns due to phenotypic switching; and, finally, excitable media and sparse structures thanks to competition and cooperation. Conversely, evolutionary processes constrain the morphological space to those most fit and stable in a complex ecology.

particular steps regarding the apparition and later recruitment of physicochemical processes into development remain unclear, yet this kind of approach can offer real insights about such crucial events in the evolution of our biosphere.

Acknowledgments

We thank the members of the Complex Systems Lab for useful discussions. This work has been supported by grants from the James McDonnell Foundation, the Botín Foundation, and the Santa Fe Institute.

References

Adami, C., Ofria, C., & Collier, T. C. (2000). Evolution of biological complexity. *Proceedings of the National Academy of Sciences of the United States of America*, 97, 4463–4468.

Alberch, P. (1980). Ontogenesis and morphological diversification. *American Zoologist*, 20, 653–667.

Anderson, P. (1972). More is different. *Science, 177*, 393–396.

Babloyantz, A. (1977). Self-organization phenomena resulting from cell–cell contact. *Journal of Theoretical Biology*, 68, 551–561.

Bonner, J. T. (2001). *First signals: The evolution of multicellular development.* Princeton, NJ: Princeton University Press.

Chuang, J. S. (2012). Engineering multicellular traits in synthetic microbial populations. *Current Opinion in Chemical Biology*, 16, 370–378.

Coen, E., Rolland-Lagan, A.-G., Matthews, M., Bangham, J. A., & Prusinkiewicz, P. (2004). The genetics of geometry. *Proceedings of the National Academy of Sciences of the United States of America*, 101, 4728–4735.

Cummings, F. (2006). On the origin of pattern and form in early metazoans. *International Journal of Developmental Biology*, 50, 193–208.

Darmon, E., & Leach, D. R. (2014). Bacterial genome instability. *Microbiology and Molecular Biology Reviews*, 78(1), 1–39.

Doursat, R. (2008). Organically grown architectures: Creating decentralized, autonomous systems by embryomorphic engineering. In R. P. Würtz (Ed.), *Organic computing* (pp. 167–200). Berlin, Germany: Springer-Verlag.

Duran-Nebreda, S., & Solé, R. V. (2015). Emergence of multicellularity in a model of cell growth, death and aggregation under size-dependent selection. *Journal of the Royal Society Interface, 12*(102).

Eggenberger, P. (1997). Evolving morphologies of simulated 3d organisms based on differential gene expression. In P. Husbands & I. Harvey (Eds.), *Fourth European Conference on Artificial Life* (pp. 205–213). Cambridge, MA: MIT Press.

Forgacs, G., & Newman, S. A. (2005). *Biological physics of the developing embryo.* Cambridge, UK: Cambridge University Press.

Gally, D. L., Bogan, J. A., Eisenstein, B. I., & Blomfield, I. C. (1993). Environmental regulation of the fim switch controlling type 1 fimbrial phase variation in *Escherichia coli* K-12: Effects of temperature and media. *Journal of Bacteriology*, 175, 6186–6193.

Glazier, J. A., & Graner, F. (1993). Simulation of the differential adhesion driven rearrangement of biological cells. *Physical Review E: Statistical Physics, Plasmas, Fluids, and Related Interdisciplinary Topics*, 47, 2128–2154.

Goodwin, B. C. (1994). *How the leopard got its spots.* Princeton, NJ: Princeton University Press.

Graner, F., & Glazier, J. A. (1992). Simulation of biological cell sorting using a two-dimensional extended Potts model. *Physical Review Letters*, 69, 2013–2016.

Gray, P., & Scott, S. K. (1990). *Chemical oscillations and instabilities: Non-linear chemical kinetics.* Oxford, UK: Oxford University Press.

Hammerschmidt, K., Rose, C. J., Kerr, B., & Rainey, P. B. (2014). Life cycles, fitness decoupling and the evolution of multicellularity. *Nature*, 515, 75–79.

Hardin, G. (1968). The tragedy of the commons. *Science, 162*, 1243–1248.

Henderson, I. R., Owen, P., & Nataro, J. P. (1999). Molecular switches—The ON and OFF of bacterial phase variation. *Molecular Microbiology*, 33, 919–932.

Hudson, R. E., Aukema, J. E., Rispe, C., & Roze, D. (2002). Altruism, cheating, and anitcheater adaptations in cellular slime molds. *American Naturalist*, 160, 31–43.

Kaandorp, J. A., Blom, J. G., Verhoef, J., Filatov, M., Postma, M., & Müller, W. E. G. (2008). Modelling genetic regulation of growth and form in a branching sponge. *Proceedings. Biological Sciences*, 275, 2569–2575.

Lejeune, O., & Tlidi, M. (1999). A model for the explanation of vegetation stripes (tiger bush). *Journal of Vegetation Science*, 10(2), 201–208.

Lenski, R. E., Ofria, C., Collier, T. C., & Adami, C. (1999). Genome complexity, robustness and genetic interactions in digital organisms. *Nature*, 400, 661–664.

Lenski, R. E., & Travisano, M. (1994). Dynamics of adaptation and diversification: A 10,000-generation experiment with bacterial populations. *Proceedings of the National Academy of Sciences of the United States of America*, 91, 6808–6814.

Libby, E., & Ratcliff, W. C. (2014). Ratcheting the evolution of multicellularity. *Science*, 346, 426–427.

Libby, E., & Rainey, P. B. (2013). A conceptual framework for the evolutionary origins of multicellularity. *Physical Biology*, 10, 035001.

Maharbiz, M. M. (2012). Synthetic multicellularity. *Trends in Cell Biology*, 22, 617–623.

Meinhardt, H., & Gierer, A. (2000). Pattern formation by local self-activation and lateral inhibition. *BioEssays*, 22, 753–760.

Murray, J. D. (1981). A pre-pattern formation mechanism for animal coat markings. *Journal of Theoretical Biology*, 88, 161–199.

Newman, S. A., & Bhat, R. (2008). Dynamical patterning modules: Physico-genetic determinants of morphological development and evolution. *Physical Biology*, 5, 015008.

Newman, S., & Comper, W. D. (1990). "Generic" physical mechanisms of morphogenesis and pattern formation. *Development*, 110, 1–18.

Newman, S. A., Forgacs, G., & Müller, G. B. (2006). Before programs: The physical origination of multicellular forms. *International Journal of Developmental Biology*, 50, 289–299.

Niklas, K. J. (1997). *The evolutionary biology of plants*. Chicago: University of Chicago Press.

Queller, D. C. (2000). Relatedness and the fraternal major transitions. *Philosophical Transactions of the Royal Society of London. Series B, Biological Sciences*, 355, 1647–1655.

Ratcliff, W. C., Denison, R. F., Borrello, M., & Travisano, M. (2012). Experimental evolution of multicellularity. *Proceedings of the National Academy of Sciences of the United States of America*, 109, 1595–1600.

Schuster, P. (1996). How does complexity arise in evolution? *Complexity*, 2, 22–30.

Sheth, R., Marcon, L., Bastida, M. F., Junco, M., Quintana, L., Dahn, R., et al. (2012). Hox genes regulate digit patterning by controlling the wavelength of a Turing-type mechanism. *Science*, 338, 1476–1480.

Solé, R. V. (2009). Evolution and self-assembly of protocells. *International Journal of Biochemistry & Cell Biology*, 41, 274–284.

Solé, R. V., & Bascompte, J. (2006). *Self-organization in complex ecosystems. (MPB-42*. Princeton, NJ: Princeton University Press.

Solé, R. V., Munteanu, A., Rodriguez-Caso, C., & Macia, J. (2007). Synthetic protocell biology: From reproduction to computation. *Philosophical Transactions of the Royal Society of London. Series B, Biological Sciences*, 362, 1727–1739.

Steinberg, M. S. (1975). Adhesion-guided multicellular assembly: A commentary upon the postulates, real and imagined, of the differential adhesion hypothesis, with special attention to computer simulations of cell sorting. *Journal of Theoretical Biology, 55*, 431–432.

Turing, A. M. (1952). The chemical basis of morphogenesis. *Philosophical Transactions of the Royal Society of London. Series B, Biological Sciences*, 237, 37–72.

Veening, J. W., Smits, W. K., & Kuipers, O. P. (2008). Bistability, epigenetics, and bet-hedging in bacteria. *Annual Review of Microbiology*, 62, 193–210.

IV METAZOA AND RELATED THEORY

9

What Are the Genomes of Premetazoan Lineages Telling Us about the Origin of Metazoa?

Iñaki Ruiz-Trillo

Metazoans: Complex Multicellularity, Development, and Explosive Diversification

Multicellularity has been acquired several times independently within the tree of life, both in prokaryotes and in eukaryotes. In eukaryotes, multicellularity evolved in fungi, animals, slime molds, charophyte algae (and their descendants, the land plants), as well as green, red, and brown algae, and several other taxa (King, 2004; Grosberg and Strathmann, 2007; Ruiz-Trillo et al., 2007; Rokas, 2008a). Undoubtedly, however, the multicellularity of metazoans or animals, together with that of embryophytes (plants), is among the most complex ever attained by eukaryotes.

A key feature of both metazoan and embryophyte complexity is their orchestrated embryonic development (Valentine et al., 1991). The regulated coordination of development processes and the diversity of cell types situates these two lineages as pivotal to understanding the origin of complex multicellularity, the origin of development, and the evolution of diverse cell types. In this review we will focus on the origin of metazoan multicellularity.

A major question with regard to the origin of metazoan multicellularity is how the genes involved in multicellular functions were acquired. In particular, animals seem to share a complex "genetic toolkit" that differentiates them from the rest of eukaryotes (including plants and fungi) (King, 2004; Rokas, 2008a, 2008b). This metazoan toolkit includes specific genes involved in cell signaling, cell adhesion, and cell differentiation. Thus, the questions are when and how did this toolkit appear.

The fossil record seems to suggest that the origin of animals was explosive (Valentine, 2004). In particular, paleontologists have shown that almost all extant animal phyla appeared suddenly in the fossil record, at the onset of the Cambrian, around 543 million years ago. This sudden appearance of animals is known as the "Cambrian explosion," and there are several hypotheses concerning its explosive origin (Valentine, 2004). Some authors suggest that the Cambrian explosion was due to environmental reasons, namely, an increase in the oxygen levels in the atmosphere. Others see ecology as the main cause, a result of an increasing arms race between prey and predators, giving rise to larger and more complex exoskeletons. Finally, some researchers have a preference for genetic explanations, such as a period of genetic innovation, including genes for body patterning and gene regulation, which provided the raw material for a huge divergence of new animal body plans. Regardless of the relative importance of each cause, we can assume that the explosive origin of metazoans was probably due to a combination of environmental, ecological, and genetic factors.

Independently of what the fossil record says, animals have one of the most complex multicellular body plans ever attained by eukaryotes. Moreover, they are highly conspicuous and extremely diverse, and, uniquely with the land plants, they exhibit a complex orchestrated embryonic development. Indeed, animals changed the face of the Earth (Conway Morris, 1998). Thus, understanding how animals emerged from their single-celled protist ancestors is an important scientific endeavor.

Comparative Genomics, a Powerful Tool to Unravel the Origin of Metazoa

Regardless of the relative importance of ecology and the environment in the origin of animals, it is clear that genetics played an important role as well. How else could the great diversity and complex body plans of animals emerge from their relatively simple single-celled ancestors? In this regard, one of the major questions is whether the main genetic components of multicellularity (i.e., genes involved in cell adhesion, cell signaling, and cell differentiation) were novel innovations of the metazoan lineage or whether they were already present in the unicellular ancestor.

Comparative genomics is, undoubtedly, the most powerful methodology we have today to address the aforementioned question and to provide clues on how the metazoan genetic toolkit evolved. A key issue here is what to compare. Comparing the genomes of early branching animals (such as cnidarians, sponges, and ctenophores) with the genomes of the more complex bilaterians provides clues into early animal evolution but not into the specific unicellular-to-multicellular transition that gave rise to metazoans. To specifically address this transition, the genomes of early-branching animals should be compared with the genomes of animal's closest unicellular relatives. However, who are the animals' closest unicellular relatives?

The Unicellular Relatives of Animals

To identify the closest unicellular relatives of animals, a well-resolved phylogenetic tree of eukaryotes is needed. Although a robust eukaryotic tree of life is still under construction for some of its nodes, molecular phylogenies have clearly shown that animals and fungi shared a more recent common ancestor than plants and other eukaryotes (Baldauf and Palmer, 1993). The clade that comprises animals and fungi is known as the opisthokonts

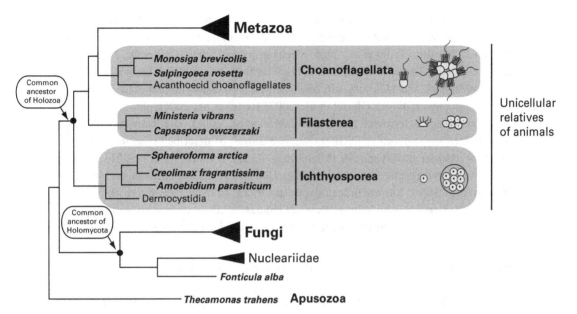

Figure 9.1
Schematic representation of the Opisthokonta clade. Phylogenetic relationships based on molecular analyses (Torruella et al., 2012). Note that the opisthokonts are divided into Holozoa, which comprises Metazoa and their relatives, and Holomycota, which comprises Fungi and their relatives. Choanoflagellata, Filasterea, and Ichthyosporea are the closest unicellular relatives of Metazoa. A schematic life cycle of colonial choanoflagellates, such as *S. rosetta*, filastereans (*C. owczarzaki*) and ichthyosporeans (*S. arctica*) are depicted in the right part of the figure. Note as well that Apusozoa are the sister group to Opisthokonta. Taxa from which genome sequence is available are shown in bold. Extracted from Figshare (http://figshare.com/articles/Opisthokont_tree/1286886) under CC-BY license.

(see figure 9.1) (Cavalier-Smith and Chao, 2003; Adl et al., 2005). Interestingly, recent molecular data have also shown that this clade includes, as well, other unicellular lineages, such as the choanoflagellates, the nucleariids, the ichthyosporeans, and the filastereans (Cavalier-Smith and Chao, 2003; Medina et al., 2003; Ruiz-Trillo et al., 2004; Steenkamp and Baldauf, 2004; Ruiz-Trillo et al., 2006; Steenkamp et al., 2006; Ruiz-Trillo et al., 2008; Torruella et al., 2012; for a review, see Paps and Ruiz-Trillo, 2010).

The choanoflagellates, a group of bacteriovore flagellate protists, have been traditionally proposed to be close relatives of animals based on morphology (King, 2004, 2005). Indeed, the morphology of these free-living flagellates closely resembles a cell type, the choano-cyte, present in sponges, one of the earliest, if not the earliest-emerging, metazoan phyla. Molecular phylogenies have confirmed that choanoflagellates are indeed the sister group to animals and, thus, their closest relatives (Cavalier-Smith and Chao, 2003; Ruiz-Trillo et al., 2006; Steenkamp et al., 2006; Carr et al., 2008; Ruiz-Trillo et al., 2008; Torruella et al., 2012). They are, however, not alone. At least two additional opisthokont lineages are more closely related to Metazoa than to Fungi, thus being, as well, close relatives of animals. These are the Filasterea and the Ichthyosporea (see figure 9.1) (Cavalier-Smith and Chao, 2003; Ruiz-Trillo et al., 2006; Steenkamp et al., 2006; Carr et al., 2008; Ruiz-Trillo et al., 2008; Torruella et al., 2012). The Filasterea has only two known taxa, *Capsaspora owczarzaki* and *Ministeria vibrans*. The first is an endosymbiont of the freshwater snail *Biomphalaria glabrata* (Hertel et al., 2002; Hertel et al., 2004), and the latter is a marine free-living bacteriovore (Shalchian-Tabrizi et al., 2008). The Ichthyosporea (also known as Mesomycetozoea or DRIPs) comprise dozens of described taxa, most of them parasites or associated with animals (Mendoza et al., 2002; Glockling et al., 2013). Therefore, the new phylogenetic framework situates Choanoflagellata, Filasterea, and Ichthyosporea as the closest known unicellular relatives of animals (see figure 9.1). Thus, comparing their genomes with the genomes of animals is key to understanding the genetic content and the genomic structure of the last common unicellular ancestor of Metazoa.

First Genome Data of a Close Unicellular Relative of Metazoa

The first unicellular relative of animals to have its genome sequenced was a choanoflagellate. In a seminal paper, King and coworkers described the genome sequence of the choanoflagellate *Monosiga brevicollis* (King et al., 2008). By comparing its genome with the genomes of metazoans, the researchers provided important clues into the origin of genes involved in multicellularity. For example, they found that *M. brevicollis* had genes involved in cell adhesion, such as cadherines and C-type lectins, as well as many protein tyrosine kinases, which were previously thought to be animal specific. Importantly, they showed that there were indeed many different genes and pathways that seemed to be specific to Metazoa and potentially important for the origin of animals. Among those potentially metazoan-specific genes and pathways were several signaling pathways (Wnt, TGFbeta, and, to a certain extent, Notch), as well as several transcription factors (such as T-box genes) and some genes involved in cell adhesion (e.g., integrins) (King et al., 2008).

Thus, the genome of *M. brevicollis* showed that choanoflagellates already had some of the genes playing key roles in multicellularity and development in animals. However, that data pointed quite strongly to the fact that some of the most important signaling pathways and transcriptional regulation genes were indeed innovations of the animal lineage. Overall, the genome of *M. brevicollis* was key to defining the potential "*metazoan genetic starter kit*," which appeared to be a rather complex toolkit, and, more importantly, partly specific to Metazoa. In a way, the genome of *M. brevicollis* supported an explosive origin of Metazoa due, in part, to a significant innovation and expansion of the genetic toolkit. Animals, one would think, got their complex body plans thanks to an explosive genetic innovation that allowed their subsequent diversification. However, a genome from a single taxon of a single lineage is clearly not enough to have a clear picture of animal origins. More data were needed.

The Genome of the Filasterean *Capsaspora owczarzaki*

Choanoflagellates are quite diverse, and some of them can form colonies. Thus, the non-colonial *M. brevicollis* was hardly representative of all choanoflagellate diversity (Ruiz-Trillo et al., 2007). Therefore, a colonial choanoflagellate, such *Salpingoeca rosetta*, was an obvious additional target for genome sequencing. Besides choanoflagellates, there are two additional unicellular lineages closely related to Metazoa (Filasterea and Ichthyosporea) from which whole genome data were not yet available. The UNICORN initiative aimed to obtain genome sequence data from those additional taxa (Ruiz-Trillo et al., 2007). Specifically, it aimed to obtain the whole genome sequence of the colonial choanoflagellate *Salpingoeca rosetta*, the two so-far described filastereans (the symbiont *Capsaspora owczarzaki* and the marine free-living *Ministeria vibrans*), and two representative ichthyosporeans (*Sphaeroforma arctica* and *Amoebidium parasiticum*). We here focus in the findings from the genome sequence of the filasterean *Capsaspora owczarzaki*, which has been completely sequenced and analyzed (Suga et al., 2013).

C. owczarzaki is an amoeba endosymbiont of the freshwater snail *Biomphalaria glabrata*. However, it is cultured without the host and, in contrast to choanoflagellates, in axenic conditions, making it a tractable organism in the lab. Interestingly, we recently described three different stages in the life cycle of *C. owczarzaki*: a filopodial stage that remains attached to the substrate, a cystic stage without filopodia, and an aggregative stage in which cells actively come together to form a multicellular structure; see videos at http://youtu.be/0Uyhor_nDts (filopodial) and http://youtu.be/83HB8srWQw4 and http://youtu.be/OvI6BvBucrc (aggregative) (Sebé-Pedrós, Irimia, et al., 2013). This is indeed the first description of an aggregative multicellular behavior among close relatives of Metazoa.

The genome of *C. owczarzaki* has a size of 28 Mb and comprises 8,657 predicted genes in total (Suga et al., 2013), with many more genes involved in cell-adhesion, transcriptional regulation, and intracellular signaling than previously expected. Interestingly, some of those genes are not present in either the unicellular (*M. brevicollis*) or the colonial (*S. rosetta*) choanoflagellates (Suga et al., 2013). The lack of those genes in choanoflagellates implies that some of the genes key to animal multicellularity were secondarily lost in choanoflagellates—or at least in the two sampled taxa—and emphasizes the importance of having a good taxon sampling when inferring evolutionary events.

The Unicellular Ancestor of Metazoa Was Genetically Complex

The most significant finding of the genome of *C. owczarzaki* is its diverse repertoire of genes involved in multicellular functions, which challenges previous views on the origin of animals and the nature of their unicellular ancestor. However, what are those genes? Cell adhesion is a crucial function in multicellular systems. The genome of *M. brevicollis* had already shown that the unicellular ancestor already had cadherins and some genes with protein domains typical of extracellular matrix components (King et al., 2008). Cadherins are also present in the genome of *S. rosetta*, and even *C. owczarzaki* has a single cadherin gene (Nichols et al., 2012) (see figure 9.2). This means that cadherins, which are the most important cell–cell adhesion system in animals, were already present before the divergence of filastereans, choanoflagellates, and metazoans (Nichols et al., 2012). The genome of *M. brevicollis* also apparently suggested, however, that the most important cell–extracellular matrix adhesion system of animals, the integrin adhesion system, was animal specific (King et al., 2008). Surprisingly, however, the genome of *C. owczarzaki* showed that a complete integrin adhesome was already present before the divergence of opisthokonts and was then secondarily lost in both fungi and choanoflagellates (Sebé-Pedrós et al., 2010). This implies that the unicellular ancestor of Metazoa already had most of the components of cell adhesion in animals: cadherins, integrins, and protein domains of extracellular matrix, as well as Sarcoglycan and Syntrophin (Suga et al., 2013) (see figure 9.2). What those genes were doing in the ancestor, or in the extant animal relatives, remains unsettled. However, it has been proposed that cadherins might work in prey capture in choanoflagellates (Abedin and King, 2008) and integrins might play a role in the formation of the multicellular aggregates of *C. owczarzaki* (Sebé-Pedrós, Irimia, et al., 2013). In any case, it is clear that the transition from unicellular to multicellular at the onset of Metazoa did not require too much innovation regarding the genes involved in cell adhesion functions.

Cell signaling is another important function for multicellular organisms. Protein tyrosine kinases (PTKs) play important roles in cell-to-cell communication, as well as differentiation in metazoans (Lim and Pawson, 2010; Miller, 2012). Because of their absence in fungi and other eukaryotes, PTKs were considered animal specific. The genome of *M. brevicollis* already showed that this taxon had more than a hundred PTKs, as many as some metazoans (Manning et al., 2008; Pincus et al., 2008). Later on, it was shown that both filastereans and ichthyosporeans also had an elaborate tyrosine signaling network, which originated in the holozoan lineage, that is the lineage that comprises Metazoa, Choanoflagellata, Filasterea, and Ichthyosporea (Suga et al., 2012; Suga et al., 2014). An important finding is the peculiar pattern of conservation between cytoplasmic and receptor tyrosine kinases. While cytoplasmic tyrosine kinases are relatively well conserved among metazoans, choanoflagellates, filastereans, and ichthyosporeans, the receptor tyrosine kinases diversified extensively, and independently, in each of those groups (Suga et al., 2012; Suga et al., 2014). This pattern suggest that even though the complex tyrosine signaling network already appeared at the onset of the Holozoa, the extracellular signal receptors (receptor PTKs) had a high evolutionary plasticity until they were co-opted to function in cell-to-cell communication in animals.

Other signaling pathways were also already present in the unicellular ancestor of Metazoa, such as the MAPK kinases, and the G-protein coupled receptor and the Hippo signaling pathways (Sebé-Pedrós et al., 2012; Suga et al., 2013; de Mendoza et al., 2014).

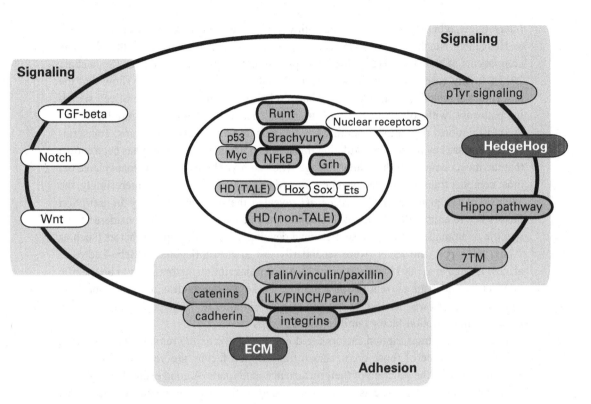

Figure 9.2
Schematic representation of a cell of the putative unicellular ancestor of Metazoa. Key cell-adhesion, cell-signaling, and transcriptional regulation genes/pathways are depicted. Genes and pathways shown in gray were already present in the unicellular ancestor of Metazoa. Label code is indicated in the legend. Light gray means present in both choanoflagellates (*S. rosetta* and *M. brevicollis*) and the filasterean *C. owczarzaki*. Light gray within wide black means present in the *C. owczarzaki* and secondarily lost in choanoflagellates. Dark gray means present in choanoflagellates, but not in *C. owczarzaki*. Genes and pathways in white are, under the current taxon sampling, specific to Metazoa. Adapted from (Suga et al., 2013).

Interestingly, other important signaling pathways that are key to animal development, such as Hedgehog, Notch, Wnt, or TGFbeta, seem to be specific to Metazoa, with only some components of Notch and Hedgehog being present in the unicellular relatives (Suga et al., 2013). In any case, the genomes of choanoflagellates and *C. owczarzaki* show that the unicellular ancestor of Metazoa already had a relatively complex repertoire of cell signaling pathways, with some pathways arising at the onset of Metazoa (see figure 9.2).

Transcriptional regulation is another key function in multicellular organisms. Transcription factors are key players in regulating transcription in eukaryotes, and it has been shown that the most complex multicellular lineages (i.e., embryophytes and metazoans) have the most complex transcription factor repertoires (de Mendoza et al., 2013). Interestingly, the analysis of the genome of *C. owczarzaki* revealed the presence of many transcription factors key to animal development, including some that were previously considered to be specific to animals. Some of those metazoan developmental transcription factors (such as T-box, NFKappa, and Runx) were secondarily lost in choanoflagellates (Sebé-Pedrós et al., 2011; Suga et al., 2013) (see figure 9.2). A few transcription factors seem to be specific to Metazoa under the current taxon sampling. These are the nuclear receptors, the ETS and the Smads, as well as several specific classes within bZIP, Homedomain, bHLH, and HMGbox transcription factor families (Sebé-Pedrós et al., 2011).

What all those transcription factors are doing in *C. owczarzaki* remains unknown, but we know that some of those premetazoan transcription factors are very well conserved and able to rescue the function of their metazoan counterparts. A good example is the gene Brachyury, a member of the class of T-box genes, identified in *C. owczarzaki* (Sebé-Pedrós et al., 2011). The homolog of Brachyury in *C. owczarzaki* is functionally conserved with respect to the metazoan Brachyurys, to the point that *C. owczarzaki* Brachyury is able to mimic the function of the endogenous *Xenopus* Brachyury (Sebé-Pedrós, Ariza-Cosano, et al., 2013). *C. owczarzaki* Brachyury, however, behaves differently at a molecular level than the metazoan Brachyury homologs. *C. owczarzaki* Brachyury lacks the target specificity of metazoan homologs, with a molecular phenotype resembling other types of T-box genes more than Brachyury itself (Sebé-Pedrós, Ariza-Cosano, et al., 2013). This was argued to be due to the fact that the subfunctionalization of T-box classes was established just at the onset of Metazoa, probably through the development of a more complex network with new interactions with other cofactors (Sebé-Pedrós, Ariza-Cosano, et al., 2013).

Regulation of Gene Expression in Premetazoan Lineages

An important component of metazoan multicellularity, and indeed any other complex multicellularity, is the regulation of gene expression, which is the basis of cell differentiation. The question is whether the tight gene regulation seen in animals is something exclusive to them. Two recent analyses in unicellular relatives of animals, the choanoflagellate *S. rosetta* and the filasterean *C. owczarzaki*, have shed some light on this question, by showing that those premetazoan lineages regulate gene expression between different developmental stages (Sebé-Pedrós, Irimia, et al., 2013; Fairclough et al., 2013).

The life cycle of the choanoflagellate *S. rosetta* features five different stages, including unicellular stages and rosette-shaped colonies of diverse number of cells (Fairclough et al., 2010; Dayel et al., 2011). Analysis of the transcriptome of *S. rosetta* revealed that different stages have distinct patterns of gene expression. Good examples are the four *S. rosetta* septins that are upregulated in colonies relative to single cells. This may be an indication that those proteins, which play roles in cytokinesis in animals, may be regulating the incomplete cytokinesis taken place during colony formation (Fairclough et al., 2013).

The life cycle of *C. owczarzaki* in culture differs significantly from that of *S. rosetta*. *C. owczarzaki* goes through three very different life stages, including one "multicellular" stage that is formed by cell aggregation. First, *C. owczarzaki* cells are filopodiated amoebas crawling over the substrate. Subsequently, the cells retract their filopodia and detach from the substrate, becoming a cyst. An alternate route, that happens randomly, is the formation of cell aggregates by active cell aggregation and not by clonal division (Sebé-Pedrós, Irimia, et al., 2013).

The analysis of the transcriptome data from each of the three different cell stages demonstrates that half of the genes of *C. owczarzaki* are differentially regulated, with each stage having a specific transcriptomic profile. Interestingly, the filopodial stage appears to be significantly enriched in genes involved in signaling functions, such as tyrosine kinase activity and G-protein-coupled receptor activity, as well as in transcription factors (Sebé-Pedrós, Irimia, et al., 2013). Moreover, this stage also has genes involved in protein synthesis and DNA replication significantly upregulated.

C. owczarzaki cystic cells, on the other hand, showed significant downregulation of genes associated with myosin transport, translation, DNA replication, and metabolic activities. However, genes involved in vesicle transport, autophagy, the ubiquitin pathway, and synaptic cell–cell communication were significantly upregulated in the cystic cells (Sebé-Pedrós, Irimia, et al., 2013). This suggests that protein turnover is probably occurring at the cystic stage. An important point is that the components of the integrin adhesome were significantly upregulated in the aggregative stage. Because of this, the authors suggested that the integrin adhesome and associated tyrosine kinases were playing an important role in the formation of the *C. owzarzaki* aggregates (Sebé-Pedrós, Irimia, et al., 2013). If this is ever experimentally confirmed, it would be a nice example of how an adhesion system that functioned in the formation of aggregates would later on be co-opted to work as a cell–extracellular matrix system within a multicellular organism.

Alternative Splicing in Premetazoan Lineages

Alternative splicing (AS) is a posttranscriptional process in which multiple transcripts are produced from a single gene (Irimia and Blencowe, 2012). Thus, AS contributes to increase the diversity of the transcriptome and proteome. AS has been reported in a wide range of eukaryotic groups, including plants, apicomplexans, diatoms, amoebae, animals, and fungi. Two types of AS can occur: intron retention and exon skipping. Intron retention is the most widespread form of AS found in nonmetazoan eukaryotic species (McGuire et al., 2008; Keren et al., 2010; Nilsen and Graveley, 2010). In contrast, metazoans have a more complex AS than the rest of eukaryotes, mainly through exon skipping. Moreover, in metazoans exon skipping often involves some cell type–specific networks of co-regulated exons that belonging to functionally related genes or pathways (Irimia and Blencowe, 2012). Analysis of AS in *C. owczarzaki* showed a significant amount of differential intron retention (Sebé-Pedrós, Irimia, et al., 2013). Moreover, some genes were shown to have exon skipping in *C. owczarzaki*. Interestingly, those genes with differential exon skipping were significantly enriched in protein kinase activity, suggesting a role in the regulation of cell signaling in *C. owczarzaki* (Sebé-Pedrós, Irimia, et al., 2013). The presence in *C. owczarzaki* of an exon network linked to a specific function (in this case, cell signaling), despite previous suggestions that it was unique to metazoans, emphasizes a premetazoan origin of this type of regulation.

Innovations at the Onset of Metazoa

The genomes of close unicellular relatives of Metazoa have, undoubtedly, challenged previous ideas. Among the most important revelation is that the unicellular ancestor of Metazoa was genetically much more complex than previously thought (Suga et al., 2013). Therefore, there was not such a drastic explosion of genetic innovation at the onset of Metazoa although it is true that some genes and pathways (especially some signaling pathways) seem, under the current taxon sampling, to be specific to Metazoa. In any case, the majority of genes and pathways involved in cell adhesion, cell signaling, and transcriptional regulation were already present in the unicellular ancestor. Thus, rather than, or in addition to, gene innovation, it seems other evolutionary mechanisms were playing important roles in the origin of Metazoa. What are those mechanisms?

The new data seem to support that four mechanisms, besides gene innovation, were playing some roles in the onset of Metazoa. These are exon shuffling, gene duplication, co-option of ancestral genes, and an increase in regulation. Exon shuffling and gene duplication did play a role in the diversification and the expansion of more complex repertoires and networks. This is the case, for example, of protein tyrosine signaling, ubiquitin signaling, and transcription factors. Co-option, that is, appropriation of ancestral genes into new roles, was probably the mechanism by which integrins and cadherins were integrated into multicellular systems. Finally, a more complex regulation via increasing the complexity of the regulatory networks was also an important feature of animal origins.

Conclusions

It is clear that understanding such an important evolutionary transition as the origin of metazoan animals requires complementary approaches including genomics, systematics, cell biology, and developmental biology. Moreover, it needs to be approached from both the perspective of early-branching animals and from the perspective of the animals' closest unicellular relatives. The latter has indeed provided important insights into this question as shown in this chapter. Thanks to this, we now know that the unicellular ancestor of Metazoa was genetically quite complex and was equipped with the capacity to tightly regulate transitions between different cell stages.

The diversity and complexity of life cycles, cell morphologies, and cell behaviors of the relatives of animals is astonishing: from the colonial bacteriovore choanoflagellates to the free-living or symbiotic filastereans that have the capacity to form multicellular aggregates, or the enigmatic ichthyosporeans with a syncytial development. But rather than regret this diversity and see this only as a challenge, we, evolutionary biologists, should rejoice and study all those biological entities, because only by understanding them will we be able to shed more light on and comprehend how multicellular animals evolved.

Acknowledgments

Funding is acknowledged from both the European Research Council (ERC-2007-StG-206883 and ERC-2012-Co -616960) and the Ministerio de Economía y Competitividad (MINECO). I thank the editors for their patience and the two reviewers for their insightful suggestions.

References

Abedin, M., & King, N. (2008). The premetazoan ancestry of cadherins. *Science*, 319, 946–948.

Adl, S. M., Simpson, A. G., Farmer, M. A., Andersen, R. A., Anderson, O. R., Barta, J. R., et al. (2005). The new higher level classification of eukaryotes with emphasis on the taxonomy of protists. *Journal of Eukaryotic Microbiology*, 52, 399–451.

Baldauf, S. L., & Palmer, J. D. (1993). Animals and fungi are each other's closest relatives: Congruent evidence from multiple proteins. *Proceedings of the National Academy of Sciences of the United States of America*, 90, 11558–11562.

Carr, M., Leadbeater, B. S., Hassan, R., Nelson, M., & Baldauf, S. L. (2008). Molecular phylogeny of choanoflagellates, the sister group to Metazoa. *Proceedings of the National Academy of Sciences of the United States of America*, 105, 16641–16646.

Cavalier-Smith, T., & Chao, E. E. (2003). Phylogeny of Choanozoa, Apusozoa, and other Protozoa and early eukaryote megaevolution. *Journal of Molecular Evolution*, 56, 540–563.

Conway Morris, S. (1998). *The crucible of creation.* Oxford, UK: Oxford University Press.

Dayel, M. J., Alegado, R. A., Fairclough, S. R., Levin, T. C., Nichols, S. A., McDonald, K., et al. (2011). Cell differentiation and morphogenesis in the colony-forming choanoflagellate *Salpingoeca rosetta*. *Developmental Biology*, 357, 73–82.

de Mendoza, A., Sebé-Pedrós, A., & Ruiz-Trillo, I. (2014). The evolution of the GPCR signaling system in eukaryotes: Modularity, conservation, and the transition to metazoan multicellularity. *Genome Biology and Evolution*, 6, 606–619.

de Mendoza, A., Sebé-Pedrós, A., Šestak, M. S., Matejcic, M., Torruella, G., Domazet-Loso, T., et al. (2013). Transcription factor evolution in eukaryotes and the assembly of the regulatory toolkit in multicellular lineages. *Proceedings of the National Academy of Sciences of the United States of America*, 110, E4858–E4866.

Fairclough, S. R., Chen, Z., Kramer, E., Zeng, Q., Young, S., Robertson, H. M., et al. (2013). Premetazoan genome evolution and the regulation of cell differentiation in the choanoflagellate *Salpingoeca rosetta*. *Genome Biology*, 14, R15.

Fairclough, S. R., Dayel, M. J., & King, N. (2010). Multicellular development in a choanoflagellate. *Current Biology*, 20, R875–R876.

Glockling, S. L., Marshall, W. L., & Gleason, F. H. (2013). Phylogenetic interpretations and ecological potentials of the Mesomycetozoea (Ichthyosporea). [Internet]. *Fungal Ecology*, 6, 237–247.

Grosberg, R. K., & Strathmann, R. R. (2007). The evolution of multicellularity: A minor major transition? *Annual Review of Ecology Evolution and Systematics*, 38, 621–654.

Hertel, L. A., Barbosa, C. S., Santos, R. A., & Loker, E. S. (2004). Molecular identification of symbionts from the pulmonate snail *Biomphalaria glabrata* in Brazil. *Journal of Parasitology*, 90, 759–763.

Hertel, L. A., Bayne, C. J., & Loker, E. S. (2002). The symbiont *Capsaspora owczarzaki*, nov. gen. nov. sp., isolated from three strains of the pulmonate snail *Biomphalaria glabrata* is related to members of the Mesomycetozoea. *International Journal for Parasitology*, 32, 1183–1191.

Irimia, M., & Blencowe, B. J. (2012). Alternative splicing: Decoding an expansive regulatory layer. *Current Opinion in Cell Biology*, 24, 323–332.

Keren, H., Lev-Maor, G., & Ast, G. (2010). Alternative splicing and evolution: Diversification, exon definition and function. *Nature Reviews. Genetics*, 11, 345–355.

King, N. (2004). The unicellular ancestry of animal development. *Developmental Cell*, 7, 313–325.

King, N. (2005). Choanoflagellates. *Current Biology*, 15, R113–R114.

King, N., Westbrook, M. J., Young, S. L., Kuo, A., Abedin, M., Chapman, J., et al. (2008). The genome of the choanoflagellate *Monosiga brevicollis* and the origin of metazoans. *Nature*, 451, 783–788.

Lim, W. A., & Pawson, T. (2010). Phosphotyrosine signaling: Evolving a new cellular communication system. *Cell*, 142, 661–667.

Manning, G., Young, S. L., Miller, W. T., & Zhai, Y. (2008). The protist, *Monosiga brevicollis*, has a tyrosine kinase signaling network more elaborate and diverse than found in any known metazoan. *Proceedings of the National Academy of Sciences of the United States of America*, 105, 9674–9679.

McGuire, A. M., Pearson, M. D., Neafsey, D. E., & Galagan, J. E. (2008). Cross-kingdom patterns of alternative splicing and splice recognition. *Genome Biology*, 9, R50.

Medina, M., Collins, A. G., Taylor, J. W., Valentine, J. W., Lipps, J. H., Amaral-Zettler, L., et al. (2003). Phylogeny of Opisthokonta and the evolution of multicellularity and complexity in Fungi and Metazoa. *International Journal of Astrobiology*, 2, 203–211.

Mendoza, L., Taylor, J. W., & Ajello, L. (2002). The class Mesomycetozoea: A heterogeneous group of microorganisms at the animal–fungal boundary. *Annual Review of Microbiology*, 56, 315–344.

Miller, W. T. (2012). Tyrosine kinase signaling and the emergence of multicellularity. *Biochimica et Biophysica Acta*, 1823, 1053–1057.

Nichols, S. A., Roberts, B. W., Richter, D. J., Fairclough, S. R., & King, N. (2012). Origin of metazoan cadherin diversity and the antiquity of the classical cadherin/β-catenin complex. *Proceedings of the National Academy of Sciences of the United States of America*, 109, 13046–13051.

Nilsen, T. W., & Graveley, B. R. (2010). Expansion of the eukaryotic proteome by alternative splicing. *Nature*, 463, 457–463.

Paps, J., & Ruiz-Trillo, I. (2010). Animals and their unicellular ancestors. In *Encyclopedia of life sciences (ELS)*. Chichester, UK: John Wiley & Sons. .10.1002/9780470015902.a0022853

Pincus, D., Letunic, I., Bork, P., & Lim, W. A. (2008). Evolution of the phospho-tyrosine signaling machinery in premetazoan lineages. *Proceedings of the National Academy of Sciences of the United States of America*, 105, 9680–9684.

Rokas, A. (2008 a). The molecular origins of multicellular transitions. *Current Opinion in Genetics & Development*, 18, 472–478.

Rokas, A. (2008 b). The origins of multicellularity and the early history of the genetic toolkit for animal development. *Annual Review of Genetics*, 42, 235–251.

Ruiz-Trillo, I., Burger, G., Holland, P. W. H., King, N., Lang, B. F., Roger, A. J., et al. (2007). The origins of multicellularity: A multi-taxon genome initiative. *Trends in Genetics*, 23, 113–118.

Ruiz-Trillo, I., Inagaki, Y., Davis, L. A., Sperstad, S., Landfald, B., & Roger, A. J. (2004). *Capsaspora owczarzaki* is an independent opisthokont lineage. *Current Biology*, 14, R946–R947.

Ruiz-Trillo, I., Lane, C. E., Archibald, J. M., & Roger, A. J. (2006). Insights into the evolutionary origin and genome architecture of the unicellular opisthokonts *Capsaspora owczarzaki* and *Sphaeroforma arctica*. *Journal of Eukaryotic Microbiology*, 53, 1–6.

Ruiz-Trillo, I., Roger, A. J., Burger, G., Gray, M. W., & Lang, B. F. (2008). A phylogenomic investigation into the origin of Metazoa. *Molecular Biology and Evolution*, 25, 664–672.

Sebé-Pedrós, A., Ariza-Cosano, A., Weirauch, M. T., Leininger, S., Yang, A., Torruella, G., et al. (2013). Early evolution of the T-box transcription factor family. *Proceedings of the National Academy of Sciences of the United States of America*, 110, 16050–16055.

Sebé-Pedrós, A., de Mendoza, A., Lang, B. F., Degnan, B. M., & Ruiz-Trillo, I. (2011). Unexpected repertoire of metazoan transcription factors in the unicellular holozoan *Capsaspora owczarzaki*. *Molecular Biology and Evolution*, 28, 1241–1254.

Sebé-Pedrós, A., Irimia, M., Del Campo, J., Parra-Acero, H., Russ, C., Nusbaum, C., et al. (2013). Regulated aggregative multicellularity in a close unicellular relative of Metazoa. *eLife*, 2, e01287.

Sebé-Pedrós, A., Roger, A. J., Lang, F. B., King, N., & Ruiz-Trillo, I. (2010). Ancient origin of the integrin-mediated adhesion and signaling machinery. *Proceedings of the National Academy of Sciences of the United States of America*, 107, 10142–10147.

Sebé-Pedrós, A., Zhen, Y., Ruiz-Trillo, I., & Pan, D. (2012). Premetazoan origin of the hippo signaling pathway. *Cell Reports*, 1, 13–20.

Shalchian-Tabrizi, K., Minge, M. A., Espelund, M., Orr, R., Ruden, T., Jakobsen, K. S., et al. (2008). Multigene phylogeny of Choanozoa and the origin of animals. *PLoS One*, 3, e2098.

Steenkamp, E. T., & Baldauf, S. L. (2004). Origin and evolution of animals, fungi and their unicellular allies (Opisthokonta). In R. P. Hirt & D. S. Horner (Eds.), *Organelles, genomes and eukaryote phylogeny: An evolutionary synthesis in the age of genomics* (pp. 109–129). Boca Raton, FL: CRC Press.

Steenkamp, E. T., Wright, J., & Baldauf, S. L. (2006). The protistan origins of animals and fungi. *Molecular Biology and Evolution, 23, 93–106.*

Suga, H., Chen, Z., de Mendoza, A., Sebé-Pedrós, A., Brown, M. W., Kramer, E., et al. (2013). The Capsaspora genome reveals a complex unicellular prehistory of animals. *Nature Communications,* 4, 2325.

Suga, H., Dacre, M., de Mendoza, A., Shalchian-Tabrizi, K., Manning, G., & Ruiz-Trillo, I. (2012). Genomic survey of premetazoans shows deep conservation of cytoplasmic tyrosine kinases and multiple radiations of receptor tyrosine kinases. *Science Signaling,* 5, ra35.

Suga, H., Torruella, G., Burger, G., Brown, M. W., & Ruiz-Trillo, I. (2014). Earliest holozoan expansion of phosphotyrosine signaling. *Molecular Biology and Evolution,* 31, 517–528.

Torruella, G., Derelle, R., Paps, J., Lang, B. F., Roger, A. J., Shalchian-Tabrizi, K., et al. (2012). Phylogenetic relationships within the Opisthokonta based on phylogenomic analyses of conserved single-copy protein domains. *Molecular Biology and Evolution,* 29, 531–544.

Valentine, J. W. (2004). *On the origin of phyla.* Chicago: University of Chicago Press.

Valentine, J. W., Tiffney, B. H., & Sepkoski, J. J. (1991). Evolutionary dynamics of plants and animals: A comparative approach. *Palaios,* 6, 81–88.

10 Sponges as the Rosetta Stone of Colonial-to-Multicellular Transition

Maja Adamska

Summary

Sponges are one of the simplest multicellular animals and are traditionally viewed as the oldest surviving animal clade. Similarities between choanocytes (the defining cell type for sponges) and choanoflagellates (single-cell and colonial protists) have long suggested an evolutionary link between them. This notion is supported by contemporary phylogenies which universally recover choanoflagellates as the sister group of animals, and in the majority of cases also place sponges as the earliest evolved animal lineage. Choanocytes combine functions which in many other animals are segregated between somatic cells and germ cells: in addition to capturing of food particles they are a source of gametes. Moreover, choanocytes express genes which are generally expressed in endodermal (e.g., gut) and stem cells of other animals. These characteristics of choanocytes, combined with the phylogenetic position of sponges, allow formulation of a highly plausible evolutionary scenario in which sponges form a link between simple colonial protists and the complex "true animals."

Introduction

Bodies of vertebrates, insects, and cephalopods are composed of dozens of types of differentiated cells, which are precisely organized and integrated to act in concert and generate complex forms and behaviors. Emergence of this complexity remains an intensively studied and disputed subject, which is of interest to not only evolutionary scientists but also the general public, as the perceived difficulty of evolving sophisticated body plans from humble single-cellular ancestors unfortunately fuels the "intelligent design" views.

While we cannot go back in time to directly observe the emergence of the first multicellular animals (metazoans), we can gain insight into these events by comparative analysis of the currently living metazoans and their nonmetazoan relatives. Such analysis aims at the reconstruction of the steps leading to metazoan complexity and identification of features (morphological, genomic, developmental, or ecological) which were a prerequisite for animal multicellularity, and those which were its consequences.

Using three complementary angles—phylogenetic, morphological, and gene expression studies—this chapter reviews what is currently understood about the evolutionary steps leading to emergence of complex animal forms. The combined evidence supports the notion that sponges (Porifera), which are one of the simplest animal forms living today, well deserve to be seen as a link between colonial protists and the "true" complex multicellular animals (eumetazoans). This "sponge-centric" perspective allows drawing of an evolutionary scenario in which all steps of evolution of animal complexity were gradual and driven by natural selection.

Phylogenetic Relationships between Animals and Their Nearest Relatives

The first step in performing the comparisons between animals and their relatives is obviously the establishment of phylogenetic relationships between nonmetazoan and metazoan clades. It is particularly important to identify clades which bracket the colonial-to-multicellular transition: the last nonanimals and the first animals to branch off the phylogenetic tree of life. The key concept of the "tree of life" was already introduced by Charles Darwin in the seminal book *On the Origin of Species by Natural Selection*, which included a simple diagram of relationships between multiple species in a theoretical genus (Darwin, 1859). In the following years, Ernst Haeckel published a series of significantly more elaborate (but not always accurate) trees representing his views on global relationships between all living phyla, as well as within specific groups (e.g., Haeckel, 1866). These early attempts were based solely on morphology; it was only the advent of broadly accessible DNA sequencing, coupled with development of computational power, which in the past two decades resulted in explosive proliferation of published phylogenetic trees based on increasingly large sequence data sets (see, e.g., Telford and Copley, 2011; Dunn et al., 2014; Telford, 2013).

There appears to be no doubt now that all bilaterally symmetrical animals, such as vertebrates, insects, nematodes, or mollusks, share a common ancestor; this well-supported monophyletic clade is known as Bilateria. On the other hand, the phylogenetic relationships between bilaterians and the nonbilaterian clades remain a hotly disputed subject (reviewed by Edgecombe et al., 2011; Dohrmann and Wörheide, 2013). These nonbilaterian clades include Cnidaria (e.g., jellyfish, hydroids, corals, and sea anemones, such as *Nematostella vectensis*, shown in figure 10.1A, B), Placozoa (which are represented by one currently recognized species, *Trichoplax adhaerens*; Schierwater, 2005; figure 10.1C), Ctenophora (comb jellies, e.g., *Mnemiopsis leidyi*, figure 10.1D), Porifera (e.g., bath sponges, Venus' flower basket sponges, and calcareous sponges such as *Sycon ciliatum*, figure 10.1E–J). While early trees based on morphological characters often combined these four clades into a monophyletic (meaning sharing a unique common ancestor) clade, referred to as Radiata or Diblastica (Diploblastica), this concept is rarely supported in recent analyses (e.g., Pick et al., 2010; Nosenko et al., 2013; but see Schierwater et al., 2009).

The current consensus view on the position of Cnidaria is that they are the sister group of Bilateria (see figure 10.1), to the extent that some authors postulate that they should indeed be included within that clade (Baguñà et al., 2008). This view is confounded by the fact that some recent phylogenetic analyses support existence of the Coelenterata clade, which would be composed of Cnidaria and Ctenophora as sister groups (Philippe et al., 2009). In contrast, several other analyses indicate that Ctenophora might be the earliest branching metazoans (Dunn et al., 2008; Ryan et al., 2013; Moroz et al., 2014). Such basal position of ctenophores has a potentially strong influence on our understating of early events in animal evolution. However, other studies using the same or comparable data sets demonstrate that recovered position of ctenophores is highly dependent on experimental variables such as gene and outgroup sampling and methodology used (Pick et al., 2010; Nosenko et al., 2013). Unfortunately, even sequencing of additional ctenophore species might not help to resolve this problem, as it is likely that all extant Ctenophora taxa are very closely related, being a result of expansion which followed a bottleneck dated as recent as the K–T boundary—only 66 million years ago (Podar et al., 2001). The phylogenetic position of Placozoa is almost equally as difficult to ascertain as that of Ctenophora, as they are recovered as a sister group to Bilateria, Cnidaria, Bilateria + Cnidaria, or Porifera but—and this is critical for the current considerations—virtually never as the most basal metazoan branch (Srivastava et al., 2008, Pick et al., 2010; Nosenko et al., 2013).

On the other hand, sponges (Porifera), which were traditionally considered to be the earliest evolved multicellular animals, are also recovered as the basal metazoans in majority of current phylogenies (reviewed by Wörheide et al., 2012). Several of the published phylogenies have in fact implied that sponges do not form a monophyletic clade, but that some sponges (Calcarea and Homoscleromorpha) are more closely related to the remaining animals than they are to the siliceous sponges (Demospongia and Hexactinellida) (e.g., Zrzavy et al., 1998; Borchiellini et al., 2001; Peterson and Butterfield, 2005; Sperling et al., 2009; Erwin et al., 2011). If this were true, the quest for the first metazoan (the "Urmetazoan") would be completed, as paraphyletic sponges at the base of the animal tree of life formally imply a sponge as the last common metazoan ancestor. However, it appears that whether sponges are recovered as mono- versus paraphyletic, similarly to whether sponges or ctenophores are recovered as the basal metazoan phylum, strongly depends on selections of genes, outgroups, and model parameters (Nosenko et al., 2013; reviewed by Dohrmann and Wörheide, 2013). In summary, the currently available methodologies do not appear to resolve the branching order of nonbilaterian clades, with sponges and ctenophores competing for the title of the earliest evolved metazoans.

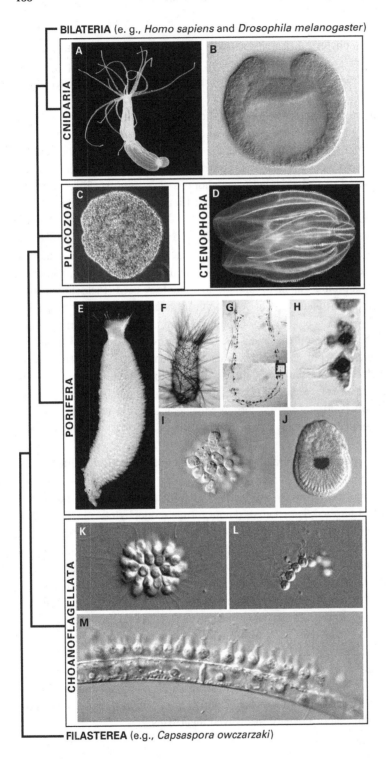

Figure 10.1
Animals and their relatives. Consensus phylogenetic relationships are indicated on the left. Right: (A) adult, and (B) gastrula stage embryo of sea anemone *Nematostella vectensis*; (C) placozoan *Trichoplax adhaerens*; (D) sea walnut *Mnemiopsis leidyi*; (E–J) calcareous sponge *Sycon ciliatum* (E, adult; F, live juvenile; G, histological section through a juvenile; H, magnification of choanocytes framed on G; I, isolated living choanocytes; J, larva); (K) rosette, and (L) chain colony of *Salpingoeca rosetta*; (M) *Salpingoeca* sp. on filamentous algae. Photographs A, B, C, D, and M were kindly provided by C. Sinigaglia, G. Richards, B. Schierwater, M. Adamski, and Y. Tsukii, respectively; E is reprinted from Fortunato et al., 2014; F–H and J are reprinted from Leininger et al., 2014; K–L are reprinted from Dayel et al., 2011.

On the other hand, there is full agreement regarding the identity of the nearest nonmetazoan relatives of animals, with choanoflagellates—which are single-cell and colonial protists—consistently recovered in this position (e.g., Ruiz-Trillo et al., 2008; King et al., 2008; Torruella et al., 2012; Carr et al., 2008; Shalchian-Tabrizi et al., 2008; figure 10.1K–M). The next branch (a sister group to Choanoflagellata + Metazoa) is currently occupied by two species, *Capsaspora owczarzaki* and *Ministeria vibrans*, and the importance of this finding is excellently covered in chapter 9 (Ruiz-Trillo, this volume).

Comparative Morphology of Early Metazoans and Their Relatives

The concept of gradual evolution implies that the first multicellular animals were not much different than their recent colonial ancestors, but it is important to keep in mind that choanoflagellates, sponges, and ctenophores have been evolving independently since they split over 600 million years ago, and the organisms we observe today might bear little resemblance to their ancestors. Yet, characters shared by choanoflagellates with either of the basally branching metazoans might have also been present in the first animals.

Modern-day ctenophores are relatively complex and highly effective predators, equipped with nervous system and striated muscles, and have a unique body plan with biradial symmetry and rows of combs built of fused cilia (figure 10.1D). Intriguingly, the development of both the nervous system and the muscles in ctenophores was suggested to utilize a somewhat different genetic toolkit than the one shared by Cnidaria and Bilateria, with an implication that both of these complex organ systems evolved independently in ctenophores and the cnidarian/bilaterian ancestor (Ryan et al., 2013; Moroz et al., 2014; but see also Ryan, 2014; Marlow and Arendt, 2014). The combination of highly contentious phylogenetic position and the derived body plan clearly complicates use of ctenophores in the quest for reconstruction of the first multicellular animals.

On the other hand, sponges and choanoflagellates display such a striking similarity of cell morphology and function that some nineteenth-century authors originally considered sponges to be colonial choanoflagellates (Clark, 1868, reviewed by Maldonado, 2004). The position of sponges as bona fide metazoans is now unquestionable, but the similarity of the form and function of choanoflagellates and the defining cell type of sponges, the choanocyte, is particularly striking in the context of the well-established relationship of choanoflagellates to metazoans (reviewed by Maldonado, 2004; King, 2004; Richter and King, 2013; but see Mah et al., 2014). Choanoflagellates and choanocytes are characterized by presence of a collar built of microvilli, which surrounds a single flagellum at the apical surface of the cell (see figure 10.1F, G, H, J, K, L). Beating of the flagellum results in movement of water carrying food particles such as bacteria or microalgae, which are captured by the collar (e.g., Dayel and King, 2014), although other parts of the cell are also able to participate in the food capture (Leys and Eerkes-Medrano, 2006). While homology of the choanoflagellate and choanocyte cell morphology should not be taken for granted, recent in-depth investigations have not identified any features of the two cell types which could challenge this broadly accepted homology, although minor differences between the two systems have been found (Mah et al., 2014).

Choanoflagellates, which are solitary and colonial, are usually sedentary, with their basal part attached to the substrate (see figure 10.1M and Leadbeater, 2008), or form free-swimming colonies which can be spherical (with the collars pointing outside), or chain-like (see figure 10.1K, L; Dayel et al., 2011). Choanocytes form the innermost epithelium of the sponge bodies, with the flagella and collars pointing into lumens of choanocyte chambers. In the simplest (asconoid) sponge body organization—for example a calcareous sponge juvenile, as illustrated in figure 10.1F, G—a single choanocyte chamber is connected with the surrounding environment by incurrent openings (ostia, pores) and the water is expelled through a single excurrent opening (osculum). In the most complex (leuconoid) sponge body forms, the choanocyte chambers are connected with ostia and oscula by a system of canals built of flat, collarless (but sometimes flagellated) cells called endopinacocytes. Similar cells, exopinacocytes, form the outermost epithelium of sponges, and the two epithelial layers (choanoderm and pinacoderm) are separated by mesohyl, which contains a variety of nonepithelial cells (e.g., sclerocytes responsible for secretion of skeletal elements), as well as gametes and embryos in viviparous sponges (reviewed by Simpson, 1984; Ereskovsky, 2010; Adamska et al., 2011).

This two-layered body plan, with the inner epithelium responsible for the feeding function, and a single major opening at the apical side, is visually and functionally similar to the body plan of the cnidarian polyps (see figure 10.1). This similarity led Haeckel to be convinced that the body plans of sponges and corals were homologous, and inspired him to propose that all multicellular animals were derived from a hypothetical two-layered, free swimming "gastrea" (Haeckel 1870, 1874; see also Leys and Eerkes-Medrano, 2005). The gastrea itself would be derived from an even simpler form, a single-layered blastula or blastea, by invagination akin to that occurring during gastrulation of many cnidarians (reviewed by Byrum and Martindale, 2004; see figure 10.1B).

The gastrea theory appears to be firmly settled in textbooks, starting from L. Hyman's (1940) *Invertebrates*, and the hypothetical blastea is now implicitly or explicitly equated with the free swimming colony of choanoflagellates (e.g., Nielsen, 2008; Richter and King,

2013). It has been suggested that some cells of the spherical colony would become free from the task of propelling and/or feeding the colony and sink to the inside to divide or produce gametes (see Nielsen, 2008), with a clear advantage of division of labor in a colony composed of genetically identical cells. However, the next evolutionary step—from "advanced choanoblastea" to a two-layered body organization with gut—is rather difficult to envisage. Evolutionary theory posits that each of the subsequent steps has an advantage over the previous one, or at least is not disadvantageous (see chapter 12, Newman, this volume, for a nonadaptationist view on the emergence of animal multicellularity). The choanoblastea is envisaged as a free swimming colony composed of cells which individually digest relatively small prey (e.g., bacteria or microalgae which are smaller than its own cells). What kind of advantage could be immediately gained by such an organism by "invagination" of its part—in effect production of a gut, which would be useful for digestion of larger prey? How would prey be captured?

Nielsen (2008) suggested that the "advanced choanoblastea" settled and became benthic while continuing to feed in the same way its ancestors fed (e.g., on bacteria), and this simple multicellular organism is referred to as an "ancestral sponge." The ancestral sponge would have a biphasic life cycle including pelagic larvae. Subsequent evolutionary steps would lead to the emergence of sponge body plans with their elaborate canal structures, such as those seen in modern sponges. Importantly, Nielsen rejects Haeckel's idea that the adult sponge body plan is homologous to the body plans of other animals, and instead derives gastrea from a neotenic larva of a sponge (Nielsen, 2008; see also Maldonado, 2004). However, such an event is as difficult to imagine as evolutionary transformation of a blastea to gastrea—raising the same questions regarding the feeding mode of this new organism.

From Choanoflagellates to Endodermal Cells in Four Easy Steps

I would like to propose a slightly modified scenario in which all subsequent steps provide selective advantages over the previous one (see figure 10.2). As suggested by Valentine (2003) the *first step* is a benthic colonial choanoflagellate-like protist. It is important to note that the majority of choanoflagellate species are indeed sedentary (Leadbeater, 2008; Carr et al., 2008). This organism would have a biphasic life cycle, with a pelagic dispersal stage in a familiar rosette form. Some modern choanoflagellates, such as *Saplingoeca rosetta*, alternate between sedentary and pelagic forms, although in this species the sedentary forms are single-celled (Dayel et al., 2011).

The *second step* is similar to Nielsen's (2008) "ancestral sponge," a benthic multicellular organism using choanoflagellate-like cells (i.e., choanocytes) for feeding, but which has already evolved an additional cell type (i.e., pinacocytes) responsible for attachment to the substrate. Gametes are produced between the two cell layers, choanoderm and pinaco-

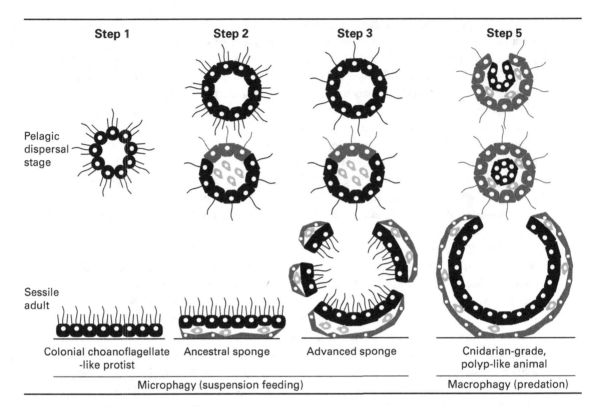

Figure 10.2
Proposed evolutionary scenario: from colonial protists to multicellular animals. Choanoflagellates, choanocytes, and endodermal lineage cells (which are considered here to be homologous) are black; pinacoderm and ectoderm lineage (which are considered here to be "younger" that the choanoderm–endoderm lineage, and also homologous to each other) is dark gray; additional cells are light gray. A variety of possible pelagic dispersal stages (including larvae and asexually produced propagules) are suggested above the sexually reproducing benthic stages. Steps 2 and 3 correspond to those described by Nielsen (2008).

derm. The pelagic dispersal stages of this animal either could be simple (like a rosette of choanocytes) or could already evolve directional swimming ability and/or sensory cells (true larvae). Modern sponges produce a variety of larval types (Leys and Ereskovsky, 2006; Ereskovsky, 2010), the simplest of which are those of calcareous sponges, such as amphiblastulae, which are composed of two major cell types: anterior ciliated micromeres and posterior macromeres (see figure 10.1J). Upon metamorphosis, these cells differentiate into choanocytes and pinacocytes, respectively—the two basic sponge "building blocks" (Amano and Hori, 1993). Extant choanoflagellates are also able to produce a small variety of differentiated cells (Dayel et al., 2011; Fairclough et al., 2010, 2013), and it is likely that their ancestors could produce even more cell types, although in contrast to true multicellularity, the cell differentiation is/was temporal rather than spatial (e.g., Mikhailov et al., 2009).

The "ancestral sponges" were flat and likely tiny; as in large colonies the cells in the middle (or downstream of the current) would have limited access to the food particles in comparison to those in the perimeter of the colony. In the *third step*, three-dimensional organization (and larger body size) could be generated by elevation of the edges of the colony to form a bowl or vase-shaped structure; pores between cells would allow water with nutrients to reach all choanocytes. The resulting organisms might have looked very much like juveniles of calcareous sponges, with the outer pinacocyte layer and the inner choanocyte chamber, and supporting skeletal elements (see figure 10.1F, G). The first sponges have likely inherited the ability to form silica-based skeletal elements from their choanoflagellate-like ancestors. The capacity to build calcium-based skeletons is also widespread among eukaryotes, including many protists and sponges. Subsequent modifications of the body plans of these first sponges can be visualized as many ways of folding of two-layered openwork fabric (choanocytes on the inside, pinacocytes on the outside, porocytes connecting the two layers). This basic design gave rise to a variety of forms and allows some siliceous sponges—for example, the giant barrel sponge *Xestospongia muta*—to reach massive sizes (e.g., 2 m in diameter).

According to Haeckel's views, the choanoderm and pinacoderm would be the direct evolutionary ancestors of the endoderm (gut) and the ectoderm (epidermis) of cnidarian polyps, and the osculum would directly correspond to the polyp's mouth (which acts as a single opening of the blind gut). Using this perspective, it appears surprisingly easy to derive polyp-like animals from the simplest (asconoid) sponges in the *fourth step*—the only difference is the feeding mode, which switches from filter feeding and intracellular digestion to capture of larger prey and extracellular digestion. The capacity of sponges to evolve extracellular digestion directly from intracellular digestion is illustrated by carnivorous sponges, which are capable of feeding on crustaceans (see, e.g., Vacelet and Duport, 2004). These sponges have absent/reduced choanocyte chambers and capture prey passively using extended filaments with Velcro-like spicules. However, macroscopic organisms often enter/inhabit choanocyte chambers of filter-feeding sponges, and it is easy to imagine that a lineage of sponges evolved the capacity to utilize these organisms as a food source. In subsequent steps, the ability to capture prey actively could be acquired by formation of nerves and muscles, and followed by loss of filter-feeding capacity. Such changes would result in emergence of cnidarian-grade organization and opening the way to evolution of the amazing diversity of the eumetazoan body plans.

In this theoretical scenario the choanocytes are thus homologous not only to choanoflagellates but also to the endodermal cells of the eumetazoans. Importantly, in cnidarians not only are the endodermal cells responsible for feeding but they are also the source of stem cells and gametes (e.g., Martin and Archer, 1986; Martin et al., 1997; Wittlieb et al.,

2006). This dual role is also shared by choanocytes, which have remarkable plasticity of function: they can proliferate, can produce gametes, and are able to transdifferentiate into a variety of other cell types (e.g., Funayama, 2013; Funayama et al., 2010; Nakanishi et al., 2014). While neither the shared position within the body nor the shared functions are sufficient to prove the proposed homologies, additional support is given by comparative analysis of developmental gene expression between sponges and eumetazoans.

Developmental Gene Expression and Homology of Sponge and Cnidarian Body Plans

In the past decade significant progress has been made toward elucidation of the relationship between cnidarian and bilaterian body plans (reviewed by Technau and Steele, 2011). One of the successful approaches used in these investigations was the comparative expression analysis of conserved developmental regulatory genes, which serves as the first step toward identification of gene regulatory networks governing development of homologous tissues and structures (Davidson and Erwin, 2006; see also Wagner, 2007). Results of these studies strongly support the notion that the oral end of cnidarian polyps (which is derived from the blastopore) is homologous to the posterior end of the bilaterians (reviewed by Guder et al., 2006; Petersen and Reddien, 2009; Holstein, 2012). At the same time, the cnidarian endoderm appears to be homologous to the bilaterian endoderm and mesoderm (Technau and Scholz, 2003; Martindale et al., 2004; Roettinger et al., 2012).

Comparative expression analysis can now be extended to sponges, as sequencing of several sponge genomes and transcriptomes demonstrated that the overwhelming majority of genes involved in the development of cnidarians and bilaterians are also present in this phylum (e.g., Nichols et al., 2006; Adamska et al., 2010; Larroux et al., 2008; Srivastava et al., 2010; Fortunato et al., 2014). In fact, sponge embryos and larvae share one of the most conserved features of animal development, which is posterior expression of Wnt genes (Adamska et al., 2007). Additionally, it has been demonstrated that the canonical Wnt pathway is involved in patterning of the adult body plan of sponges in a manner strikingly similar to its role in cnidarians, as in both cases increase in Wnt signaling results in multiplication of body axes (Windsor and Leys, 2010; Müller et al., 2007).

Using calcareous sponge *Sycon ciliatum* as a model system, we recently investigated expression of a broad range of genes which in cnidarians and bilaterians are responsible for establishment of axial polarity and for determination of the endodermal layer (Leininger et al., 2014). Expression of the analyzed marker genes is fully consistent with homology of the sponge osculum with the cnidarian mouth. In particular, multiple Wnt and TGF-beta ligands are expressed uniquely or are enriched around the osculum, and a battery of genes involved in endoderm development (e.g., *beta-catenin*, *gata*, *brachyury*) are expressed in the larval micromeres and in differentiated choanocytes. At the same time, several genes responsible for gastrulation, including *forkhead*, *snail*, and *strabismus*, are absent from sponges. This is particularly interesting in light of the fact that specification of the endoderm (which is dependent on canonical Wnt signaling through *beta-catenin*) is uncoupled from gastrulation movements (which are dependent on noncanonical Wnt signaling through *strabismus*) in the cnidarian *Nematostella vectensis* (Wikramanayake et al., 2003; Kumburegama et al., 2011). Thus, it appears likely that while the endoderm is derived from the choanoderm, gastrulation by invagination evolved after sponges separated from the lineage leading to cnidarians and bilaterians. Notably, genes which are involved in endoderm formation and specification of the stem cells, such as *vasa* and *pl10* (Extavour et al., 2005; Roettinger et al., 2012), are also expressed in the choanoderm lineage in *Sycon*, in line with the dual role of the choanocytes.

Finally, a recent phylostratigraphic analysis demonstrated that genes associated with endoderm development are, as a group, significantly older than those expressed in developing ectoderm and mesoderm. Based on these findings, the authors "propose that the endoderm program dates back to the origin of multicellularity, whereas the ectoderm originated as a secondary germ layer freed from ancestral feeding functions" (Hashimshony et al., 2014, p. 219).

In conclusion, multiple lines of evidence support the notion that the ancestry of endoderm can be traced back, via choanocytes, to colonial (choanoflagellate-like) ancestors of all multicellular animals. Further research will be needed to fully appreciate the changes associated with emergence of the beautiful and complex animal bodies from their humble protistan beginnings.

Acknowledgments

I would like to thank Karl J. Niklas and Stuart A. Newman for the invitation to participate in this exciting project, Fabian Rentzsch for sharing his knowledge on cnidarian development, Chiara Sinigaglia for figure 10.1A, Gemma Richards for figure 10.1B, Bernd Schierwater for figure 10.1C, Yuuji Tsukii for figure 10.1M, Marcin Adamski for figure 10.1D and the never-ending discussions on animal evolution and phylogeny, and the Sars International Centre for Marine Molecular Biology for supporting research in my group.

References

Adamska, M., Degnan, S. M., Green, K. M., Adamski, M., Craigie, A., Larroux, C., et al. (2007). Wnt and TGF-beta expression in the sponge *Amphimedon queenslandica* and the origin of metazoan embryonic patterning. *PLoS One*, 2, e1031.

Adamska, M., Degnan, B. M., Green, K., & Zwafink, C. (2011). What sponges can tell us about the evolution of developmental processes. *Zoology (Jena)*, 114, 1–10.

Adamska, M., Larroux, C., Adamski, M., Green, K., Lovas, E., Koop, D., et al. (2010). Structure and expression of conserved Wnt pathway components in the demosponge *Amphimedon queenslandica*. *Evolution & Development*, 12, 494–518.

Amano, S., & Hori, I. (1993). Metamorphosis of calcareous sponges: II. Cell rearrangement and differentiation in metamorphosis. *Invertebrate Reproduction & Development*, 24, 13–26.

Baguñà, J., Martinez, P., Paps, J., & Riutort, M. (2008). Back in time: A new systematic proposal for the Bilateria. *Philosophical Transactions of the Royal Society of London. Series B, Biological Sciences*, 363, 1481–1491.

Borchiellini, C., Manuel, M., Alivon, E., Boury-Esnault, N., & Vacelet, J. (2001). Sponge paraphyly and the origin of Metazoa. *Journal of Evolutionary Biology*, 14, 171–179.

Byrum, C. A., & Martindale, M. Q. (2004). Gastrulation in the Cnidaria and Ctenophora. In C. D. Stern (Ed.), *Gastrulation: From cells to embryo* (pp. 33–50). Cold Spring Harbor, NY: CSH Press.

Carr, M., Leadbeater, B. S., Hassan, R., Nelson, M., & Baldauf, S. L. (2008). Molecular phylogeny of choano-flagellates, the sister group to Metazoa. *Proceedings of the National Academy of Sciences of the United States of America*, 105, 16641–16646.

Clark, H. (1868). On the *Spongiae ciliatae* as *Infusoria flagellata*; or observations on the structure, animality and relationship of *Leucosolenia botryoides* Bowerbank. *Annals & Magazine of Natural History*, 4, 133–142, 188–215, 250–264.

Darwin, C. (1859). *On the origin of species by means of natural selection, or the preservation of favoured races in the struggle for life*. London: J. Murray.

Davidson, E. H., & Erwin, D. H. (2006). Gene regulatory networks and the evolution of animal body plans. *Science*, 311, 796–800.

Dayel, M. J., Alegado, R. A., Fairclough, S. R., Levin, T. C., Nichols, S. A., McDonald, K., et al. (2011). Cell differentiation and morphogenesis in the colony-forming choanoflagellate *Salpingoeca rosetta*. *Developmental Biology*, 357, 73–82.

Dayel, M. J., & King, N. (2014). Prey capture and phagocytosis in the choanoflagellate *Salpingoeca rosetta*. *PLoS One*, 9(5), e95577.

Dohrmann, M., & Wörheide, G. (2013). Novel scenarios of early animal evolution—Is it time to rewrite text-books? *Integrative and Comparative Biology*, 53, 503–511.

Dunn, C. W., Giribet, G., Edgecombe, G. D., & Hejnol, A. (2014). Animal phylogeny and its evolutionary implications. *Annual Review of Ecology Evolution and Systematics*, 45, 371–395.

Dunn, C. W., Hejnol, A., Matus, D. Q., Pang, K., Browne, W. E., Smith, S. A., et al. (2008). Broad phylogenomic sampling improves resolution of the animal tree of life. *Nature*, 452, 745–749.

Edgecombe, G. D., Giribet, G., Dunn, C. W., Hejnol, A., Kristensen, R. M., Neves, R. C., et al. (2011). Higher-level metazoan relationships: Recent progress and remaining questions. *Organisms, Diversity & Evolution*, 11, 151–172.

Ereskovsky, A. (2010). *The comparative embryology of sponges*. Springer Netherlands.

Erwin, D. H., Laflamme, M., Tweedt, S. M., Sperling, E. A., Pisani, D., & Peterson, K. J. (2011). The Cambrian conundrum: Early divergence and later ecological success in the early history of animals. *Science*, 334, 1091–1097.

Extavour, C. G., Pang, K., Matus, D. Q., & Martindale, M. Q. (2005). *vasa* and *nanos* expression patterns in a sea anemone and the evolution of bilaterian germ cell specification mechanisms. *Evolution & Development*, 7, 201–215.

Fairclough, S. R., Chen, Z., Kramer, E., Zeng, Q., Young, S., Robertson, H. M., et al. (2013). Premetazoan genome evolution and the regulation of cell differentiation in the choanoflagellate *Salpingoeca rosetta*. *Genome Biology*, 14(2), R15.

Fairclough, S. R., Dayel, M. J., & King, N. (2010). Multicellular development in a choanoflagellate. *Current Biology*, 20, R875–R876.

Fortunato, S., Adamski, M., Mendivil, O., Leininger, S., Liu, J., Ferrier, D. E. K., et al. (2014). Calcisponges have a ParaHox gene and dynamic expression of dispersed NK homeobox genes. *Nature*, 514, 620–623.

Funayama, N. (2013). The stem cell system in demosponges: Suggested involvement of two types of cells: Archeocytes (active stem cells) and choanocytes (food-entrapping flagellated cells). *Development Genes and Evolution*, 223(1–2), 23–38.

Funayama, N., Nakatsukasa, M., Mohri, K., Masuda, Y., & Agata, K. (2010). Piwi expression in archeocytes and choanocytes in demosponges: Insights into the stem cell system in demosponges. *Evolution & Development*, 12, 275–287.

Guder, C., Philipp, I., Lengfeld, T., Watanabe, H., Hobmayer, B., & Holstein, T. W. (2006). The Wnt code: Cnidarians signal the way. *Oncogene*, 25, 7450–7460.

Haeckel, E. (1866). *Generelle Morphologie der Organismen: Allgemeine Grundzüge der organischen Formen-Wissenschaft, mechanisch begründet durch die von Charles Darwin reformirte Descendenztheorie.* Berlin, Germany: Verlag von Georg Reimer.

Haeckel, E. (1870). On the organization of sponges and their relationship to the corals. *Annals & Magazine of Natural History*, 5, 1–13, 107–120.

Haeckel, E. (1874). Die Gastrae Theorie, die phylogenetische Classification des Thierreichs und die Homologie der Keimblatter. *Jenaische Zeitschrift für Naturwissenschaft*, 8, 1–55.

Hashimshony, T., Feder, M., Levin, M., Hall, B. K., & Yanai, I. (2014). Spatiotemporal transcriptomics reveals the evolutionary history of the endoderm germ layer. *Nature.* doi:.10.1038/nature13996

Holstein, T. W. (2012). The evolution of the Wnt pathway. *Cold Spring Harbor Perspectives in Biology*, 4(7), a007922.

Hyman, L. H. (1940). *Invertebrates: Protozoa through Ctenophora* (Vol. 1). New York: McGraw-Hill.

King, N. (2004). The unicellular ancestry of animal development. *Developmental Cell*, 7, 313–325.

King, N., Westbrook, M. J., Young, S. L., Kuo, A., Abedin, M., Chapman, J., et al. (2008). The genome of the choanoflagellate *Monosiga brevicollis* and the origin of metazoans. *Nature*, 451, 783–788.

Kumburegama, S., Wijesena, N., Xu, R., & Wikramanayake, A. H. (2011). Strabismus-mediated primary archenteron invagination is uncoupled from Wnt/β-catenin-dependent endoderm cell fate specification in *Nematostella vectensis* (Anthozoa, Cnidaria): Implications for the evolution of gastrulation. *EvoDevo*, 2(1), 2.

Larroux, C., Luke, G. N., Koopman, P., Rokhsar, D. S., Shimeld, S. M., & Degnan, B. M. (2008). Genesis and expansion of metazoan transcription factor gene classes. *Molecular Biology and Evolution*, 25, 980–996.

Leadbeater, B. S. C. (2008). Choanoflagellate evolution: The morphological perspective. *Protistology*, 5, 256–267.

Leininger, S., Adamski, M., Bergum, B., Guder, C., Liu, J., Laplante, M., et al. (2014). Developmental gene expression provides clues to relationships between sponge and eumetazoan body plans. *Nature Communications*, 5, 3905.

Leys, S. P., & Eerkes-Medrano, D. (2005). Gastrulation in calcareous sponges: In search of Haeckel's gastraea. *Integrative and Comparative Biology*, 45, 342–351.

Leys, S. P., & Eerkes-Medrano, D. I. (2006). Feeding in a calcareous sponge: Particle uptake by pseudopodia. *Biological Bulletin*, 211, 157–171.

Leys, S. P., & Ereskovsky, A. V. (2006). Embryogenesis and larval differentiation in sponges. *Canadienne De Zoologie*, 84, 262–287.

Mah, J. L., Christensen-Dalsgaard, K. K., & Leys, S. P. (2014). Choanoflagellate and choanocyte collar–flagellar systems and the assumption of homology. *Evolution & Development*, 16, 25–37.

Maldonado, M. (2004). Choanoflagellates, choanocytes, and animal multicellularity. *Invertebrate Biology*, 123, 1–22.

Marlow, H., & Arendt, D. (2014). Evolution: Ctenophore genomes and the origin of neurons. *Current Biology*, 24, R757–R761.

Martin, V. J., & Archer, W. E. (1986). Migration of interstitial cells and their derivatives in a hydrozoan planula. *Developmental Biology*, 116, 486–496.

Martin, V. J., Littlefield, C. L., Archer, W. E., & Bode, H. R. (1997). Embryogenesis in hydra. *Biological Bulletin*, 192, 345–363.

Martindale, M. Q., Pang, K., & Finnerty, J. R. (2004). Investigating the origins of triploblasty: "Mesodermal" gene expression in a diploblastic animal, the sea anemone *Nematostella vectensis* (phylum, Cnidaria; class, Anthozoa). *Development.*, 131, 2463–2474.

Mikhailov, K. V., Konstantinova, A. V., Nikitin, M. A., Troshin, P. V., Rusin, L. Y., Lyubetsky, V. A., et al. (2009). The origin of Metazoa: A transition from temporal to spatial cell differentiation. *BioEssays*, 31, 758–768.

Moroz, L. L., Kocot, K. M., Citarella, M. R., Dosung, S., Norekian, T. P., Povolotskaya, I. S., et al. (2014). The ctenophore genome and the evolutionary origins of neural systems. *Nature*, 510, 109–114.

Müller, W., Frank, U., Teo, R., Mokady, O., Guette, C., & Plickert, G. (2007). wnt signaling in hydroid development: Ectopic heads and giant buds induced by GSK-3beta inhibitors. *International Journal of Developmental Biology*, 51, 211–220.

Nakanishi, N., Sogabe, S., & Degnan, B. M. (2014). Evolutionary origin of gastrulation: Insights from sponge development. *BMC Biology*, 12, 26.

Nichols, S. A., Dirks, W., Pearse, J. S., & King, N. (2006). Early evolution of animal cell signaling and adhesion genes. *Proceedings of the National Academy of Sciences of the United States of America*, 103, 12451–12456.

Nielsen, C. (2008). Six major steps in animal evolution: Are we derived sponge larvae? *Evolution & Development*, 10, 241–257.

Nosenko, T., Schreiber, F., Adamska, M., Adamski, M., Eitel, M., Hammel, J., et al. (2013). Deep metazoan phylogeny: When different genes tell different stories. *Molecular Phylogenetics and Evolution*, 67, 223–233.

Petersen, C. P., & Reddien, P. W. (2009). Wnt signaling and the polarity of the primary body axis. *Cell*, 139, 1056–1068.

Peterson, K. J., & Butterfield, N. J. (2005). Origin of the Eumetazoa: Testing ecological predictions of molecular clocks against the Proterozoic fossil record. *Proceedings of the National Academy of Sciences of the United States of America*, 102, 9547–9552.

Philippe, H., Derelle, R., Lopez, P., Pick, K., Borchiellini, C., Boury-Esnault, N., et al. (2009). Phylogenomics revives traditional views on deep animal relationships. *Current Biology*, 19, 706–712.

Pick, K. S., Philippe, H., Schreiber, F., Erpenbeck, D., Jackson, D. J., Wrede, P., et al. (2010). Improved phylogenomic taxon sampling noticeably affects non-bilaterian relationships. *Molecular Biology and Evolution*, 27, 1983–1987.

Podar, M., Haddock, S. H., Sogin, M. L., & Harbison, G. R. (2001). A molecular phylogenetic framework for the phylum Ctenophora using 18S rRNA genes. *Molecular Phylogenetics and Evolution*, 21, 218–230.

Richter, D. J., & King, N. (2013). The genomic and cellular foundations of animal origins. *Annual Review of Genetics*, 47, 509–537.

Roettinger, E., Dahlin, P., & Martindale, M. Q. (2012). A framework for the establishment of a cnidarian gene regulatory network for "endomesoderm" specification: The inputs of ß-catenin/TCF signaling. *PLOS Genetics*, 8, e1003164.

Ruiz-Trillo, I., Roger, A. J., Burger, G., Gray, M. W., & Lang, B. F. (2008). A phylogenomic investigation into the origin of Metazoa. *Molecular Biology and Evolution*, 25, 664–672.

Ryan, J. F. (2014). Did the ctenophore nervous system evolve independently? *Zoology (Jena)*, 117, 225–226.

Ryan, J. F., Pang, K., Schnitzler, C. E., Nguyen, A. D., Moreland, R. T., Simmons, D. K., et al. (2013). The genome of the ctenophore *Mnemiopsis leidyi* and its implications for cell type evolution. *Science*, 342, 1242592.

Schierwater, B. (2005). My favorite animal, *Trichoplax adhaerens*. *BioEssays*, 27, 1294–1302.

Schierwater, B., Eitel, M., Jakob, W., Osigus, H. J., Hadrys, H., Dellaporta, S. L., et al. (2009). Concatenated analysis sheds light on early metazoan evolution and fuels a modern "urmetazoon" hypothesis. *PLoS Biology*, 7(1), 36–44.

Shalchian-Tabrizi, K., Minge, M. A., Espelund, M., Orr, R., Ruden, T., Jakobsen, K. S., et al. (2008). Multigene phylogeny of Choanozoa and the origin of animals. *PLoS One*, 3(5), e2098.

Simpson, T. L. (1984). *The cell biology of sponges*. New York: Springer.

Sperling, E. A., Peterson, K. J., & Pisani, D. (2009). Phylogenetic-signal dissection of nuclear housekeeping genes supports the paraphyly of sponges and the monophyly of Eumetazoa. *Molecular Biology and Evolution*, 26, 2261–2274.

Srivastava, M., Begovic, E., Chapman, J., Putnam, N. H., Hellsten, U., Kawashima, T., et al. (2008). The *Trichoplax* genome and the nature of placozoans. *Nature*, 454, 955–960.

Srivastava, M., Simakov, O., Chapman, J., Fahey, B., Gauthier, M. E., Mitros, T., et al. (2010). The *Amphimedon queenslandica* genome and the evolution of animal complexity. *Nature*, 466, 720–726.

Technau, U., & Scholz, C. B. (2003). Origin and evolution of endoderm and mesoderm. *International Journal of Developmental Biology*, 47, 531–539.

Technau, U., & Steele, R. E. (2011). Evolutionary crossroads in developmental biology: Cnidaria. *Development.*, 138, 1447–1458.

Telford, M. J. (2013). Evolution: The animal tree of life. *Science*, 339, 764–766.

Telford, M. J., & Copley, R. R. (2011). Improving animal phylogenies with genomic data. *Trends in Genetics*, 27, 186–195.

Torruella, G., Derelle, R., Paps, J., Lang, B. F., Roger, A. J., Shalchian-Tabrizi, K., et al. (2012). Phylogenetic relationships within the Opisthokonta based on phylogenomic analyses of conserved single-copy protein domains. *Molecular Biology and Evolution*, 29, 531–544.

Vacelet, J., & Duport, E. (2004). Prey capture and digestion in the carnivorous sponge *Asbestopluma hypogea* (Porifera: Demospongiae). *Zoomorphology*, 123, 179–190.

Valentine, J. W. (2003). Architectures of biological complexity. *Integrative and Comparative Biology*, 43, 99–103.

Wagner, G. P. (2007). The developmental genetics of homology. *Nature Reviews. Genetics*, 8, 473–479.

Wikramanayake, A. H., Hong, M., Lee, P. N., Pang, K., Byrum, C. A., Bince, J. M., et al. (2003). An ancient role for nuclear Beta-catenin in the evolution of axial polarity and germ layer segregation. *Nature*, 426, 446–450.

Windsor, P. J., & Leys, S. P. (2010). Wnt signaling and induction in the sponge aquiferous system: Evidence for an ancient origin of the organizer. *Evolution & Development*, 12, 484–493.

Wittlieb, J., Khalturin, K., Lohmann, J. U., Anton-Erxleben, F., & Bosch, T. C. (2006). Transgenic Hydra allow in vivo tracking of individual stem cells during morphogenesis. *Proceedings of the National Academy of Sciences of the United States of America*, 103, 6208–6211.

Wörheide, G., Dohrmann, M., Erpenbeck, D., Larroux, C., Maldonado, M., Voigt, O., et al. (2012). Deep phylogeny and evolution of sponges (phylum Porifera). *Advances in Marine Biology*, 61, 1–78.

Zrzavy, J., Mihulka, S., Kepka, P., Bezdek, A., & Tietz, D. (1998). Phylogeny of the Metazoa based on morphological and 18S ribosomal DNA evidence. *Cladistics*, 14, 249–285.

11

A Scenario for the Origin of Multicellular Organisms: Perspective from Multilevel Consistency Dynamics

Kunihiko Kaneko

Biological systems consist of a hierarchy, from molecules to cells to multicellular organisms. To investigate such hierarchical systems, one often analyzes units at a lower level first and then endeavors to study the higher-level whole system as a collection of lower-level units. This bottom-up approach from the lower-level units to a higher-level ensemble, however, is not ideal for the understanding of biologically complex systems because the state of each lower-level unit is often defined by a higher-level unit. Indeed, lower-level units change their state depending on the state of an ensemble to which the unit belongs. Hence, the mutually dependent dynamics between lower and higher levels of hierarchical systems must be seriously considered in studies of complex-systems biology (Kaneko, 2006).

Consider, for example, a single cell within a multicellular organism. This single cell has diverse, internal components (degrees of freedom) and can change its chemical composition or gene expression pattern. The extent of each cellular component's change in abundance depends on the surrounding cells, through the exchange of certain molecules across the cell membrane. Indeed, a cell in isolation and a cell in a community often show quite distinct characteristics, which is known as a *community effect* (Gurdon et al., 1993). Hence, a higher-level system (cell ensemble) constrains the behavior of the lower-level units (cells) that it comprises.

In addition, a biological unit generally can reproduce itself under suitable conditions, and, thus, the ensemble of cells grows as a total. If the cell ensemble is stable, the growth rate of each cell type has to be balanced to maintain the population distribution of the various cell types that compose the ensemble. This "stationary growth" cannot be achieved without cell–cell interactions, as these interactions provide information from an ensemble to each cell.

Temporal change of the intracellular state is represented by dynamical systems, as envisioned by Waddington (1957), and has been investigated extensively since then (Goodwin, 1963; Rosen, 1970; Kauffman, 1993). To investigate multicellular dynamics, we need to study an ensemble of such dynamical systems interacting with each other, which constitutes multilevel dynamics. As cells grow and their number increases, the stability of multilevel dynamics must be satisfied to achieve consistency between levels.

In considering multicellularity, it is often debated how the fitness at a cellular level is aligned with that at a multicellular level since conflict between the two levels might be an obstacle for evolution of multicellular organisms. However, before discussing fitness alignment, the mechanism by which a stable state is achieved both intracellularly and in multicellular dynamical systems must first be addressed. We have proposed complex-systems biology to deal with such multilevel consistency, where we are interested in how multilevel robustness is achieved in dynamical systems rather than optimization of their fitness (Kaneko, 2006, Kaneko and Furusawa, 2008).

Let us consider a cell with diverse components that grows and divides. As the number of cells increases, resources will become limited, and cells interact with each other by various molecules as they become more crowded. In this situation, is multicellularity a necessary course? Here, as a requisite for multicellularity, we postulate that multiple types of cells are inevitably differentiated from a single cell type, and these cells grow and coexist together to form an aggregate (ensemble) with division of labor.

The basic questions to be addressed in this chapter are as follows:

1. *Diversification:* How can different cell types coexist, instead of only the fittest cell type's surviving?

2. *Differentiation:* How is cell type diversification achieved among cells that share common intrinsic characteristics (i.e., having the same genotype and differentiating from the same cell)?

3. *Recursive production as a multicellular organism:* From an ensemble of cells developed from a single cell, a single (or few) cell(s) is detached to form the next generation. How is such recursive production of cell ensembles through a bottleneck in cell number achieved?

4. *Developmental and evolutionary robustness:* Can the above cycle of differentiation, development, and reproduction acquire robustness against perturbations? Is cancer a result of the loss of such robustness, and an unavoidable consequence of complex cell ensembles?

Coexistence of Diverse Cell Types

Now, consider multilevel dynamics (i.e., intracellular dynamics with chemical reactions, interactions, and growth) (see figure 11.1). Can diverse cell types sharing common resources coexist? Mathematically, these multilevel dynamics are formulated as described below.

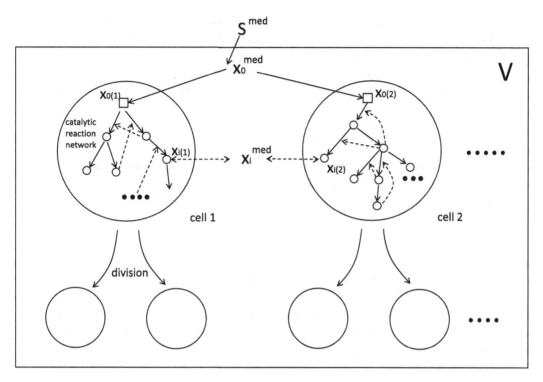

Figure 11.1
Schematic representation of multilevel dynamics: Intracellular catalytic reaction dynamics, cell–cell interactions through medium (med), and cell growth. V denotes the volume of medium, against the unit volume of a cell.

Let us consider a cell consisting of K components. The concentration of the i-th chemical component of the m-th cell is denoted by $x_i(m;t)$. In each cell, there are intracellular reaction dynamics that change the concentration $x_i(m;t)$ (given by a certain function) $f_i^m(\{x_j(m)\})$, while these cells interact with each other with an influence described by the function $h_i([x_i(\ell)])$. The cell volume grows at the rate $\mu^m(t)$, such that the concentration is diluted by the rate $\mu^m x_i(m)$. Then, in general, the temporal change of each component is given by

$$\frac{dx_i(m)}{dt} = f_i^m(\{x_j(m)\}) + h_i([x_i(\ell)]) - \mu^m x_i(m), \tag{11.1}$$

where the growth rate is given as a function of the components (phenotype), $\mu^m = g(\{x_j(m)\})$. If the total system as a whole exhibits stationary growth, $\mu^m = \mu^\ell$ has to be satisfied (i.e., every cell must exhibit balanced growth) (Kaneko et al., 2015).

The above is what is termed an "individual-based model" and, with this simulation, we can trace the dynamics of each cell individually. When a population of cells shares the same rule of dynamics (genotype), it is relevant to extract the dynamics of the population of each type, that is, the higher-level dynamics. If each cellular state attains a stationary state, each $x_i(m)$ reaches a fixed-point solution with $f_i^m(\{x_j^*(m)\}) + h_i([x_i^*(\ell)]) = \mu^m x_i^*(m)$. Assuming that each cell type of the network exhibits the same phenotypic state x^*, and that the interaction term depends on the number density of each cell type, the stationary state of the multilevel dynamics must satisfy

$$f_i^\alpha(\{x_j^*(\alpha)\}) + h_i([x_i^*(\beta)], \rho^\beta) = \mu^\alpha x_i^*(\alpha),$$

where α, β,... represents the index of each type. Here, ρ^α denotes the population fraction of each type as $\rho^\alpha = M^\alpha / M_{tot}$, where M is the number of each cell type and M_{tot} is the total population. By using the growth rate μ^α, the population fraction changes as

$$\frac{d\rho^\alpha}{dt} = \left(\mu^\alpha - \sum_v \mu^v \rho^v\right)\rho^\alpha \tag{11.2}$$

for all cell types $v = 1,..,\alpha,...,M$. Now, if all of the cell types coexist as a stationary distribution, then the growth rate of each cell type must be equal: $\mu^\alpha = \mu^\beta$.

For example, intracellular catalytic reactions as shown in figure 11.1 are considered (for related models, see Kaneko and Yomo, 1999; Furusawa and Kaneko, 2001, 2003, 2012b) in which chemical resources are transformed to cellular components with the aid of catalysts (typically enzymes) that are also synthesized by the catalytic reactions given by

$$\frac{dx_i}{dt} = \sum_{j,\ell=0}^{N-1} W_{ji\ell} x_j x_\ell^n - \sum_{j',\ell'=0}^{N-1} W_{ij'\ell'} x_i x_{\ell'}^n - x_i \sum_{j=0}^{N-1} F_j \quad (+R_i), \tag{11.3}$$

where $W_{ij\ell}$ is 1 if reaction $i + n\ell \to j + n\ell$ occurs; otherwise, it is 0. In this reaction, ℓ works as a catalyst and, for a simple catalytic reaction $n = 1$, while $n = 2$ if the catalyst works as a dimer, and so forth. The third term with the sum of nutrient uptake flux gives a constraint of $\sum_{i=0}^{N-1} x_i = 1$, because of the volume growth. The last term is added only for the nutrients, and it represents their transport into a cell from the environment.

On the other hand, the simplest form of interaction is the exchange of some biomolecules through the media (see figure 11.1) as

$$h_i([x_i(m)]) = D(x_i^{med} - x_i(m)), \tag{11.4}$$

where x_i^{med} is the concentration of the chemical in the media, which is changed accordingly as

$$dx_i^{med} / dt = -\sum_m D(x_i^{med} - x_i(m)) \Big/ V, \tag{11.5}$$

with V being the ratio of the media volume to that of the cell. Here, a resource chemical x_q^{med} (say $q = 0$) is transported from the media into a cell m as a function $R_q(\{x_j(m)\}, x_q^{med})$, for simple diffusion $R_q \propto (x_q^{med} - x_\mu(m))$, and for active transport mediating some chemical P, $R_q \propto x_p x_q^{med}$. The concentration of the resource chemical in the media is, thus, decreased, while there is supply from the outside S:

$$dx_q^{med} / dt = -x_q^{med} R_q(\{x_j(m)\}, x_q^{med}) / V + D_{out}(S - x_q^{med}). \tag{11.6}$$

Finally, the growth rate is determined as a function of $\mu^m = g([x_i(m)])$. For example, if the growth is determined by the transport of L species of chemical resources, it is given by

$$g([x_i(m)]) = \sum_q^L R_q(\{x_j(m)\}, x_q^{med}). \tag{11.7}$$

We have simulated the above model by considering a variety of reaction networks to determine whether cell types with different reaction networks can coexist (Kaneko, 2015). For example, we consider a model with catalytic reaction networks consisting of $K = 100$ chemical species under $L = 4$ chemical resources. Initially, we set up 100 types of cells with different networks and numerically studied the time evolution of the population of these cells types. We examined how many types coexist by discarding those types whose density was less than 0.001 and decreasing.

In the above model, the inverse of the medium-cell volume ratio V works as a parameter to control cell–cell interactions and competition for resources. If V is large, cells exist in dilute conditions, such that each cell grows almost independently. With a decrease in V, cells begin to exhibit stronger interactions. If the interaction is weak, only a few cell types (less than the number of resource types) coexist. Within each cell, only a few chemical components remain (i.e., simple replicators are formed). On the other hand, coexistence of multiple cell types with limited resources (strong interaction) is sometimes achieved (see figure 11.2A and 11.2B).

In ecology, the upper bound of the number of coexisting species is known as Gause's limit (Gause, 1934). According to this limit, the number of possible species that coexist is bounded by the number of given resource species. This limit, however, is not always reached. If cell–cell interactions are disregarded and only growth rates at a single-cell level are compared, then only the cell type with the highest growth rate survives under the Darwinian selection process. For example, if one cell type can utilize all resources better than the other cell types, it will dominate.

Indeed, if cell–cell interactions and competition among resources are weak, only the cell type with the highest growth rate remains, and the number of coexisting cell types is below Gause's limit. In this case, the number of chemical species that exist in a cell is

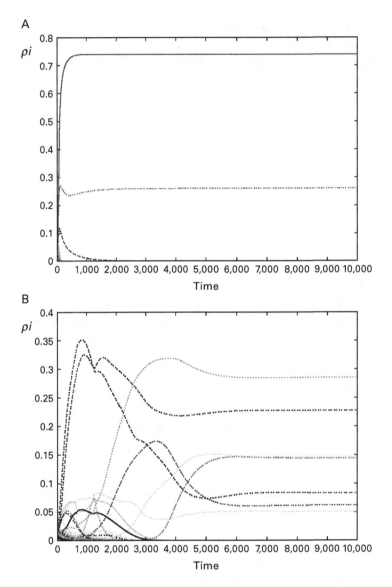

Figure 11.2
Time series of the population ratio ρ^j in our models in equations 11.1–11.7, plotted against the time. Starting from 100 species, and those with $\rho^j < 10^{-4}$ are discarded. (A) $V = 10^4$; two species remain, each of which contains two chemical components besides the resources; (B) $V = 2$; seven species remain, which contain almost all (100) chemical components.

often much smaller than all of the possible chemical species. Indeed, only a simple auto-catalytic replicator (a single autocatalytic loop $X_0 + X_\ell \rightarrow 2X_\ell$ or mutually catalytic loop $X_0 + X_m \rightarrow X_\ell + X_m$, $X_0 + X_\ell \rightarrow X_m + X_\ell$) often remains.

In contrast, from numerical simulations of models with equations 11.2–11.7 with small V (i.e., under limited resources with strong cell–cell interactions), several cell types can coexist. Indeed, more cell types than allowed by Gause's limit often coexist. Since chemicals that are not considered resources are exchanged by cells and sometimes play a role in the growth of other cells, the number of effective chemical species relevant for growth can be larger than that of the resource species. Especially under strong competition for resources and interactions, many more species coexist than the resource number. In this case, there are many components in each cell, such that diversity in components is also achieved.

These numerical results are interpreted as follows. First of all, dominance of simple replicators under sufficient nutrient supply is easily understood. Consider a catalytic reaction process for cell growth $X_i + X_j \rightarrow X_k + X_j$. Then, the rate of this reaction is proportional to the product of concentrations $x_i x_j$. In a cell of a given volume, the number of molecules within the cell is limited. Let us assume this number to be N. Now, assume that these N molecules are equally distributed into K_{eff} species. Then, the concentration of each component is proportional to $1/K_{eff}$, and, thus, the reaction rate for each of the above reactions is estimated to be proportional to K_{eff}^{-2}. Considering that the number of reaction paths is expected to be proportional to K_{eff}, the total number of reaction events will be proportional to $1/K_{eff}$. In this sense, the total reaction rate for growth is higher for smaller K_{eff} (i.e., for a smaller number of components). This tendency is further increased if the catalytic reaction process is of a higher order than $X_i + nX_j \rightarrow X_k + nX_j$, with the rate proportional to $x_i x_j^n$. In this case, the reaction speed is estimated to be $1/K_{eff}^n$, and acceleration by the concentration of a few components is increased. (If the reaction from resource chemical S, $S + nX_j \rightarrow X_k + nX_j$ is dominant and the resource abundance does not decrease with $1/K_{eff}$, the speed will be $1/K_{eff}^{n-1}$, instead, and in this case, $n > 1$ is necessary for the argument below.)

If the concentrations of each remaining chemical species are not equally distributed but are nonuniform, this tendency for acceleration will be relaxed. However, the concentrations of fewer relevant molecules will in any case speed up the growth rate. Therefore, with sufficient resource chemicals, simple replicators generally undergo higher growth speed; thus, only the fittest cell type (or fittest for each chemical resource) remains, such that the possible number of coexisting cell types is highly restricted.

Why, then, is coexistence of many cell types possible for smaller V? Theoretically, there are two possible mechanisms, one due to strong cell–cell interactions and the other due to resource limitation. We will first consider a consequence of strong interactions.

As an illustration, consider the simplest situation with the coexistence of two cell types sharing a single resource. A catalyst needed for growth is missing within a given cell type

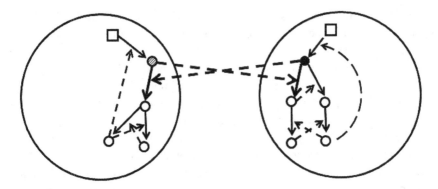

Figure 11.3
Schematic representation of symbiotic growth of two types of cells.

but is synthesized by the other cell type and transported. The latter cell type also lacks a catalyst, which is synthesized by the former. In this sense, the two cell types form a symbiotic relationship (see figure 11.3).

With strong cell–cell interactions, the energetic cost of transporting chemicals synthesized by other cells is low. Instead of paying the cost to produce all the necessary components by the cell itself, use of chemicals synthesized by cells of other types may be advantageous for growth. If the cost of transport is not so large, utilization of the surplus made by cells of other types, which are needed to produce the necessary catalyst, will be advantageous for cell growth. If this use is unidirectional (i.e., one type takes advantage of the other, but not vice versa), the growth rate of the two types cannot be balanced and, ultimately, one type would become extinct; then, the other type could not survive on its own either. Hence, the use of surplus should be bidirectional. This situation is similar to the idea of "comparative advantage" in economics put forth by Ricardo (1817). The idea is that instead of producing every good necessary in a single country, the mutual use of surplus from another country is advantageous if the transport cost is not too high.

As cell number increases, the available resources needed for growth will become limited. With this limitation of resources, the transport of resources from the media to a cell becomes the rate-limiting step for cell growth. Reaction by the optimal autocatalytic reaction loop that was adopted in the high-resource case then progresses only occasionally, so that there is sufficient time for other diverse reactions to occur. Then, the optimization of reaction rates is no longer important. Because of the slow transport of resources, there will be sufficient time to transform resources and perform successive intracellular reactions. Storing chemical components by distributing them to diverse components will no longer be disadvantageous. Thus, diverse components begin to coexist under limited resources.

In some sense, the transition to increase the diversity of components is regarded as a change from r-selection to K-selection (Pianka, 1970). Indeed, in ecology, r-selection favors higher growth rate under sufficient resources, while K-selection favors carrying capacity, to increase survivability under a limited resource condition. As a high growth rate is not possible, a cell holds on to all chemicals, instead of concentrating only a few chemical species, in an attempt to survive with limited resources. Here, this change from r- to K-selection emerges as a result of reaction dynamics under limited resources.

Combining this diversification of components with the mutual dependency discussed already, diversification in both cell type and composition will result. Mutual dependency can be much more complex among cells with strong interactions; thus, more cell types can coexist by growing together. Since exchangeable components may work as resources for some cells, they can be included among the resources that determine the limit for coexisting cell types, and, thus, more cell types may coexist than that given by the afore-mentioned Gause's limit.

Once this coexistence with complex mutual dependency with diverse components is initiated, it is difficult to determine which cell type is optimal (i.e., the fittest). If there were multiple cell types with similar growth rates, it would take an enormous amount of time for the fittest to win, and, before that occurs, the criteria for fitness could be altered because of changes in the population. Thus, under limited resources and strong cell–cell interactions, the coexistence of diverse cell types with diverse components is expected. Indeed, simulation results support this expectation. (The results are schematically represented in figure 11.4.) An increase in diversity with a decrease in resources has also been analyzed in the transition from exponential to nonexponential growth (Kamimura and Kaneko, 2015).

Note here that the resource limitation and strong interactions are a natural outcome of the increase in cell number due to cell growth and division, which leads to cell crowding.

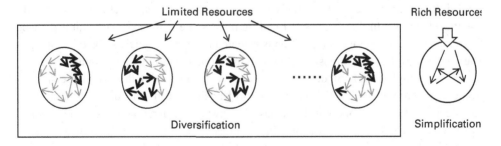

Figure 11.4
Schematic representation of the result on cell-type diversification. Several diverse cell types coexist with diverse components under limited resources while a single fittest cell with few components is selected under rich resources.

In this sense, the emergence of cell aggregates of diverse types with diverse components is a necessity unless the cells have access to unlimited resources.

Cell Differentiation by Isologous Diversification

According to the argument in the previous section, a primitive form of multicellular aggregate, consisting of diverse cell types with differentiation of their roles, is achieved as a result of resource limitation and cell–cell interactions. Still, these different cell types generally embody different reaction networks (i.e., they do not share a common genotype). With this arrangement, an aggregate of cells with different genes emerges, but this cannot be regarded as a multicellular organism in the usual sense. Indeed, a biofilm consisting of diverse bacteria genotypes (Stewart and Franklin, 2008) might belong to this class of primitive multicellular aggregate, while diverse genotypes can also coexist in wild slime mold (Sathe et al., 2010). In plants also, cells expressing different genes often coexist (Gill et al., 1995). These examples may be regarded as retaining some signature of the primitive form of cell aggregates with hetero-genes.

In a standard multicellular organism, however, each individual (cell) shares an identical network (i.e., the same genes). Of course, each cell type of the present multicellular organisms adopts a different subpart of the network according to epigenetic modifications, but these modifications are also a result of a developmental process from a single cell type with a given gene regulation network. Hence, cell types have to be diversified through development from a single cell. Indeed, we previously proposed isologous diversification (Kaneko and Yomo, 1999), in which cells with identical networks (dynamics) differentiate into several types, taking advantage of intracellular interactions (see figure 11.5).

To be specific, let us take a cellular system whose intracellular dynamics allow for oscillation of the concentration of some components (see figure 11.1). With cell division and an increase in cell number, the oscillation starts to desynchronize. This leads to temporal differentiation of cells, but, after taking the temporal average, cells are not yet differentiated. With a further increase in cell number, some cells begin to exhibit distinct behavior, where the average composition is deviated from the original oscillatory cell type. These cells have biased compositions, with a greater abundance of some components and an almost complete lack of other components, compared with the original cell type (Kaneko and Yomo, 1994, 1999; Furusawa and Kaneko, 2001). This differentiation does not occur for every network, but, for a class of networks, differentiation to a few distinct types is achieved. The mechanism for this is understood in terms of dynamical-systems theory (see the "Pluripotency Provided by Oscillatory Expression State and Differentiation via Cell–Cell Interactions" section below). Depending on the ratio of cells of each type, the influence by cell–cell interactions changes, so that the differentiation ratio is changed, which causes a change in the cell number distribution. Through this mutual relationship

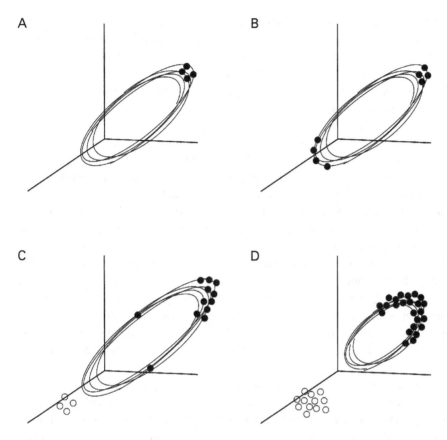

Figure 11.5
Schematic representation of isologous diversification represented in the state space of chemical compositions, where circles represent cells, and the locus shows the time course of the chemical concentrations. (A) Synchronized oscillation, (B) desynchronization by cells with an increase in their number, (C) differentiation of some cells to lose oscillation, and (D) balanced increase of two cell types.

between each intracellular dynamic and cell-to-cell interaction, a cell ensemble with robust distribution is achieved.

Indeed, this isologous diversification theory matches well with the argument in the last section and complements the missing part for multicellular organisms (i.e., differentiation of cells sharing the same genes developed from a single cell). As shown in the previous section, diverse components exist in a cell under resource limitation, and they form a complex reaction network. With this complexity, oscillatory dynamics are often generated. A type of oscillation called heteroclinic oscillation often appears, in which the concentration of some components changes from an almost null level to a high level and back to the almost null level, as is commonly observed in nonlinear population dynamics. This type of temporal evolution often appears in chemical concentration dynamics when the resource is limited. In this case, some components become extinct because of noise. With this extinction process, reactions now occur only in a subnetwork of cells from the original network. In addition, under strong cell–cell interactions, chemicals are exchanged with each other to compensate for the extinct components. Thus, mutual dependency of distinct cell types for growth is achieved, starting from cells with diverse components and complex networks.

For example, we numerically studied the models with equations 11.1 and 11.3–7 in the last section by taking an individual-based model and starting from a single cell and keeping the reaction network identical. With this, a population of cells of identical genotype (i.e., identical reaction networks) was studied. For a class of networks (taking the catalytic degree $n = 2$), we observed the above differentiation starting from irregular oscillatory dynamics. An example of this differentiation process is shown in figure 11.6 while the increase in the degree of cell diversification against the decrease in V (i.e., increasing competition for resources) is shown in figure 11.7. The localization index σ, characterizing the degree to which chemicals are concentrated on fewer components, is also plotted. Here, σ is defined as $\sigma = \sum_{j=1}^{K} x_j^2 / \sum_{j=1}^{K} x_j$; if the concentration is equally distributed with respect to all components, it is $1/K$, and, if it is concentrated in two components equally, it is $1/2$. As shown with the decrease in V and achievement of cell differentiation, σ increases, implying that the chemical abundance is concentrated in fewer components for differentiated cells.

Note that we have not imposed external fitness conditions to promote the differentiation of roles for growth. Rather, differentiation emerges as a result of a generic process for sufficiently complex intracellular reaction dynamics under conditions of resource limitation and strong cell–cell interactions. The state in which this differentiation in dynamical systems works to differentiate roles to allow for growth remains (Yamagishi et al., 2015). In this sense, the multicellular organism is regarded as a biological consolidation of such generic dynamics in terms of Stuart A. Newman (Newman et al., 2006).

Recursive Production of a Multicellular Colony through a Bottleneck

In considering the origin of multicellularity, we need to further consider how a cellular aggregate of given size is formed from which the next generation emerges. The argument so far is insufficient, as all cells interact with each other globally, and resources are distributed in any neighborhood evenly, so that there is neither a spatial structure nor an appropriate size limit. In reality, cells are located in three-dimensional space, where cell–cell interactions and resource transport from the exterior is a local process (Forgacs and Newman, 2005).

Hence, we studied a system with intracellular dynamics located on a spatial domain (Furusawa and Kaneko, 1998, 2000). Resource chemicals diffuse in from the medium and

A

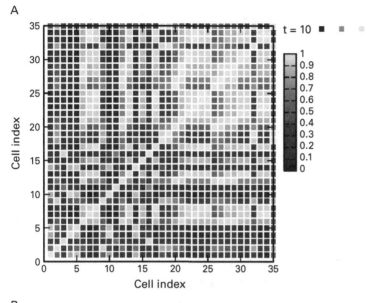

B

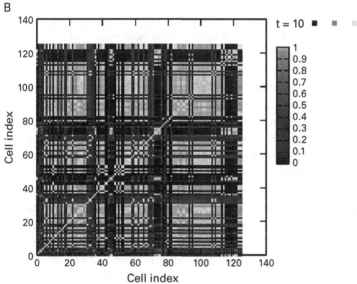

Figure 11.6
Similarity between a pair of cells that exist at each stage. Similarity between cells m and n is plotted by a color code against cell index n and m on two axes. Similarity is computed by

$$\frac{\sum_i x^i(m) x^i(n)}{\sqrt{\sum_i x^i(m)^2 \sum_i x^i(n)^2}},$$

which takes unity if the two cells have identical compositions $x^i(m) = x^i(n)$. (A) At the stage when cell number increased up to 35 and (B) the stage with a cell number of 120.

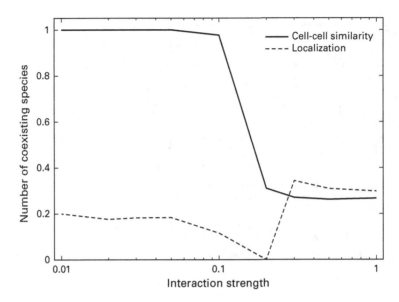

Figure 11.7
Average of cell–cell similarity (solid line) and localization index σ (dashed line) over cells, plotted as a function of interaction strength $1/V$. The similarity is defined as in figure 11.6, and the localization index is defined in the text.

are consumed by cells while some chemicals diffuse back and forth between cells and the medium. The diffusion constant in the medium is not large, such that the concentration change progresses locally, and a spatial gradient in concentration can be generated. With the differentiation of cell types, the concentrations of these diffusible chemicals can be spatially inhomogeneous and form a spatial gradient. Then, depending on the location, intracellular reactions can be different, which leads to the differentiation of cell types. Hence, differentiation of cells and spatial pattern formation reinforce each other, so that a robust pattern of cell types is generated. To achieve balanced cell growth as discussed in the previous section, the aggregate grows to maintain the ratio of each cell type, and the spatial pattern is preserved with proper scaling to the total size of the aggregate.

Here, it should be noted that a system undergoing differentiation will benefit by continued growth, as the distinct cell types can differentiate in their use of resources and help each other's growth. Indeed, for a network that leads to oscillatory dynamics and differentiation, the growth speed of the cell ensemble is not much decreased as cell number increases, even if the growth speed of individual cells is low. This is in strong contrast to cells with neither oscillatory dynamics nor differentiation, where the growth speed is drastically reduced as the cell number grows because of competition for the same resources. As numerically verified (Furusawa and Kaneko, 2000), the growth speed of an ensemble of cells is much faster in a system with differentiation than that with homogeneous cells.

In other words, K-selection at a single-cell level enables higher growth as an ensemble and satisfies r-selection at a multicellular level.

However, under limited resources, it is difficult to continue this growth while maintaining cell types and patterns. Note that resources enter through the boundary surface of the cell aggregate, while the resources needed by cells for their maintenance and growth increase with cell number (i.e., in proportion to the volume of the cell aggregate). Hence, the resource per cell decreases as the volume-to-surface ratio increases with the increased size of the cell aggregate. Ultimately, the growth–division process of cells will stop because of lack of surface area for sufficient resource uptake. To continue growing, the cell aggregate must be disintegrated to form smaller clusters at some stage in order to reduce the volume-to-surface ratio.

Hence, whether cells remain located nearby or apart is important for shaping a colony as a unit of a multicellular organism. One typical source of detachment of cells is cell motion, either by a specific directed motion following cell-to-cell force or by random motion by noise. As we are interested in the origin of multicellular organisms, an advanced mechanism for ordered motion might not exist, while random motion naturally exists. Hence, we consider the latter case here.

On the other hand, if the degree of random motion is large, cells will be disintegrated before they form an aggregate. There are two possibilities to maintain aggregate integrity. Either the adhesion between cells prohibits their independent motion, which can cause separation, or cell division is much faster than detachment by random motion. Assuming these two mechanisms for integrity, there are two corresponding possibilities for disintegration. One is the loss of cell adhesion and the other is cell death, which work at a certain stage of increasing cell number. We now discuss these two scenarios:

1. *Dependence of cell adhesion strength upon cell type (Furusawa and Kaneko, 2002):* In general, cell adhesion depends on cellular state. After differentiation to new cell types, some that have weaker adhesion may appear which, through random motion, may become detached from the aggregate (see figure 11.8). When separated, cells lose their interaction with other cells in the aggregate and can take in more resources. Hence, an original cell type is regained, which grows and divides to form a new colony. With this process, reproduction of a multicellular aggregate is achieved.

2. *Division of minority cells that keep symbiotic growth with the surrounding majority cells (see figure 11.9):* Let us consider the case in which the growth and death of cells is balanced to form an aggregate. For example, two types of cells (A and B) mutually help each other's growth, while also dying at some rate. Let us assume that cell type A has higher rates of reproduction and death than cell type B. Then, typically, there is just one cell of type B surrounded by many other cells of type A. If the two cell types mutually help the growth of each other, both must exist together, as also discussed in the "Coexistence of Diverse Cell Types" section. However, if a type A cell is distant from type B cells,

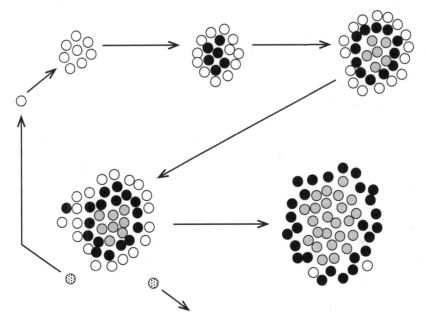

Figure 11.8
Schematic representation of the recursive production of a multicellular colony. After cell differentiation with an increase in cell number, a cell (dotted one) separates from the original colony by losing its adhesion to form a novel colony of the next generation.

the type A cell cannot survive, because it needs chemical(s) diffused from type B cells for its maintenance and growth. Hence, even without explicit cell adhesion, the cells form an aggregate. Then, with a much slower rate than the type A cell, the type B cell reproduces, and the two B cells are slowly separated by random motion. Each B cell is surrounded by type A cells to form a dumbbell-like structure. Finally, the two aggregates are split to form two colonies (see figure 11.9). By taking advantage of symbiotic relationships and the importance of the minority cell, recursive production of multicellular aggregates is possible.

Note that the above scenario was originally proposed for the growth–division process of a protocell consisting of two types of catalytic molecules (Kamimura and Kaneko, 2010). Although the levels are different between molecule–cell and cell–multicellularity, the result therein can be applied, and the condition for the growth–division process of the multicellular aggregate is obtained directly from the argument in the aforementioned publication.

In both cases 1 and 2, the next generation of multicellular aggregates is triggered by one (or few) cell(s). In case 1, the detached cell losing cell adhesion provides the next generation, while in case 2, the next generation starts from the division of a minority cell.

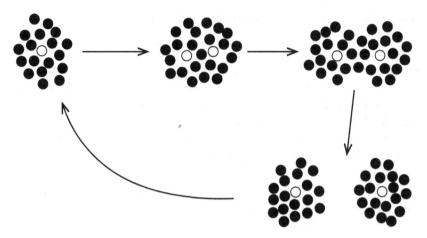

Figure 11.9
Schematic representation of the recursive production of a multicellular colony by the division of a minority cell at the center. The type A cell (black) and type B cell (white) form a mutually symbiotic relationship.

Hence, in both cases, the next generation of multicellular aggregate is not generated from many cells in the aggregate but is originated by one (or few) cell(s) (i.e., the next generation is started via a bottleneck to a single cell).

Production from a specific bottlenecked cell is important for evolvability. In the context of cell reproduction from catalytic molecules, we proposed a minority control theory based on simulations of reproducing cells (Kaneko and Yomo,1992), where replication of a molecule that is in the minority in number (e.g., DNA, whose number is much smaller than proteins) is correlated with cell reproduction. This molecule controls the cellular behavior and is essential for evolvability since the change in minority molecules exhibits dominant influence over cellular behavior.

Similarly, multicellularity from a bottlenecked single cell is important for evolvability and survivability. If a majority cell contributed to the offspring, the average character over cells would be transferred to the next offspring. In general, deleterious mutations are more frequent than advantageous ones, so that functions will be deteriorated on average. Thus, if the number of transferred cells is large and the average character is transferred, the fitness of offspring would decrease in each generation. On the other hand, for offspring generated from a single (or few) cell(s), the mutation is less frequent, so that the offspring will not exhibit decreased survivability. For deleterious mutations, the corresponding offspring will just die out while, for rare advantageous mutations, the fitness will be increased. Hence, the generation of offspring via a single (or few) bottlenecked cell(s) offers evolvability. This type of formation of the next generation from a specific cell is consistent with the separation between germ and somatic cells, where the somatic cells (majority in number) cannot produce the next generation.

Pluripotency Provided by Oscillatory Expression State and Differentiation via Cell–Cell Interactions

Based on the argument of isologous diversification in the "Cell Differentiation by Isologous Diversification" section, we proposed a possible mechanism for robust differentiation, in which stem cells irreversibly lose their pluripotency (Suzuki et al., 2011; Furusawa and Kaneko, 2012a). It is summarized as follows.

By starting from a cell with oscillatory gene expression dynamics, desynchronized irregular oscillation appears through cell–cell interactions with an increase in cell number. With a further increase in cell number, some cells switch to a novel state. Mathematically, this switching of state is understood as bifurcation (Goto and Kaneko, 2013), in which the state falls onto a different attractor upon a change in a relevant parameter, which is introduced by changes in cell–cell interactions. With the appearance of differentiated cells, the cell–cell interactions leading to the switch to the original cell type are relaxed, so that the original oscillatory state is stable. Then, cells of the original oscillatory type are reproduced by cell division. If the ratio of the original type is increased, differentiation occurs again. Hence, the present mechanism offers both differentiation and autonomous regulation of the cell type distribution since the fate of differentiation or self-renewal depends on the distribution of each cell type (Furusawa and Kaneko, 2001; Nakajima and Kaneko, 2008).

According to the present theory, the pluripotent state is characterized by (1) diversity of expressed genes, (2) larger cell–cell variation, and (3) oscillation in gene expression. The first two are generally accepted while the third point is still controversial. There is a recent report by Kobayashi et al. (2009), however, in which the concentration of the protein HeS1 shows oscillation in embryonic stem cells, but no oscillation in the differentiated cells.

If one accepts the present picture that the pluripotent state is characterized by an oscillatory state involving the interplay between several genes with mutual activation and suppression, and that the differentiated state loses activity of some genes, it is expected that pluripotency can be recovered by overexpressing the genes that lost activity in the differentiated cells. This regaining of pluripotency was predicted theoretically (Furusawa and Kaneko, 2001) while Takahashi and Yamanaka (2006) succeeded in inducing pluripotent cells by overexpressing four genes.

Finally, we offer speculation on the cancer state based on the above dynamical-systems framework (Kaneko, 2011). Consider the evolution of a multicellular organism, achieved by intra- and intercellular reaction (gene-expression) dynamics. The dynamics to achieve differentiation and growth are generally subject to noise since the number of relevant molecules in a cell is not too large and, thus, the chemical reaction process is stochastic in nature. Then, the dynamic process has to be robust to noise. As shown already (Kaneko, 2007), robustness to noise leads to robustness to mutation through evolution, to achieve a normal developmental process. Next, as the multicellular organism increases its size, each cell experiences greater limitation in nutrient conditions, and development involves stronger cell–cell interactions. With this evolution, the complexity of multicellular organisms increases, and more cell types are differentiated successively. For such complex dynamics involving expression of many genes, there will exist spurious, attracting states other than those adopted by normal cell types. These aberrant attracting states do not necessarily achieve differentiation of roles due to cell–cell interactions. In general, they will not have a stabilizing relationship with other cells. In this sense, they are "selfish," and consistency between growth of each cell and a cell society will be lost. Furthermore, these aberrant states are not the object of selection, as they do not appear in normal developmental processes. Hence, they are not necessarily robust to noise. Here, because of lack of robustness to noise, the phenotypes will be heterogeneous across cells. By recalling the above correlation between the robustness to noise and to mutation, these aberrant states will not be robust to mutation either. Then, by mutation, the state could increase robustness to acquire stability. Thus, mutations are accumulated, so that cell types increase robustness to noise and mutation.

Here, the aberrant state is generated by environmental perturbation, rather than mutation. The state loses robustness in phenotype and harmony with other cells (see also Soto and Sonnenschein, 2004; Bissell and Radisky, 2001, for relevance of cell–cell interaction to carcinogenesis). This state easily accumulates mutations. With these characteristics, one may be tempted to assign this aberrant state (attractor) as a cancer stem cell and to assume that the accumulation of mutations leads to cancer cells. In this picture, mutation is not a cause of cancer but rather a result of the appearance of cancer (stem) cells.

Summary: Possible Scenario for the Origin of Multicellularity and Future Questions

Life systems generally consist of different levels with hierarchies, from molecules and cells to organisms. States at each level change dynamically over time while consistency between different levels is achieved for a stationary biosystem. Based on the consistency between growth of an ensemble of cells and of a single cell incorporating intracellular reaction dynamics, we propose a possible scenario for the origin of multicellular organisms, as is summarized as follows:

1. *As cell number increases because of cell division, cells become crowded and resources available for growth become limited. With resource limitation and cell–cell interactions, cellular states are diversified:* As cells grow and divide, their number increases. Then, cells compete for limited resources. In this situation with slow growth, diverse components are contained and coexist within a cell. On the other hand, as cells become crowded in a confined space, the degree of cell–cell interaction is increased. Under this condition,

diverse cell types with different chemical compositions begin to coexist. At this stage, cells are diversified in both genotype and phenotype coexist.

2. *Cell–cell interaction leads to differentiation of roles for symbiotic growth:* Under these strong interactions, cellular states are no longer rigidly determined and change over time. From a single genotype, diverse phenotypes can be generated. For some genotypes, different cell types attain a symbiotic relationship by achieving a division of labor. Indeed, there are several examples of intracellular reaction dynamics that support this "isologous diversification." Even though genotypes giving rise to such reaction dynamics may be rare, colonies of cells without such reaction dynamics cease to grow and die out while those that achieve differentiation continue to grow. Thus, such genotypes that produce isologous diversification and symbiotic relationships, albeit rare, will remain.

3. *By taking advantage of state-dependent cell adhesion, a multicellular colony is recursively produced via a detached cell of a specific type:* These cells form a cell colony by forming adhesions with each other. With differentiated cell types, some cell types may lose their adhesion with neighboring cells. Thus, one or few cell(s) will be detached from the original colony, and a new colony of the next generation is generated from the detached cell(s). Then, the next generation of multicellular colony will be produced.

4. *This primitive form of germ–soma separation allows for evolution, and loss of cell differentiation potency is established from a cell type with an oscillatory expression state:* Once this prototypical multicellular organism is achieved, genetic evolution will fix the cell differentiation scheme, as a system that reproduces itself from a minority component within has greater evolvability in general. Genetic changes in minority cells producing the next generation are a major influence for selection. Thus, germ–somatic differentiation is established.

5. *Differentiation mediated by cell–cell interactions is established and results in regulation in the proportion of each cell type:* With the interplay between intracellular reaction and cell-cell interaction, both the phenotype of each cell type and the number ratio of cells of each type is robust to developmental perturbation. Evolution further enhances developmental robustness to noise and mutation while the origin of cancer cells is proposed as acquisition of a state with decreased robustness. Note that this evolutionary process is a result of multilevel dynamics, a combination of intra- and intercellular dynamics. As these developmental dynamics are highly noisy, evolution occurs to enhance robustness to noise. Diverse, distinct cell types emerge from a single cellular state, containing diverse components, and have nonstationary dynamics. These cell types achieve differentiation of roles, and they sustain their phenotype and grow collectively. Evolution stabilizes this differentiation process. Thus, a developmental process from a stem cell with pluripotency is established, where differentiated cell types take rigid and fixed states with a loss in pluripotency. Now, the present form of cell differentiation is achieved.

This viewpoint on multicellularity may open up the possibility of several research projects. On the one hand, multicellularity as a generic process under limited resources and crowded cell conditions allows for a constructive-biology approach to multicellularity. Indeed, there have been recent approaches to construct basic characteristics of multicellular organisms using bacteria (Rainey and Rainey, 2003) and yeast (Ratcliff et al., 2012), where differentiation of cells is achieved to allow growth as an ensemble of cells. It will be interesting to examine the validity of our theoretical scenario in such constructive experiments, or to propose an alternative multilevel dynamics scenario that corresponds with experimental observations.

On the other hand, our multilevel dynamics provides an integrated picture of intracellular dynamics, cell–cell interactions, differentiation, morphogenesis, and growth. It provides experimentally verifiable predictions, including a possible relationship between pluripotency and oscillatory dynamics, autonomous regulation of cell-type ratio, and robust pattern formation. Also, the hypothesis on the nonrobust cancer attractor and the accumulation of mutations as its consequence should be examined in the future.

Of course, several questions remain on the origin and evolution of multicellular organisms, which have to be understood in terms of consistency in multilevel dynamics. They include the following: (1) Which form of multicellular states is possible (logically and physio-chemically) as a multilevel stable state allowing for growth? (2) Which condition allows for isogenic multicellularity rather than heterogenic symbiosis? (3) Is there a complementary relationship between diversification in intracellular components and cell types? (4) Which type of development from single (stem) cells enables the achievement recursive production of multicellular organisms? (5) Is there evolution–development congruence as a result of robust multilevel dynamics? Or, put differently, which phenotype is evolutionarily possible? (6) Is cancer an unavoidable occurrence in multicellular systems beyond some level of complexity?

Acknowledgment

The author would like to thank Tetusya Yomo, Chikara Furusawa, Tetsuhiro Hatakeyama, Nen Saito, and Jumpei Yamagishi for illuminating discussion. This work was supported in part by a platform for Dynamic Approaches to the Living System from MEXT, Japan, and the Dynamical Micro-Scale Reaction Environment Project, Japan Science and Technology Agency.

References

Bissell, M. J., & Radisky, D. (2001). Putting tumours in context. *Nature Reviews. Cancer*, 1, 46–54.

Forgacs, G., & Newman, S. A. (2005). *Biological physics of the developing embryo*. Cambridge, UK: Cambridge University Press.

Furusawa, C., & Kaneko, K. (1998). Emergence of multicellular organisms with dynamic differentiation and spatial pattern. *Artificial Life*, 4, 79–93.

Furusawa, C., & Kaneko, K. (2000). Origin of complexity in multicellular organisms. *Physical Review Letters*, 84(26), 6130.

Furusawa, C., & Kaneko, K. (2001). Theory of robustness of irreversible differentiation in a stem cell system: Chaos hypothesis. *Journal of Theoretical Biology*, 209, 395–416.

Furusawa, C., & Kaneko, K. (2002). Origin of multicellular organisms as an inevitable consequence of dynamical systems. *Anatomical Record*, 268, 327–342.

Furusawa, C., & Kaneko, K. (2003). Zipf's law in gene expression. *Physical Review Letters*, 90(8), 088102.

Furusawa, C., & Kaneko, K. (2012 a). A dynamical-systems view of stem cell biology. *Science*, 338, 215–217.

Furusawa, C., & Kaneko, K. (2012 b). Adaptation to optimal cell growth through self-organized criticality. *Physical Review Letters*, 108(20), 208103.

Gause, G. F. (1934). *The struggle for existence*. Baltimore, MD: Williams and Willkins.

Gill, D. E., Chao, L., Perkins, S. L., & Wolf, J. B. (1995). Genetic mosaicism in plants and clonal animals. *Annual Review of Ecology and Systematics*, 26, 423–444.

Goodwin, B. C. (1963). *Temporal organizations in cells*. London: Academic Press.

Goto, Y., & Kaneko, K. (2013). Minimal model for stem-cell differentiation. *Physical Review E: Statistical, Nonlinear, and Soft Matter Physics*, 88(3), 032718.

Gurdon, J. B., Lemaire, P., & Kato, K. (1993). Community effects and related phenomena in development. *Cell*, 75, 831–834.

Kamimura, A., & Kaneko, K. (2010). Reproduction of a protocell by replication of a minority molecule in a catalytic reaction network. *Physical Review Letters*, 105(26), 268103.

Kamimura, A., & Kaneko, K. (2015). Transition to diversification by competition for resources in catalytic reaction networks. *Journal of Systems Chemistry*, 6, 5.

Kaneko, K. (2006). *Life: An introduction to complex systems biology*. Berlin, Germany: Springer.

Kaneko, K. (2007). Evolution of robustness to noise and mutation in gene expression dynamics. *PLoS One*, 2, e434.

Kaneko, K. (2011). Characterization of stem cells and cancer cells on the basis of gene expression profile stability, plasticity, and robustness. *BioEssays*, 33, 403–413.

Kaneko K. (2015). Manuscript in preparation.

Kaneko, K., & Furusawa, C. (2008). Consistency principle in biological dynamical systems. *Theory in Biosciences*, 127, 195–204.

Kaneko, K., Furusawa, C., & Yomo, T. (2015). Universal relationship in gene-expression changes for cells in steady-growth state. *Physical Review X*, X5, 011014.

Kaneko, K., & Yomo, T. (1994). Cell division, differentiation and dynamic clustering. *Physica D. Nonlinear Phenomena*, 75(1), 89–102.

Kaneko, K., & Yomo, T. (1999). Isologous diversification for robust development of cell society. *Journal of Theoretical Biology*, 199, 243–256.

Kauffman, S. A. (1993). *The origins of order: Self-organization and selection in evolution*. Oxford, UK: Oxford University Press.

Kobayashi, T., Mizuno, H., Imayoshi, I., Furusawa, C., Shirahige, K., & Kageyama, R. (2009). The cyclic gene Hes1 contributes to diverse differentiation responses of embryonic stem cells. *Genes & Development*, 23, 1870–1875.

Nakajima, A., & Kaneko, K. (2008). Regulative differentiation as bifurcation of interacting cell population. *Journal of Theoretical Biology*, 253, 779–787.

Newman, S. A., Forgacs, G., & Muller, G. B. (2006). Before programs: The physical origination of multicellular forms. *International Journal of Developmental Biology*, 50(2–3), 289.

Pianka, E. R. (1970). On *r* and *K* selection. *American Naturalist*, 104, 592–597.

Rainey, P. B., & Rainey, K. (2003). Evolution of cooperation and conflict in experimental bacterial populations. *Nature*, 425, 72–74.

Ratcliff, W. C., Denison, R. F., Borrello, M., & Travisano, M. (2012). Experimental evolution of multicellularity. *Proceedings of the National Academy of Sciences of the United States of America*, 109, 1595–1600.

Ricardo, D. (1817). *Principles of political economy and taxation*. London: John Murray.

Rosen, R. (1970). *Dynamical system theory in biology*. New York: John Wiley & Sons.

Sathe, S., Kaushik, S., Lalremruata, A., Aggarwal, R. K., Cavender, J. C., & Nanjundiah, V. (2010). Genetic heterogeneity in wild isolates of cellular slime mold social groups. *Microbial Ecology*, 60, 137–148.

Soto, A. M., & Sonnenschein, C. (2004). The somatic mutation theory of cancer: Growing problems with the paradigm? *BioEssays*, 26, 1097–1107.

Stewart, P. S., & Franklin, M. J. (2008). Physiological heterogeneity in biofilms. *Nature Reviews. Microbiology*, 6, 199–210.

Suzuki, N., Furusawa, C., & Kaneko, K. (2011). Oscillatory protein expression dynamics endows stem cells with robust differentiation potential. *PLoS One*, 6(11), e27232.

Takahashi, K., & Yamanaka, S. (2006). Induction of pluripotent stem cells from mouse embryonic and adult fibroblast cultures by defined factors. *Cell*, 126, 663–676.

Waddington, C. (1957). *The strategy of the genes*. London: George Allen and Unwin.

Yamagishi J., Saito N, and Kaneko K. (2015). Manuscript in preparation.

12 Multicellularity, the Emergence of Animal Body Plans, and the Stabilizing Role of the Egg

Stuart A. Newman

Evolutionary scenarios for the origin of multicellularity typically focus on the selective advantages of such forms relative to the ways of life of their unicellular progenitors. But while it would be difficult to rationalize the survival of multicellular forms if they were physiologically or ecologically inadequate relative to unicellular ones, there is no logical need for multicellular forms to be competitively superior to unicellular forms in order to endure. After all, both kinds of organisms can thrive in the same or similar environments; in particular, unicellular holozoans and metazoans, which are derived from common ancestors, do coexist in the modern world.

Similar considerations apply to complex body plans and organ forms, the evolutionary and developmental consequences of multicellularity. While three-layered body plans eventually came to vastly predominate over two-layered ones, each of these types has prevailed in one setting or another. The emergence of triploblasts from diploblasts, or segmented from unsegmented body plans, was, in fact, independent of any of these forms' eventual success.

The default impulse to conceptualize the emergence of biological novelties such as multicellularity or segmentation as the outcome of a competitive game stems from an assumption deeply embedded in Darwinism, in both its original and Modern Synthesis versions. Natural selection is a theory in which any major transition or large-scale change is held to come about gradually, with each incrementally different form contributing more progeny to the population, or going extinct, based on its relative fitness. The underlying assumption is the Malthusian one, that the old and new variants reside in the same ecological niche, and that the resources they are vying for are scarce (Callebaut, 2007).

Ecological niches, however, are no longer thought of as preexisting features of the natural world that organisms strive to adapt to and occupy. Instead, organisms are acknowledged to play active roles in constructing, defining, and potentially inventing their modes of life (Odling-Smee et al., 2003). Devaluing this tent of Darwinism, however, brings to the forefront the question of the origin of novelties (Müller and Newman, 2005), which in the standard account is merely an epiphenomenon of the struggle for existence.

To the nineteenth-century formulators of the theory of natural selection, the abrupt emergence of novel forms (such as those which Darwin called "sports") seemed miraculous and could not be reconciled with materialist science. New forms supplanted old ones because they were better at meeting some challenge (the doctrine of adaptationism), but they could only do so in marginal steps, because any organism with a markedly changed phenotype could not persist in the niche in which it arose (the doctrine of gradualism). Now that we know, however, that existing niche-adapted forms can thrive in environments entirely different from the ones in which they evolved (Sax et al., 2013), and that novel forms can depart from their point of origin and establish themselves in more suitable settings (Dittrich-Reed and Fitzpatrick, 2013; Rieseberg et al., 1999; Rieseberg et al., 2003), there is no need to appeal to competition to account for the survival of novel forms, even if abruptly appearing.

An implication of this new view is that traits can become adaptive after the fact, when their carriers find ways of exploiting them to their advantage (Palmer, 2004). With the link between phenotypic innovation and adaptation thus broken on the theoretical level, new explanations for generation of novelties are called for. In fact, as noted by one of its founding scientists, Ernst Mayr (Mayr, 1960), the Modern Synthesis never addressed this problem: the variations purported to be selected were so minor ("insensibly fine gradations" in Darwin's terminology; Darwin, 1859) that they did not seem to require explanation (Newman and Linde-Medina, 2013; Salazar-Ciudad, 2006). Current understanding of developmental mechanisms, however, indicates that morphological reorganization can indeed occur abruptly since the underlying material processes are plastic, typically exhibiting nonlinear dynamics and discrete alternative states (Forgacs and Newman, 2005).

Such physically based plasticity was inevitably active during the emergence of the first multicellular forms. This helps us to address the following apparent paradox: although morphogenesis and pattern formation (once embryo multicellularity has been achieved), in all present-day animals are coordinated by the activities of a highly conserved set of "toolkit" genes, the earliest stages of development (in seeming conflict with the assumption of shared mechanism of body plan origination), begin in eggs that are morphologically highly diverse, containing clade-specific arrangements of ooplasmic determinants.

In this chapter I propose to resolve this puzzle by presenting a scenario for the origin of multicellularity in the Metazoa, for the surprisingly (in view of the standard model) predictable array of morphological motifs in the various animal phyla resulting from this major evolutionary transition, and for the evolutionary basis for an interpolation of morphologically and biochemically varied egg stages early in the life cycles of these forms. The explanatory framework I propose is largely nonadaptationist and nongradualist, with physico-genetic-based plasticity of cell aggregates the driving force of morphological innovation. Events that occurred at the later-evolved egg stage of development in this scenario served to refine, channel, and stabilize developmental pathways.

Multicellular Holozoans and Protometazoans

The Metazoa—animals—belong to a phylogenetic group, the opisthokonts, which also includes the fungi and modern unicellular and transiently colonial forms. Most present-day metazoans become multicellular when cells clonally derived from a zygote or fertilized egg remain attached to one another by one or more members of the cadherin family of homophilic Ca^{2+}-dependent cell adhesion molecules (CAMs) (Halbleib and Nelson, 2006). (Complex plants and fungi achieved a multicellular state using different means; Niklas and Newman, 2013.) During subsequent development the cadherins typically change in abundance and type in those embryonic tissues that remain *epithelioid* (defined by cells being directly attached to each other). When cells disaggregate to form more loosely organized *mesenchyme* in those animals which also contain this tissue type, cohesion is maintained by the presence on the cell surfaces of integrin-family membrane proteins bound to secreted extracellular matrix (ECM) molecules (reviewed in Forgacs and Newman, 2005).

A number of critical differences distinguish the behaviors of clusters of unicellular nonfungal opisthokonts (holozoans) from those of their metazoan cousins. *Monosiga brevicollis*, a choanoflagellate (the holozoan group comprising the closest living relatives of the metazoans), is exclusively unicellular, despite having several cadherin genes (Abedin and King, 2008). A different choanoflagellate, *Salpingoeca rosetta*, does achieve a multi-cellular state, but one distinct from that of almost all metazoan embryos. It retains cyto-plasmic bridges after division (Dayel et al., 2011), and the resulting colonies do not depend on cadherins to remain intact (Dayel and King, 2014). *Capsaspora owczarzaki*, a member the filastereans, a holozoan group more distantly related to the metazoans, also exhibits a multicellular life-cycle stage, one mediated by aggregation (Sebé-Pedrós et al., 2013; see also Ruiz-Trillo, this volume). Again, cell–cell attachment is probably independent of cadherins, in this case possibly being mediated by the organism's integrins. In contrast, during animal development integrins are not generally mobilized until gastrulation (Bökel and Brown, 2002).

As discussed below, morphogenesis in early-stage metazoan embryos, leading to characteristic structural motifs such as multiple tissue layers, body cavities, and body and organ primordia elongation, depends on cells simultaneously remaining attached to each other while being independently mobile. Cadherin-based adhesion potentially affords such a balance between and persistence and release of cell–cell bonds, but it requires not only the attachment-enabling presence of calcium ions, but transient interac-tions of the cytoplasmic portion of the cadherin proteins with cytoskeletal elements includ-ing β-catenin and vinculin (Halbleib and Nelson, 2006). The genome of *M. brevicollis* contains neither of these, nor do the cadherins it specifies contain the cytoplasmic domain that mediates these interactions (Abedin and King, 2008; King et al., 2008).

While the cytoplasmic bridges and associated protein complexes of *S. rosetta* mediate persistent connections between cells in this organism, this type of adhesion is not consistent with the individual cell mobility required for further morphogenesis. It is not known whether the integrin-based attachment mode used by *C. owczarzaki* exhibits the interplay of attachment and detachment needed for animal development. More likely a common ancestor of filastereans and choanoflagellates that employed a cadherin-based mode of aggregation was in the direct line leading to metazoans.

The earliest metazoan embryo-like cell clusters depended on a repurposing ("neofunctionalization") of cadherins in such an ancestor (see figure 12.1). Acquisition of the cadherin cytoplasmic domain and β-catenin, vinculin, and other components of the submembrane apparatus would have been necessary for sufficiently strong but transient adhesion (Miller et al., 2013). Whether this occurred in gradual steps (as per the standard evolutionary model), or by horizontal transfer from other unicellular lineages (Tucker, 2013), or through other mechanisms of gene innovation (Long et al., 2013), gene-based

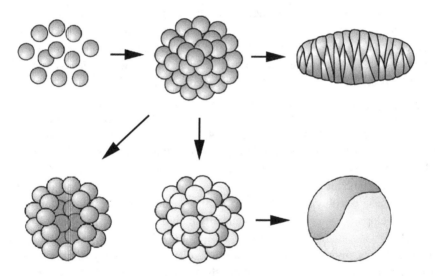

Figure 12.1
Scenario for origin and early consequences of holozoan multicellularity. Protocadherin cell surface proteins present on free-living cells acquired a homophilic adhesion function due to a rise in ambient calcium ion levels, mutations, or both. Evolution of cadherin intracellular domains produced liquid drop-like aggregates of cells that were simultaneously cohesive and independently mobile within the clusters. Lax regulation of cadherin expression led to aggregates containing subpopulations of cells with different amounts of cadherins on their surfaces, leading to sorting out of cells, and phase separation and multilayering of the respective immiscible tissues. Alternatively, induced apicobasal polarization of cells with, in some cases, nonuniform expression of cadherins on the cell surface, led to sorting of cells so that their less adhesive regions came to adjoin a central lumen, yielding hollow cell clusters. Finally, induced planar polarization of cell shape (in conjunction with apicobasal polarization of cell adhesivity) led to cell alignment and intercalation, shortening of the tissue mass along the axis parallel to the cells' long axes, and elongation of the mass parallel to cells' short axes. These self-organizational effects can occur in combination with one another in given cell clusters. See the text for further discussion.

processes would have been only part of the story since, as noted, cadherins also depend on sufficient ambient calcium ion levels to mediate cell–cell adhesion. Given that seawater concentrations of Ca^{2+} rose markedly during the Neoproterozoic (Precambrian) era (Fernandez-Busquets et al., 2009; Kazmierczak and Kempe, 2004; Petrychenko et al., 2005), the rise of the multicellular antecedents of the metazoans—protometazoans— during that period could have been relatively sudden.

Multicellular clusters formed by aggregation (as in filastereans, and in protists even more distantly related to the metazoans such the dictyostelids, and enabled in holozoans by the neofunctionalized cadherin scenario described above) are not true individuals and would thus tend toward evolutionary instability. Insofar as their constituent cells are genetically variable, they will fare differently under natural selection. There is also no way for genetically chimeric aggregates to propagate in a determinate fashion from one generation to the next. Only a mechanism that produces clonality of the constituent cells ("alignment of fitness") will ensure that multicellular forms are incipient individuals (Grosberg and Strathmann, 2007; Michod and Roze, 2001).

Later in this chapter I will describe how the evolution of an egg stage of development solved the problem of generating clonal embryos while utilizing an originally aggregative (i.e., cadherin-based) mode of cell attachment. But first I will show how the essentially universal morphological themes in animal embryos of tissue multilayering, lumen formation, and body and organ primordium elongation were consequences of precisely that mode of cell–cell adhesion.

Physico-Genetic Bases of Multilayering, Lumen Formation, and Elongation in Multicellular Aggregates

Once multicellularity was achieved in protometazoans, the new biological entities—cell clusters—had physical properties very different from those of the individual cells that composed them. Eukaryotic cells are internally structured parcels of matter surrounded by mechanically inextensible membranes. In a multicellular aggregate, if mediated by cadherins or similar CAMs, the independently mobile cells assume the role of subunits of what is, in a formal sense, a liquid droplet (Forgacs and Newman, 2005).

This interpretation is confirmed by observation of tissues of early-stage embryos, which indeed have liquid-like properties as a result of their constituent cells being independently mobile while remaining collectively cohesive (Foty et al., 1994; Steinberg, 2007b). Newly formed embryos that are clusters of cells (in contrast to some exceptions, like *Drosophila*, which start out as syncytia) are spherical and lacking in interior spaces, similarly to free liquid droplets. These "generic" morphological features result from the fact that the subunits of liquids (transiently associated molecules in the case of nonliving counterparts) constantly readjust their positions so as to minimize the surface tension of the collective. The random mobility of molecules in nonliving liquids is due to Brownian motion; in liquid-like cell aggregates it is due to undirected locomotory behavior, a default behavior of cells.

Two contiguous liquids will be immiscible if the bonds between the subunits of one of them are stronger than the bonds between its subunits and the subunits of the other liquid. The second liquid can be more, or less, cohesive than the first. If the former, it will be engulfed by the first; if the latter it will engulf it. The same rules hold for liquid-like tissues (Forgacs and Newman, 2005). Previously, it was thought that the relative quantities of CAMs on the respective cell surfaces entirely accounted for the described tissue immiscibility. Indeed, mixtures of cells differing only in levels of cadherin expression will sort out (like oil and water) into separate layers in certain experimental settings (Steinberg and Takeichi, 1994). In recent years, however, this differential adhesion hypothesis (DAH) (Steinberg, 2007a) has been superseded by the differential interfacial tension hypothesis, which recognizes a major, and sometimes decisive, role in tissue multilayering for mechanical effects exerted at the level of the cytoskeleton (Amack and Manning, 2012; Brodland, 2002; Krieg et al., 2008).

In liquid-like tissues, affinity between the cellular subunits is not due (as the DAH proposed) solely to simple physicochemical forces of adhesion. The tension the cytoskeleton exerts on the interior surface of the cell membrane influences the extent of their contact with their neighbors, affecting adhesion from the inside (Krieg et al., 2008; Maitre et al., 2012). Intracellular tension, moreover, affects the conformation of the surface proteins, enabling cadherins on one cell to bind in a specific fashion to those on an adjoining cell (Halbleib and Nelson, 2006). These effects, in conjunction with the quantities of cadherins and other CAMs on the cell surfaces (as per the DAH), regulate the strength of intercellular attachment (Amack and Manning, 2012; Heisenberg and Bellaiche, 2013).

Multilayering

In present-day metazoan embryos the development of phylum-specific body plans is initiated by the formation of immiscible layers: two in diploblasts such as ctenophores and cnidarians, and three in triploblasts such as arthropods and chordates. This multilayering is the first step in *gastrulation* (Gilbert, 2010). The allocation of cells to the different embryonic layers is well-regulated in the cell clusters of early development (e.g., morulas, blastulas, inner cell masses), often by nonuniformly distributed maternally deposited cytoplasmic factors in the eggs from which the gastrulation-competent multicellular cluster arises. In the earliest protometazoans, in contrast, such egg-patterning processes (EPPs) would not yet have evolved. Nonetheless, it seems reasonable to assume that primitive multilayered aggregates existed before the appearance of developmental signaling.

This inference is based on the plausible assumption that the levels of CAMs were variable on the surfaces of aggregation-competent populations of ancestral unicellular holozoans. This being the case, for many aggregates the self- and nonself affinities between two or more cellular subpopulations would have been different enough to promote spontaneous sorting out into distinct, or relatively distinct layers. In this way, ancient surface molecules of solitary cells, acting in an environment that permitted them to mobilize the physical force of adhesion, became not only the mediators of colony formation but, by a process of self-organization, of the generation of gastrula-like multilayered forms. The layers, due to the engulfment hierarchy imposed by differential interfacial tension, would have had spatial relationships to one another that would have been reproducible across "generations" (i.e., successive clusters derived from the original one), so long as the descendent aggregates were constituted of similar cell populations (see figure 12.1).

Lumen Formation

As noted, the default shape and topology of a cluster of cells attached to each other by CAMs will be spheroids, and lacking in internal cavities (i.e., topologically solid). However, cells can become *polarized* by expressing distinct populations of cell surface molecules along their apical–basal (A/B) axis (Karner et al., 2006a), or by exhibiting shape or cytoskeletal orientation anisotropies within the plane orthogonal to this axis ("planar cell polarity"; PCP; Karner et al., 2006b; Mlodzik, 2002). Clusters of such cells will deviate from the morphological defaults of topological solidity and sphericality (Forgacs and Newman, 2005).

Both A/B polarity and PCP are based on cell capabilities that predated metazoan evolution, but are typically elicited in animal embryos by one or another branch of the Wnt signaling pathway. The origin of the Wnt family of secreted factors is obscure—they are present in all metazoan phyla but in no unicellular opisthokonts (Ryan and Baxevanis, 2007; Suga et al., 2013). In contrast, the Frizzled (Fz) family of cell surface receptors that convey the Wnt signal to the interior of the cell in all animal phyla (Schenkelaars et al., 2015) has deeper roots in cellular evolution. Frizzled genes and proteins are found in some fungi and, even further afield phylogenetically from the metazoans, in the social amoeba *Dictyostelium discoideum* (Krishnan et al., 2012), a member of Amoebozoa, a sister clade of the group comprised of Opisthokonta + Apusozoa (Torruella et al., 2012). In all these organisms, Frizzled must have had a Wnt-independent function. The effector of the Wnt signal directly downstream from Fz is a cytoplasmic phosphoprotein known as Dishevelled (Dsh). Although Dsh itself is confined to the animals (Dillman et al., 2013), domains homologous to portions of Dsh are present in intracellular fungal proteins, where they are involved in subcellular membrane targeting functions, similarly to Dsh itself in metazoan cells (Ramanujam et al., 2012).

Notwithstanding the absence of Wnt in the ancestors of the animals, important molecular components mediating the two kinds of cell polarization had already evolved.

The cytoplasmic scaffold protein Mo25, which in multicellular animals coordinates A/B-related cell shape changes with Wnt signaling (Chien et al., 2013), has a homolog in the yeast *Schizosaccharomyces pombe*, where it is essential for polar growth and regulation of cell division (Mendoza et al., 2005). The serine/threonine kinase Lkb1, which functions in association with Mo25 in mammalian cells (Baas et al., 2004), is a downstream effector of the Wnt pathway when the latter induces A/B polarity, but can mediate this change in a Wnt-independent fashion when activated in isolated cells (reviewed in Karner et al., 2006a). This enzyme has deep phylogenetic roots, being found in *D. discoideum*, where it is essential for sporulation (Veeranki et al., 2011). Key components of A/B polarity regulation which later came to be mobilized by the Wnt pathways in animals thus predated the divergence of amoebozoans and fungi from metazoan ancestors.

When it occurs in the context of multicellular clusters, A/B polarity can generate novel morphological outcomes. If the polarity is manifested, for instance, in a specialization of the cell membrane into CAM-rich and CAM-poor portions, the cells, instead of forming a solid aggregate, will orient themselves so that their adhesive portions bind to each other while their less adhesive regions enclose an internal free space or lumen (Tsarfaty et al., 1992; Tsarfaty et al., 1994) (see figure 12.1). This topological change is analogous to what occurs on the nanoscale when phospholipid molecules in water spontaneously assemble into micelles, with their hydrophilic regions interacting with the aqueous environment on the micelle surface and their hydrophobic tails interacting with each other in the interior. On both scales, these interiorization effects are driven by the same principle of free energy minimization that causes the default morphology of a liquid droplet or an aggregate of nonpolarized cells to be spherical, and a liquid or cell aggregate of lower cohesivity to surround a more cohesive liquid or cell aggregate (Forgacs and Newman, 2005).

The described self-organizational mechanism is responsible for forming the proamniotic cavity in the egg cylinder of the early-stage mouse embryo (Bedzhov and Zernicka-Goetz, 2014). A subset of cells of the epiblast, the layer from which the entire body will derive, sorts out from the surrounding tissue and organizes into a rosette. By polar expression of anti-adhesive proteins such as podocalyxin and actin-dependent constriction, the apical portion of each of these cells loses affinity for the basolateral surfaces of its neighbors, defining a central lumen, which, when expanded by inward-directed transport of fluid, becomes the proamniotic cavity (Bedzhov and Zernicka-Goetz, 2014).

This lumen-producing effect is a generic property (or "biogeneric," since the self-organizational process involves a sophisticated cellular response function; see Newman, 2014) of clusters of prespecialized animal cells. This is indicated by the capacity of the effect to be induced by changes in the surrounding ECM even in clusters of embryonic stem cells that lack the developmental history and maternal cues of the intact embryo (Bedzhov and Zernicka-Goetz, 2014).

The identification of what appear to be small, hollow cell clusters in Chinese fossil beds of the Precambrian suggests that internal cavities or lumens were among the earliest innovations of metazoan evolution (Chen et al., 2004; Hagadorn et al., 2006). Given the inescapability of the physics of self-organization, the emergence of these forms early in multicellular evolution should not be surprising.

Tissue Elongation

Planar cell polarity, in which cells elongate along an axis orthogonal to the apicobasal axis, is, like A/B polarization, also induced by the binding of Wnts to Fz receptors, with the relay of the signal also dependent on Dsh. Unlike A/B polarization, however, which can be set into motion by the "canonical" Wnt pathway, in which the actin-binding protein β-catenin translocates from the cytoplasmic face of the plasma membrane to the cell nucleus where it serves as a transcriptional cofactor, planar polarization employs other auxiliary molecules. These include the membrane protein Van Gogh/Strabismus (Vang/Stbm; Ciruna et al., 2006) and the cytoplasmic scaffold protein Scribble (Courbard et al., 2009; also involved in A/B polarization), which mobilize the cytoskeleton in an entirely different fashion from A/B polarity. Like the latter, PCP can be activated independently of Wnt (Karner et al., 2006b) and can function cell-autonomously (Walsh et al., 2011), but in certain cases it also employs non-cell-autonomous mechanisms (Ezan and Montcouquiol, 2013). The extracellular domain of Vang/Stbm, for example, can act as a ligand for Fz receptors on cells that adjoin one another (Wu and Mlodzik, 2008).

The morphological consequences of PCP in a multicellular context can be dramatic. Elongated cells with anisotropic adhesive properties align and intercalate among one another, leading to a narrowing of the tissue mass in one direction accompanied by elongation in the orthogonal direction (figure 12.1; see Zajac et al., 2003, for computer simulations of tissue elongation based on these assumptions). This reshaping, like tissue multilayering and lumen formation, has counterparts on the molecular level: liquid crystalline polymers, which have a propensity to align with each other, form ellipsoidal nanoparticles or droplets with hyperbolic profiles rather than particles or droplets with the default spherical shapes formed by molecular subunits that do not align (Croll et al., 2006; Yang et al., 2005).

In early-stage animal embryos PCP-dependent cellular rearrangements lead to germ band extension in insects like *Drosophila* (Irvine and Wieschaus, 1994) and the phenomenon of "convergent extension," which establishes the elongated body axis during gastrulation in amphibians (Keller, 2002). Interestingly, there have been no reports of PCP-specific or -related proteins such as Vang/Stbm and Scr, or their functional domains, outside the metazoans. These genes and their associated morphological effects may be novelties in the animals, but not in all of them. Indeed, the earliest-evolving animals to exhibit the "noncanonical" (β-catenin-independent) Wnt pathway and PCP are not those with indeterminate body organization such as the simple sandwich-like placozoans or tubular and labyrinthine sponges (see Adamska, this volume), but ones that exhibit gastrulation and true body plans such as ctenophores and cnidarians (Jager et al., 2013; Kumburegama et al., 2011).

The Role of the Egg in Stabilizing the Animal Body Plan

All three of the morphogenetic consequences of holozoan multicellularity described above, multilayering, lumen formation, and tissue elongation, can take place in cell clusters that form by aggregation. Indeed, it was suggested above that the functional recruitment of the cadherins of ancestral unicellular organisms to homophilic adhesion was what got the metazoans started. An implication of this origination scenario, however, is that the earliest protometazoans (i.e., multicellular organisms with metazoan-type morphological motifs) were genetically chimeric. Even if they were formed by cells that shared genes of what would later become the developmental toolkit (e.g., cadherins, β-catenin, Wnt, Dsh, Vang/ Stbm, Scr), such cells would inevitably have exhibited allelic variation. As noted in the introduction to this chapter, this situation would not have been conducive to evolutionary stability.

Despite the presumed origin of metazoan body plan motifs as cell aggregates, tissue multilayering, lumen formation, and tissue elongation in present-day embryos takes place instead in multicellular clusters that derive developmentally from a single enlarged cell, the egg. I have termed such mid-development clusters which can be, depending on the type of organism, morulas, blastulas, blastoderms, or inner cell masses, the "morphogenetic stage" (Newman, 2011).

The scenario proposed here shifts the arena of causation in the evolution of development from gradual, undirected adaptations that potentially transformed a primordial single cell into a complex multicellular body, to the inescapable morphological outcomes of physical effects occasioned by the presence of certain toolkit genes in primordial multicellular clusters. (See also Newman and Bhat, 2008, and Newman and Bhat, 2009, for an expanded list of such physico-genetic determinants—"dynamical patterning modules"—of animal form.) The egg, notwithstanding its being the first step in the development of metazoan forms, was not the first step in their evolution in this scenario. If this hypothesis is correct, the question arises of how and why an egg stage became inserted into the life cycles of multicellular holozoans.

One common assumption that can be dispelled before taking up the origin of the egg is that heredity in animals requires germ cells or even a separate germline. Many extant animals are not, or not necessarily, derived from an egg. *Trichoplax* (the one extant placozoan), and asexually reproducing sponges and hydroids, can breed true by means of multicellular propagules or buds, which in these cases represent the morphogenetic stage of these organisms (Eitel et al., 2011; Hammel et al., 2009; Plickert et al., 2012). From this we can infer that aggregation-based protometazoans, however allelically heteroge-

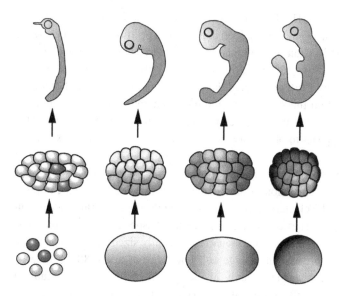

Figure 12.2
Proposed role of a proto-egg stage of development in the evolutionary consolidation of phylotype identity. In the left column a protometazoan organism arising from aggregation of unicellular holozoans, though genetically heterogeneous (indicated by different-shaded cells), produces a body (the embryonic stage is shown) that reflects the morphological motifs mediated by the toolkit genes shared by the cells in the founding population. Though the chimeric form can successfully undergo development, the morphogenetic results will vary and the form will not propagate in a reproducible fashion. In the successive columns, cell clusters (assumed to contain the same toolkit as the aggregated form) arise from the subdivision (i.e., cleavage) of an enlarged cell, the egg. Because of the molecular nonuniformities present in such cells, the cells of the morphogenetic stage, though now genetically uniform, will be differentiated by virtue of incorporating different ooplasmic determinants. Genetic uniformity will promote consistent propagation and thus evolutionary stability, while dependably generated epigenetic nonuniformity ("preparation") of the morphogenetic stage embryo will promote reliable development of subtypes (e.g., taxonomic classes) within a phylum. This is reflected in the general similarity in form of all the embryos shown, despite subtype-associated variations. (See main text and Newman, 2011, for additional details.)

neous their cells may have been, would have been able to shed single cells or small cell clusters that propagated the parental cluster's capacity to form complex morphological phenotypes, so long as the enabling toolkit genes were present in each of the cells (see figure 12.2; see also Nanjundiah, this volume, for analogous examples from the dictyostelids). While fertilization (along with meiosis) may have arisen early in the evolution of metazoans (Levin and King, 2013), it was neither required for the propagation of developmental routines nor did it solve the problem of destabilizing chimerism.

Why, then, does development almost invariably begin with large, morphologically specialized eggs? I have proposed that the egg stage of development (or proto-egg, since fertilization was not necessarily involved) arose as a consequence of nonstringent regulation of cell size and/or the confinement caused by secreted materials (Newman, 2011;

Newman, 2012; Newman, 2014). The first of these could have led to compromised cyto-kinesis in enlarged cells and thus generation of cell clusters by a primitive form of cleavage (Chen et al., 2014). Such subdivision without increase in mass, termed *palintomy*, is seen in some protozoans (Molloy et al., 2005) and in *Volvox* (Desnitskiy, 2008). The second, confinement, would have caused fully separated daughter cells to remain together within a primitive zona pellucida. Both these effects would have generated genetically uniform multicellular clusters and have thus promoted evolutionary stability of these forms. One or another of these modes appears to obligatory, and both are found to varying extents, in present-day embryos.

The first mode, the generation of cell clusters from enlarged founder cells, enabled a new kind of developmental and evolutionary stabilizing effect over and above the imposi-tion of genetic uniformity on the clusters' cells. In enlarged cells, common time-dependent metabolic and biochemical processes can take on novel spatial aspects. Calcium ion oscil-lations and transients, for example, which are seen in all typically sized cells, can generate sweeping patterns across the breadth of a large cell (Berridge et al., 1988). Similarly, proteins that would otherwise distribute evenly at the typical cell scale can assume a nonuniform profile by reaction–diffusion processes (Daniels et al., 2009).

These examples of intracellular spatiotemporal patterning of cellular activities are based on observations of present-day postfertilized eggs. However, it is important to recognize that such dynamical phenomena either happen or do not depending on the parameters (e.g., rate constants, diffusion coefficients) of the component processes and the boundary condi-tions (e.g., reactor or cell diameter) of the domain within which they occur. Ancestral enlarged cells would have been prone to spontaneously generating such intracellular pat-terns, which therefore were unlikely to have resulted from the refinements of gradual selection. The role of such EPPs (Newman, 2011) in present-day embryonic systems must be considered in light of their likelihood of having been abruptly imposed on organismal forms with preexisting developmental propensities.

Significantly, EPPs, though essential for successful development, seem to have little role in, or effect on, the characteristics of the body plan, such as the multilayering, lumen formation, and body axis elongation described above. Several lines of evidence support this conclusion. First, different species in a given animal phylum can have different EPPs. The nematodes, for example, exhibit exceptional morphological conservation despite extreme variation in EPPs. The external and internal anatomies of most of these worms (apart from size) are remarkably similar while the clade-specific ooplasmic patterns in each studied species are reproducible, precise, and, in those species that exhibit them, essential (Guedes and Priess, 1997; Prodon et al., 2004).

The egg of the nematode *Caenorhabditis elegans*, for instance, is unpolarized before it is fertilized, and reorganizes its cortical cytoplasm by flows initiated at the point of sperm entry. The resulting asymmetrical distribution of factors before the first cleavage estab-lishes the embryo's anteroposterior (A-P) axis, with the sperm entry point becoming the future embryo's posterior pole (Munro et al., 2004; Rohrschneider and Nance, 2009). In *Bursaphelenchus xylophilus*, in contrast, the sperm entry point becomes the future *anterior* pole of the embryo and the patterns of cortical cytoplasmic flow are entirely different from that in *C. elegans* (Hasegawa et al., 2004). Other nematodes (e.g., *Romanomermis culicivorax* and *Tobrilus diversipapillatus*) initiate cleavage in a symmetrical rather than asymmetrical fashion, with assignments of cell fates and A-P polarity that differ from both *C. elegans* and *B. xylophilus* (Schierenberg, 2005; Schulze and Schierenberg, 2008). Thus, although acquisition of A-P polarity is an obligatory aspect of nematode developmental anatomy, the way that it is acquired during embryogenesis, sometimes via EPPs and sometimes via morphogenetic-stage processes, seems to have little impact on the final morphological outcome.

Second, disruption or quelling of EPPs can affect the quality of developmental outcome without significantly compromising phylum-specific body plan features. In mammals, postfertilization Ca^{2+} transients are essential for egg activation and subsequent development (Kline and Kline, 1992; Swann and Yu, 2008). Calcium transients in mammalian eggs have broadly similar properties (Swann and Yu, 2008), but the onset and periodicity of the Ca^{2+} signals vary dramatically by species, with a dependence on the egg size among other parameters (Fissore et al., 1992; Macháty et al., 1997; Swann and Yu, 2008; Taylor et al., 1993). When normal patterns of Ca^{2+} oscillations were perturbed, interrupted, or circumvented by a variety of techniques, rates of embryo implantation and growth were impaired, but the major anatomical features remained intact (Kurokawa and Fissore, 2003; Miao and Williams, 2012; Ozil et al., 2006; Rogers et al., 2006; Takahashi et al., 2009).

Thus, despite the appearance of Ca^{2+} transients and cytoplasmic flows at the earliest postfertilization stages of embryogenesis, dramatic changes in the dynamics, spatiotemporal distribution, and even their presence at the egg stage, of these determinants, experimentally, within a given species, or phylogenetically, across related species, have only subtle effects on the body plans and associated motifs of the respective species. Compare this to the effects of abrogating the cell–cell interaction-mediating toolkit genes such as the cadherins or Wnts, and their related signaling pathways, at the morphogenetic stage of development. In such cases the embryo would fail to cohere, to undergo the multilayering and lumen formation essential to gastrulation, or the elongation required for body organization in most phyla.

For these reasons, I have suggested that the events that take place at the morphogenetic stage are what define phylum identity, and that anything that occurs earlier or later, however necessary for developmental success or survival, does not influence this identity (Newman, 2011). This, in turn, is proposed to resolve (in terms of developmental morphology, not solely gene expression patterns) the longstanding paradox (relative to the expectations of the Modern Synthesis) of the "embryonic hourglass" of comparative developmental biology (Duboule, 1994; Kalinka and Tomancak, 2012; Raff, 1996).

A consequence of egg stage cytoplasmic heterogeneity (manifested in spatiotemporal signaling transients and ooplasmic gradients resulting from EPPs) is that when cleavage occurs, giving rise to the cell clusters of the morphogenetic stage, the resulting population of cells, while genetically uniform, will be reproducibly heterogeneous in the cytoplasmic determinants they contain. This provides the morphogenetic stage cells with the potential to specialize epigenetically and functionally just as the embryo embarks on gastrulation and the subsequent body plan and organ generating stages of development (Van Blerkom, 2011).

Holozoan cell clusters, bound to one another via cadherin-based homophilic adhesion, were fully capable of self-organizing into complex forms containing differentiated cells, by virtue of the "interaction toolkit" and the array of transcription factors they contained prior to and coincident with the emergence of the metazoans (Newman et al., 2009). While this was all possible without an egg stage of development, the outcomes of these early forays into body plan generation would have been only approximately reproducible. The enforcement of genetic uniformity among the morphogenetic stage cells by the invention of a founder cell (i.e., proto-egg) mode of development, and the "preparation" of the cluster by the reliable imposition of cytoplasmic nonuniformities by the EPPs occurring in proto-eggs, ensured the rise of stable evolutionary and developmental trajectories from these tentative beginnings (see figure 12.2).

Conclusions

I have presented a scenario whereby the metazoans, or animals, arose in ancient populations of free-living holozoan cells by the recruitment of cell surface cadherins, which had evolved to serve single-cell functions, to a homophilic cell attachment role. This step could have been rather sudden: a simple change in ambient Ca^{2+} levels would have been sufficient to make the ancestral single cells form clumps. These aggregates would have been a first step along the path that led to the metazoans, but much more needed to happen before they were fully on their way. I will list them in order:

1. Primitive aggregates were unable to develop into anything more complex morphologically until classical cadherins, containing a cytoplasmic domain, and the associated cadherin/β-catenin/α-catenin complex which allows for the balance and interplay between adhesion and mobility that makes metazoan cell clusters behave like liquid droplets, had evolved. Metazoa is the only group that contains all these components, although related proteins arose much deeper in eukaryotic and even prokaryotic phylogeny (Miller et al., 2013). Given the antecedents, the rise of the transmembrane complex could have been gradual, but it was not necessarily so.

2. Once "liquid-drop" holozoan aggregates (protometazoans) had arisen, animal-type morphogenesis became possible. The first step might have been based on lax regulation

of cadherin expression, giving rise to differentially adherent cell subpopulations and tissue multilayering. By the physical principle of minimization of surface and interfacial tension, simple quantitative differences in cadherin abundance could lead to reproducible hierarchies of engulfment behaviors of layers constituted by sorting out of cells (Ninomiya et al., 2012; Steinberg and Takeichi, 1994). Later on, pattern-forming processes based on morphogen gradients and lateral inhibition would have made allocation of cells in a cluster to one or another adhesive state more reliable (Newman and Bhat, 2008, 2009), as would prespecification at the egg stage (Newman, 2011). But initially, self-organization based on stochastic assignment of cellular phenotypes would have been sufficient to produce gastrula-like forms.

3. Lumen formation and elongation of tissue primordia are hallmarks of embryogenesis in most animal phyla, and both effects are mediated by the Wnt pathway. The capacity of individual cells to undergo versions of apicobasal and planar cell polarization, and some components of both the canonical and noncanonical Wnt pathways, can be found well beyond the holozoans. But the global morphogenetic effects of these cell behaviors will only occur if they are induced in a global or concerted fashion. This requires a diffusible signal like Wnt itself. However there is no sign, thus far, of antecedents to this protein in either the eukaryotes or prokaryotes. Perhaps the life form in which it originally evolved is extinct. If so, Wnt may have come into existence in gradual steps and then horizontally transferred to a metazoan ancestor (Tucker, 2013; Chand et al., 2013). But other modes of new gene origination, for example, from noncoding sequences, have been documented (Long et al., 2013). However important Wnt was for animal origination and development, we simply do not know how or when it arose.

4. Protometazoans with the capacity to form multiple tissue layers and lumens, to elongate, and even to form segments, appendages, and reproducible patterns of distinct cell types by utilizing a broad array of dynamical patterning modules based on additional ancient toolkit molecules and mesoscale physical effects (Newman and Bhat, 2008, 2009) were still not genetic individuals, though they arguably had many of the defining features of animals. It took the generation of the morphogenetic stage from a single cell for protometazoans to be converted into true metazoans, or at least premetazoans. Once again, this may have been based in part on lax regulation of an ordinary cell property at these early evolutionary times, in this case cell size (Sung et al., 2013). This would have created a tendency toward a palintomic mode of division, eventually taking the form of cleavage in later-evolved forms. Alternatively, the production of an ECM by a shed cell would have promoted the continued association of daughter cells and given rise to the modern zona pellucida. These proto-eggs, actually stem-like founder cells, became true eggs when phenomena like fertilization and sequestration of the germline were invented (with the caveat, mentioned above, that a form of fertilization with weakly dimorphic gametes may have appeared at the protometazoan stage; Levin and King, 2013). But the fact that some

animals retain the capacity to reproduce asexually, by propagules (Hammel et al., 2009), and that the distinction between soma and germline appears to have evolved later than complex morphogenetic processes (Extavour and Akam, 2003), is consistent with a hypothesis asserting that the proto-egg as embryo founder cell was itself a late arrival in animal evolution.

Returning to the theoretical considerations discussed at the beginning of this chapter, and the theme of the workshop that occasioned it, it is clear that multicellularity, as a major innovation in animal evolution, and its consequences in terms of the origination of complex body plans, proceeded by a series of steps that can be rationalized on the basis of the genetic toolkits available at various points along the way and the mesoscale physical effects mobilized by the products of those genes. While the forms that "worked" were the ones that survived, this could easily have been due to occupation of novel niches after the fact rather than selection for adaptation to preexisting niches (see also Bonner, this volume).

Although gradual adaptive evolution may have operated during some phases of metazoan emergence, there was no point at which this is the only plausible scenario—many key transitions could have occurred by physical self-organization and the consequences of horizontal gene transfer. The origin of genetic individuality in the metazoans could reasonably have been a side-effect of palintomic or ECM-confined modes of cell division rather than the result of competition between unicellular and multicellular levels of selection. The Modern Synthesis has long been considered the only legitimate framework for comprehending the evolution of complexity. While it cannot be definitively falsified given the inaccessibility of the past, it does not seem to provide the only, or best, account of the phenomena discussed here.

References

Abedin, M., & King, N. (2008). The premetazoan ancestry of cadherins. *Science*, 319, 946–948.

Amack, J. D., & Manning, M. L. (2012). Knowing the boundaries: Extending the differential adhesion hypothesis in embryonic cell sorting. *Science*, 338, 212–215.

Baas, A. F., Smit, L., & Clevers, H. (2004). LKB1 tumor suppressor protein: PARtaker in cell polarity. *Trends in Cell Biology*, 14, 312–319.

Bedzhov, I., & Zernicka-Goetz, M. (2014). Self-organizing properties of mouse pluripotent cells initiate morphogenesis upon implantation. *Cell*, 156, 1032–1044.

Berridge, M. J., Cobbold, P. H., & Cuthbertson, K. S. (1988). Spatial and temporal aspects of cell signalling. *Philosophical Transactions of the Royal Society of London. Series B, Biological Sciences*, 320, 325–343.

Bökel, C., & Brown, N. H. (2002). Integrins in development: Moving on, responding to, and sticking to the extracellular matrix. *Developmental Cell*, 3, 311–321.

Brodland, G. W. (2002). The differential interfacial tension hypothesis (DITH): A comprehensive theory for the self-rearrangement of embryonic cells and tissues. *Journal of Biomechanical Engineering*, 124, 188–197.

Callebaut, W. (2007). Herbert Simon's silent revolution. *Biological Theory*, 2, 76–86.

Chand, D., de Lannoy, L., Tucker, R., & Lovejoy, D. A. (2013). Origin of chordate peptides by horizontal protozoan gene transfer in early metazoans and protists: Evolution of the teneurin C-terminal associated peptides (TCAP). *General and Comparative Endocrinology*, 188, 144–150.

Chen, J. Y., Bottjer, D. J., Oliveri, P., Dornbos, S. Q., Gao, F., Ruffins, S., et al. (2004). Small bilaterian fossils from 40 to 55 million years before the Cambrian. *Science*, 305, 218–222.

Chen, L., Xiao, S., Pang, K., Zhou, C., & Yuan, X. (2014). Cell differentiation and germ–soma separation in Ediacaran animal embryo-like fossils. *Nature*, 516, 238–241.

Chien, S. C., Brinkmann, E. M., Teuliere, J., & Garriga, G. (2013). *Caenorhabditis elegans* PIG-1/MELK acts in a conserved PAR-4/LKB1 polarity pathway to promote asymmetric neuroblast divisions. *Genetics*, 193, 897–909.

Ciruna, B., Jenny, A., Lee, D., Mlodzik, M., & Schier, A. F. (2006). Planar cell polarity signalling couples cell division and morphogenesis during neurulation. *Nature*, 439, 220–224.

Courbard, J. R., Djiane, A., Wu, J., & Mlodzik, M. (2009). The apical/basal-polarity determinant Scribble cooperates with the PCP core factor Stbm/Vang and functions as one of its effectors. *Developmental Biology*, 333, 67–77.

Croll, A. B., Massa, M. V., Matsen, M. W., & Dalnoki-Veress, K. (2006). Droplet shape of an anisotropic liquid. *Physical Review Letters*, 97, 204502.

Daniels, B. R., Perkins, E. M., Dobrowsky, T. M., Sun, S. X., & Wirtz, D. (2009). Asymmetric enrichment of PIE-1 in the *Caenorhabditis elegans* zygote mediated by binary counterdiffusion. *Journal of Cell Biology*, 184, 473–479.

Darwin, C. (1859). *On the origin of species by means of natural selection, or, The preservation of favoured races in the struggle for life*. London: J. Murray.

Dayel, M. J., Alegado, R. A., Fairclough, S. R., Levin, T. C., Nichols, S. A., McDonald, K., & King, N. (2011). Cell differentiation and morphogenesis in the colony-forming choanoflagellate *Salpingoeca rosetta*. *Developmental Biology*, 357, 73–82.

Dayel, M. J., & King, N. (2014). Prey capture and phagocytosis in the choanoflagellate *Salpingoeca rosetta*. *PLoS One*, 9, e95577.

Desnitskiy, A. G. (2008). On the problem of ecological evolution in *Volvox*. *Russian Journal of Developmental Biology*, 39, 122–124.

Dillman, A. R., Minor, P. J., & Sternberg, P. W. (2013). Origin and evolution of dishevelled. *G3 (Bethesda)*, 3, 251–262.

Dittrich-Reed, D. R., & Fitzpatrick, B. M. (2013). Transgressive hybrids as hopeful monsters. *Evolutionary Biology*, 40, 310–315.

Duboule, D. (1994). Temporal colinearity and the phylotypic progression: A basis for the stability of a vertebrate Bauplan and the evolution of morphologies through heterochrony. *Development. Supplement*, 135–142.

Eitel, M., Guidi, L., Hadrys, H., Balsamo, M., & Schierwater, B. (2011). New insights into placozoan sexual reproduction and development. *PLoS One*, 6, e19639.

Extavour, C. G., & Akam, M. (2003). Mechanisms of germ cell specification across the metazoans: Epigenesis and preformation. *Development*, 130, 5869–5884.

Ezan, J., & Montcouquiol, M. (2013). Revisiting planar cell polarity in the inner ear. *Seminars in Cell & Developmental Biology*, 24, 499–506.

Fernandez-Busquets, X., Kornig, A., Bucior, I., Burger, M. M., & Anselmetti, D. (2009). Self-recognition and Ca2+-dependent carbohydrate–carbohydrate cell adhesion provide clues to the Cambrian explosion. *Molecular Biology and Evolution*, 26, 2551–2561.

Fissore, R. A., Dobrinsky, J. R., Balise, J. J., Duby, R. T., & Robl, J. M. (1992). Patterns of intracellular Ca2+ concentrations in fertilized bovine eggs. *Biology of Reproduction*, 47, 960–969.

Forgacs, G., & Newman, S. A. (2005). *Biological physics of the developing embryo*. Cambridge, UK: Cambridge University Press.

Foty, R. A., Forgacs, G., Pfleger, C. M., & Steinberg, M. S. (1994). Liquid properties of embryonic tissues: Measurement of interfacial tensions. *Physical Review Letters*, 72, 2298–2301.

Gilbert, S. F. (2010). *Developmental biology* (9th ed.). Sunderland, MA: Sinauer Associates.

Grosberg, R. K., & Strathmann, R. (2007). The evolution of multicellularity: A minor major transition? *Annual Review of Ecology, Evolution, and Systematics*, 38, 621–654.

Guedes, S., & Priess, J. R. (1997). The *C. elegans* MEX-1 protein is present in germline blastomeres and is a P granule component. *Development*, 124, 731–739.

Hagadorn, J. W., Xiao, S., Donoghue, P. C., Bengtson, S., Gostling, N. J., Pawlowska, M., et al.. (2006). Cellular and subcellular structure of Neoproterozoic animal embryos. *Science*, 314, 291–294.

Halbleib, J. M., & Nelson, W. J. (2006). Cadherins in development: Cell adhesion, sorting, and tissue morphogenesis. *Genes & Development*, 20, 3199–3214.

Hammel, J. U., Herzen, J., Beckmann, F., & Nickel, M. (2009). Sponge budding is a spatiotemporal morphological patterning process: Insights from synchrotron radiation-based x-ray microtomography into the asexual reproduction of *Tethya wilhelma*. *Frontiers in Zoology*, 6, 19.

Hasegawa, K., Futai, K., Miwa, S., & Miwa, J. (2004). Early embryogenesis of the pinewood nematode *Bursaphelenchus xylophilus*. *Development, Growth & Differentiation*, 46, 153–161.

Heisenberg, C. P., & Bellaiche, Y. (2013). Forces in tissue morphogenesis and patterning. *Cell*, 153, 948–962.

Irvine, K. D., & Wieschaus, E. (1994). Cell intercalation during *Drosophila* germband extension and its regulation by pair-rule segmentation genes. *Development*, 120, 827–841.

Jager, M., Dayraud, C., Mialot, A., Queinnec, E., le Guyader, H., & Manuel, M. (2013). Evidence for involvement of Wnt signalling in body polarities, cell proliferation, and the neuro-sensory system in an adult ctenophore. *PLoS One*, 8, e84363.

Kalinka, A. T., & Tomancak, P. (2012). The evolution of early animal embryos: Conservation or divergence? *Trends in Ecology & Evolution*, 27, 385–393.

Karner, C., Wharton, K. A., & Carroll, T. J. (2006 a). Apical–basal polarity, Wnt signaling and vertebrate organogenesis. *Seminars in Cell & Developmental Biology*, 17, 214–222.

Karner, C., Wharton, K. A., Jr., & Carroll, T. J. (2006 b). Planar cell polarity and vertebrate organogenesis. *Seminars in Cell & Developmental Biology*, 17, 194–203.

Kazmierczak, J., & Kempe, S. (2004). Calcium build-up in the Precambrian sea: A major promoter in the evolution of eukaryotic life. In J. Seckbach (Ed.), *Origins* (pp. 329–345). Dordrecht, the Netherlands: Kluwer.

Keller, R. (2002). Shaping the vertebrate body plan by polarized embryonic cell movements. *Science*, 298, 1950–1954.

King, N., Westbrook, M. J., Young, S. L., Kuo, A., Abedin, M., Chapman, J., et al.. (2008). The genome of the choanoflagellate *Monosiga brevicollis* and the origin of metazoans. *Nature*, 451, 783–788.

Kline, D., & Kline, J. T. (1992). Repetitive calcium transients and the role of calcium in exocytosis and cell cycle activation in the mouse egg. *Developmental Biology*, 149, 80–89.

Krieg, M., Arboleda-Estudillo, Y., Puech, P. H., Kafer, J., Graner, F., Muller, D. J., & Heisenberg, C. P. (2008). Tensile forces govern germ-layer organization in zebrafish. *Nature Cell Biology*, 10, 429–436.

Krishnan, A., Almen, M. S., Fredriksson, R., & Schioth, H. B. (2012). The origin of GPCRs: Identification of mammalian like *Rhodopsin, Adhesion, Glutamate* and *Frizzled* GPCRs in fungi. *PLoS One*, 7, e29817.

Kumburegama, S., Wijesena, N., Xu, R., & Wikramanayake, A. H. (2011). Strabismus-mediated primary archenteron invagination is uncoupled from Wnt/beta-catenin-dependent endoderm cell fate specification in *Nematostella vectensis* (Anthozoa, Cnidaria): Implications for the evolution of gastrulation. *Evodevo*, 2, 2.

Kurokawa, M., & Fissore, R. A. (2003). ICSI-generated mouse zygotes exhibit altered calcium oscillations, inositol 1,4,5-trisphosphate receptor-1 down-regulation, and embryo development. *Molecular Human Reproduction*, 9, 523–533.

Levin, T. C., & King, N. (2013). Evidence for sex and recombination in the choanoflagellate *Salpingoeca rosetta*. *Current Biology*, 23, 2176–2180.

Long, M., VanKuren, N. W., Chen, S., & Vibranovski, M. D. (2013). New gene evolution: Little did we know. *Annual Review of Genetics*, 47, 307–333.

Macháty, Z., Funahashi, H., Day, B. N., & Prather, R. S. (1997). Developmental changes in the intracellular Ca^{2+} release mechanisms in porcine oocytes. *Biology of Reproduction*, 56, 921–930.

Maitre, J. L., Berthoumieux, H., Krens, S. F., Salbreux, G., Julicher, F., Paluch, E., & Heisenberg, C. P. (2012). Adhesion functions in cell sorting by mechanically coupling the cortices of adhering cells. *Science*, 338, 253–256.

Mayr E. (1960). The emergence of evolutionary novelties. In S. Tax (Ed.), *Evolution after Darwin* (pp. 349–380). Cambridge, MA: Harvard University Press.

Mendoza M, Redemann S, Brunner D. 2005. The fission yeast MO25 protein functions in polar growth and cell separation. *Eur J Cell Biol* 84: 915–926.

Miao, Y. L., & Williams, C. J. (2012). Calcium signaling in mammalian egg activation and embryo development: The influence of subcellular localization. *Molecular Reproduction and Development*, 79, 742–756.

Michod, R. E., & Roze, D. (2001). Cooperation and conflict in the evolution of multicellularity. *Heredity*, 86, 1–7.

Miller, P. W., Clarke, D. N., Weis, W. I., Lowe, C. J., & Nelson, W. J. (2013). The evolutionary origin of epithelial cell–cell adhesion mechanisms. *Current Topics in Membranes*, 72, 267–311.

Mlodzik, M. (2002). Planar cell polarization: Do the same mechanisms regulate *Drosophila* tissue polarity and vertebrate gastrulation? *Trends in Genetics*, 18, 564–571.

Molloy, D. P., Lynn, D. H., & Giamberini, L. (2005). Ophryoglena hemophaga n. sp. (*Ciliophora: Ophryoglenidae*): A parasite of the digestive gland of zebra mussels *Dreissena polymorpha*. *Diseases of Aquatic Organisms*, 65, 237–243.

Müller, G. B., & Newman, S. A. (2005). The innovation triad: An EvoDevo agenda. *Journal of Experimental Zoology. Part B, Molecular and Developmental Evolution*, 304, 487–503.

Munro, E., Nance, J., & Priess, J. R. (2004). Cortical flows powered by asymmetrical contraction transport PAR proteins to establish and maintain anterior–posterior polarity in the early *C. elegans* embryo. *Developmental Cell*, 7, 413–424.

Newman, S. A. (2011). Animal egg as evolutionary innovation: A solution to the "embryonic hourglass" puzzle. *Journal of Experimental Zoology. Part B, Molecular and Developmental Evolution*, 316, 467–483.

Newman, S. A. (2012). Physico-genetic determinants in the evolution of development. *Science*, 338, 217–219.

Newman, S. A. (2014). Why are there eggs? *Biochemical and Biophysical Research Communications, 450*, 1225–1230.

Newman, S. A, & Bhat, R. (2008). Dynamical patterning modules: Physico-genetic determinants of morphological development and evolution. *Physical Biology*, 5, 15008.

Newman, S. A., & Bhat, R. (2009). Dynamical patterning modules: A "pattern language" for development and evolution of multicellular form. *International Journal of Developmental Biology*, 53, 693–705.

Newman SA, Bhat R, Mezentseva NV. 2009. Cell state switching factors and dynamical patterning modules: complementary mediators of plasticity in development and evolution. *J Biosci* 34: 553–572.

Newman, S. A., & Linde-Medina, M. (2013). Physical determinants in the emergence and inheritance of multicellular form. *Biological Theory*, 8, 274–285.

Niklas, K. J., & Newman, S. A. (2013). The origins of multicellular organisms. *Evolution & Development*, 15, 41–52.

Ninomiya, H., David, R., Damm, E. W., Fagotto, F., Niessen, C. M., & Winklbauer, R. (2012). Cadherin-dependent differential cell adhesion in *Xenopus* causes cell sorting in vitro but not in the embryo. *Journal of Cell Science*, 125, 1877–1883.

Odling-Smee, F. J., Laland, K. N., & Feldman, M. W. (2003). *Niche construction: The neglected process in evolution.* Princeton, NJ: Princeton University Press.

Ozil, J. P., Banrezes, B., Toth, S., Pan, H., & Schultz, R. M. (2006). Ca^{2+} oscillatory pattern in fertilized mouse eggs affects gene expression and development to term. *Developmental Biology*, 300, 534–544.

Palmer, A. R. (2004). Symmetry breaking and the evolution of development. *Science*, 306, 828–833.

Petrychenko, O. Y., Peryt, T. M., & Chechel, E. I. (2005). Early Cambrian seawater chemistry from fluid inclusions in halite from Siberian evaporites. *Chemical Geology*, 219, 149–161.

Plickert, G., Frank, U., & Muller, W. A. (2012). *Hydractinia*, a pioneering model for stem cell biology and reprogramming somatic cells to pluripotency. *International Journal of Developmental Biology*, 56, 519–534.

Prodon, F., Pruliere, G., Chenevert, J., & Sardet, C. (2004). [Establishment and expression of embryonic axes: Comparisons between different model organisms]. *Médecin Sciences (Paris)*, 20, 526–538.

Raff, R. A. (1996). *The shape of life: Genes, development, and the evolution of animal form.* Chicago: University of Chicago Press.

Ramanujam, R., Yishi, X., Liu, H., & Naqvi, N. I. (2012). Structure–function analysis of Rgs1 in Magnaporthe oryzae: Role of DEP domains in subcellular targeting. *PLoS One*, 7, e41084.

Rieseberg, L. H., Archer, M. A., & Wayne, R. K. (1999). Transgressive segregation, adaptation and speciation. *Heredity*, 83(Pt 4), 363–372.

Rieseberg, L. H., Widmer, A., Arntz, A. M., & Burke, J. M. 2003). The genetic architecture necessary for transgressive segregation is common in both natural and domesticated populations. *Philosophical Transactions of the Royal Society of London. Series B, Biological Sciences*, 358, 1141–1147.

Rogers, N. T., Halet, G., Piao, Y., Carroll, J., Ko, M. S., & Swann, K. (2006). The absence of a Ca^{2+} signal during mouse egg activation can affect parthenogenetic preimplantation development, gene expression patterns, and blastocyst quality. *Reproduction*, 132, 45–57.

Rohrschneider, M. R, & Nance, J. (2009). Polarity and cell fate specification in the control of *Caenorhabditis elegans* gastrulation. *Developmental Dynamics*, 238, 789–796.

Ryan, J. F., & Baxevanis, A. D. (2007). Hox, Wnt, and the evolution of the primary body axis: Insights from the early-divergent phyla. *Biology Direct*, 2, 37.

Salazar-Ciudad, I. (2006). Developmental constraints vs. variational properties: How pattern formation can help to understand evolution and development. *Journal of Experimental Zoology. Part B, Molecular and Developmental Evolution*, 306, 107–125.

Sax, D. F., Early, R., & Bellemare, J. (2013). Niche syndromes, species extinction risks, and management under climate change. *Trends in Ecology & Evolution*, 28, 517–523.

Schenkelaars, Q., Fierro-Constain, L., Renard, E., Hill, A. L., & Borchiellini, C. (2015). Insights into Frizzled evolution and new perspectives. *Evolution & Development*, 17, 160–169.

Schierenberg, E. (2005). Unusual cleavage and gastrulation in a freshwater nematode: Developmental and phylogenetic implications. *Development Genes and Evolution*, 215, 103–108.

Schulze, J., & Schierenberg, E. (2008). Cellular pattern formation, establishment of polarity and segregation of colored cytoplasm in embryos of the nematode *Romanomermis culicivorax*. *Developmental Biology*, 315, 426–436.

Sebé-Pedrós, A., Irimia, M., Del Campo, J., Parra-Acero, H., Russ, C., Nusbaum, C., et al. (2013). Regulated aggregative multicellularity in a close unicellular relative of Metazoa. *eLife*, 2, e01287.

Steinberg, D. (2007 a). *Biomedical ethics: A multidisciplinary approach to moral issues in medicine and biology.* Hanover, NH: University Press of New England.

Steinberg, M. S. (2007 b). Differential adhesion in morphogenesis: A modern view. *Current Opinion in Genetics & Development*, 17, 281–286.

Steinberg, M. S., & Takeichi, M. (1994). Experimental specification of cell sorting, tissue spreading, and specific spatial patterning by quantitative differences in cadherin expression. *Proceedings of the National Academy of Sciences of the United States of America*, 91, 206–209.

Suga, H., Chen, Z., de Mendoza, A., Sebé-Pedrós, A., Brown, M. W., Kramer, E., et al.. (2013). The Capsaspora genome reveals a complex unicellular prehistory of animals. *Nature Communications*, 4, 2325.

Sung, Y., Tzur, A., Oh, S., Choi, W., Li, V., Dasari, R. R., et al. (2013). Size homeostasis in adherent cells studied by synthetic phase microscopy. *Proceedings of the National Academy of Sciences of the United States of America*, 110, 16687–16692.

Swann, K., & Yu, Y. (2008). The dynamics of calcium oscillations that activate mammalian eggs. *International Journal of Developmental Biology*, 52, 585–594.

Takahashi, T., Igarashi, H., Kawagoe, J., Amita, M., Hara, S., & Kurachi, H. (2009). Poor embryo development in mouse oocytes aged in vitro is associated with impaired calcium homeostasis. *Biology of Reproduction*, 80, 493–502.

Taylor, C. T., Lawrence, Y. M., Kingsland, C. R., Biljan, M. M., & Cuthbertson, K. S. (1993). Oscillations in intracellular free calcium induced by spermatozoa in human oocytes at fertilization. *Human Reproduction*, 8, 2174–2179.

Torruella, G., Derelle, R., Paps, J., Lang, B. F., Roger, A. J., Shalchian-Tabrizi, K., & Ruiz-Trillo, I. (2012). Phylogenetic relationships within the Opisthokonta based on phylogenomic analyses of conserved single-copy protein domains. *Molecular Biology and Evolution*, 29, 531–544.

Tsarfaty, I., Resau, J. H., Rulong, S., Keydar, I., Faletto, D. L., & Vande Woude, G. F. (1992). The met proto-oncogene receptor and lumen formation. *Science*, 257, 1258–1261.

Tsarfaty, I., Rong, S., Resau, J. H., Rulong, S., da Silva, P. P., & Vande Woude, G. F. (1994). The Met proto-oncogene mesenchymal to epithelial cell conversion. *Science*, 263, 98–101.

Tucker, R. P. (2013). Horizontal gene transfer in choanoflagellates. *Journal of Experimental Zoology. Part B, Molecular and Developmental Evolution*, 320, 1–9.

Van Blerkom, J. (2011). Mitochondrial function in the human oocyte and embryo and their role in developmental competence. *Mitochondrion*, 11, 797–813.

Veeranki, S., Hwang, S. H., Sun, T., Kim, B., & Kim, L. (2011). LKB1 regulates development and the stress response in *Dictyostelium*. *Developmental Biology*, 360, 351–357.

Walsh, G. S., Grant, P. K., Morgan, J. A., & Moens, C. B. (2011). Planar polarity pathway and Nance-Horan syndrome-like 1b have essential cell-autonomous functions in neuronal migration. *Development*, 138, 3033–3042.

Wu, J., & Mlodzik, M. (2008). The Frizzled extracellular domain is a ligand for Van Gogh/Stbm during nonautonomous planar cell polarity signaling. *Developmental Cell*, 15, 462–469.

Yang, Z., Huck, W. T., Clarke, S. M., Tajbakhsh, A. R., & Terentjev, E. M. (2005). Shape-memory nanoparticles from inherently non-spherical polymer colloids. *Nature Materials*, 4, 486–490.

Zajac, M., Jones, G. L., & Glazier, J. A. (2003). Simulating convergent extension by way of anisotropic differential adhesion. *Journal of Theoretical Biology*, 222, 247–259.

V   PHILOSOPHICAL ASPECTS OF MULTICELLULARITY

13 Integrating Constitution and Interaction in the Transition from Unicellular to Multicellular Organisms

Argyris Arnellos and Alvaro Moreno

Introduction and Motivation: Multicellular Organizations Exhibit Different Types and Degrees of Integration

Multicellular (MC) systems have evolved from a variety of ancestral unicellular lineages. Multicellularity has arisen independently in each of the kingdoms, and it is very likely that it arose more than once in some phyla (see Buss, 1987; Maynard Smith and Szathmáry, 1995; Bonner, 1999; Carroll, 2001; Kaiser, 2001; Medina et al., 2003, for detailed treatments of the topic). The current abundance of MC systems gives the impression that under certain conditions some key aspects of multicellularity (compared to the unicellular level) confer a selective advantage (Rokas, 2008a). Size-related aspects such as escaping predation and increasing the efficiency of food consumption (Bonner, 1999, this volume; Kaiser, 2001), aggregative or motility-related aspects conducive to better dispersal and exploration of the environment (Bonner, 1999), the spatiotemporal separation of certain biological processes and the overcoming of the limits of passive diffusion (Kaiser, 2001), the division of labor (Bonner, 2003; Michod, 2007), and the limiting of the deleterious interactions with noncooperative unicellular entities (Grosberg and Strathmann, 2007) are often mentioned as reasons for the appearance of MC phenotypes.

These "survival-related" advantages are key aspects implemented through specific biological processes characteristic of all multicellularity. In all MC systems we find routine cellular processes, such as protein synthesis, cellular transport, and metabolism, the capacity for openness to the environment for energy and nutrients acquisition, and the capacity for reproduction, as well as processes characteristic of multicellularity, such as cell–cell adhesion, cell–cell signaling pathways for intercellular communication, and transcription regulation for cellular differentiation. In general, all MC associations comprise a number of different cell types and they are characterized by specialized cell-to-cell interactions, thereby exhibiting some kind of *integration*, which in turn enables their maintenance and adaptation in the environment.

However, not all forms of multicellularity are the same. These (more or less universal) characteristics have multiple ways and degrees of realization in the various MC systems

(Niklas and Newman, 2013). Prokaryotes establish MC organizations with relatively simple morphological features, and sometimes with no clearly demarcated boundaries, using up to three or four cell types (e.g., cyanobacteria, myxobacteria, actinobacteria), and similar levels of complexity (considered as the number of distinct cell types—see Bonner, 1988; Bell and Mooers, 1997, for detailed discussions) are observed in many cases of eukaryotic multicellularity—for example, cellular slime molds (Nanjundiah, this volume). Animals, plants and fungi show a much greater diversity in cellular complexity, as well as a remarkable variety of morphologies and of underlying organizations, thus producing architecturally complex body plans with several cell types and much higher degrees of integration. Therefore, not all MC organizations achieve the same degree or type of cohesion and integration.

Notwithstanding all this diversity, the different characters of multicellularity are usually explained merely on the basis of the gradual and even measurable difference of the degree of realization of the specific biological processes characterizing all MC systems. This is mainly because of the high commonality of the biological processes characterizing the transition to different MC systems. More specifically, there is no doubt that the appearance of biological processes associated with the transition to multicellularity is correlated with a diversity of domains of proteins involved in intercellular signaling and transcription regulation. However, this characterizes the transitions to all forms of multicellularity, be it green plants compared to *Chlamydomonas* (a unicellular green alga), metazoans compared to the unicellular choanoflagellate *Monosiga brevicollis,* or comparisons between bacterial biofilms and unicellular prokaryotic lineages (see Rokas, 2008b, for a discussion and relevant references). Also, the basic elements for intercellular adhesion seem to participate (in various forms) in the life cycles of almost every unicellular organism (Niklas and Newman, 2013) while those for intercellular communication also have ancestral unicellular characters exhibiting possibly different functionalities (see King, 2004; Abedin and King, 2008, for detailed discussions). Moreover, cellular differentiation is not only a characteristic of multicellularity. Unicellular bacteria, yeast, amoeba, and algae exhibit alternative states of gene activity and morphology during their life cycles (Niklas and Newman, 2013). Last but not least, from a developmental point of view, it seems that multicellularity (independently of its many forms) is a recurrent trend of morphological evolution achieved in many different ways through a generally common scheme, that is, the mobilization of physically based patterning modules by common or unique genetic toolkits (Newman and Bhat, 2009).

In this respect, all types of MC systems (from temporary microbial aggregates and biofilms, to colonial, modular, full-fledged MC systems, and from incoherent groupings to more integrated societies of MC systems) are ensembles that manage to exhibit at the MC level all aspects characterizing their unicellular constituents. This is achieved via various forms of biological realization—but within a (more or less) common context of morphological evolution—thereby all MC systems exhibit the same characteristics and properties, however in various degrees. Actually, all these associations manage to generate and maintain relatively cohesive forms of functionally integrated organizations. In turn, the result of this integration (the various functional interactions between the cells) is externally observed as an expansion of the overall adaptive capacity of the MC associations as they tend to occupy new niches and to increase the possibilities of survival of the constituting units, and of the associations as a whole. In this respect, all MC systems, at least from a phenotypic point of view, seem to operate as *individual organisms*[1] exhibiting different degrees and types of integration.

Obviously, this explanatory framework does not bear any distinctive power with respect to the different characters of multicellularity. Firstly, the character of several composite MC systems, such as bacterial colonies, slime molds, lichens, clonal plants, and colonial invertebrates, remains unclear, and there is a great controversy whether such composite systems should be considered as individuals, organisms, parts of organisms, colonies, or just symbiotic relationships. Secondly, the different degrees and types of functional differentiation and integration exhibited by different MC systems have implications for their overall behavior in their environments. Given the rather limited degree of functional differentiation in many MC associations, their coordinated interactive capacity (the set of functional interactions the MC system can perform) is accordingly pretty low and almost negligible compared to their unicellular members. And in several other forms of multicellularity with a higher degree of functional differentiation, the interactive capacity is also rather weak—that is, it does not involve significant innovations with regard to those of the constitutive entities, let alone any capacity for diverse behaviors, and/or for their complexification. Thirdly, the more complex body plans with several cell types and relatively higher levels of anatomic and physiological complexity (i.e., tissues, organs) is not necessarily indicative of higher behavioral capacity. Certain plants can be consisted of many different cells and can grow extremely big, but their (motility-based) behavior is very simple. Finally, considering the high variability in the degree of functional differentiation between different MC systems, it is not clear which is the type and degree of integration a MC organization exhibits, and how is this reflected in its individual and/or organismal status.

Several authors have recently dealt with this problem in various ways. From a radical point of view, Dupré and O'Malley (2009) discard completely the notion of individual or/ and organism as a characteristic of biological systems. They conceive of life as a collaborative enterprise, and of metabolism as a collaborative process. Any emphasis on individual metabolisms is problematic because even large animals (being the paradigmatic biological individuals) are dependent on symbiotic associations with many other microbial or microbial-like entities, thereby constituting a very diverse group of things that both reproduce and participate in metabolic systems (wholes) on which selection acts (see also Gilbert et al., 2012). In all, seeing life in a continuum of collaboration blurs any conceptual (but also ontological) distinction between "so-called individual organisms and these larger organismal groupings of which they are parts" (Dupré and O'Malley, 2009, p. 12).

Yet, acknowledging the collective dimension of life does not mean ignoring individuality, which has, after all, played a key role in the history of biology as the locus of mechanisms, of adaptations, and of selective-evolutionary dynamics. A strong notion of an individual (metabolic) organization is conceptually coextensive to a naturalized account of other fundamental concepts of living systems like genetic information, functionality, agency, autonomy, and cognition (Ruiz-Mirazo and Moreno, 2012). But even more pragmatically, it would be difficult to make a clear-cut distinction between organisms, parts of organisms and groups of organisms, and other forms of cooperative or "ecological" networks (Ruiz-Mirazo et al., 2000; Arnellos et al., 2013). After all, the notion of organism is necessary as well for purely biological reasons, such as counting of individual biological units, comparing functional traits among different taxonomic groups, distinguishing growth from reproduction, and so on (Herron et al., 2013).

Neo-Darwinist (and adaptationist) approaches conceptualize organisms based on the notion of the "unit of selection"—that is, a biological individual on which natural selection acts because it exhibits variation, differential fitness, and heredity. A common feature of all these approaches is that the characteristic property of an organism is its *adaptation as a whole*. Queller and Strassmann (2009) suggest this can be achieved through *high actual collaboration and low actual conflict* between the unicellular constituents. In this case, *"the parts work together for the integrated whole"* (p. 3144), which should then be considered an organism. Accordingly, a blue whale, a marmoset, a sequoia tree, a giant coral, a lichen, the green algae *Volvox*, a Buchnera-aphid or a squid–vibrio symbiont, aggregations of *Myxococus xanthus* and of *Dictyostelium discoideum*, and so forth (see Queller and Strassmann, 2009, figures 1 and 3) all qualify as organisms. Folse IIIand Roughgarden (2010) suggest an MC system qualifies as an organism if it exhibits *alignment* (genetic homogeneity for the avoidance or even the elimination of noncooperative individuals) and *exportation of fitness* (the constituent parts can now be reproduced only interdependently as parts of the whole) and if it operates as an *"integrated functional agent, whose components work together in coordinated action ... thus demonstrating adaptation at the level of the whole organism"* (Ibid. p. 449). They conclude that colonies of modular organisms and genetic chimeras (such as the cellular slime mold *D. discoideum*) should not be considered organisms, while the degree of individuality possessed by plants may be said to vary continuously, thus rendering plant individuality an empirical question. Godfrey-Smith (2009) suggests a heuristic solution for the consideration of whether a particular MC system qualifies as an individual. He examines the entire issue under the umbrella of reproduction by plotting each case in a space using the three reproduction-related dimensions of bottleneck, reproductive specialization and integration.[2] Among many other rankings, based on his analysis, a buffalo herd would score zero and humans would score high in all dimensions. A sponge reproducing by fragmentation differs from a herd only in the degree of integration, while *D. discoideum* is considered to have some reproductive specialization, an intermediate level of integration, and no bottleneck. *Volvox carteri* seems to be doing well with respect to bottleneck and reproduction but lacks in integration, while the oak grown from an acorn differs from humans only in reproductive specialization (i.e., germ–soma distinctions and related phenomena).

A common ground of these neo-Darwinian and adaptationist-centered frameworks is that they reduce integration to any instantiation of other properties characteristic of multicellularity (e.g., high collaboration, low conflict, division of labor, etc.). While this explanatory strategy accounts for the *reproducible-functional* individuation[3] of MC systems, it cannot account (or can only do so in a poor way) for the strikingly different characters of multicellularity. No one would deny that higher animal multicellularity is more complex in terms of numbers of cell types than early eukaryotic multicellularity.[4] But this cannot tell us much regarding why most instances of multicellularity exhibit simple architectures with few cell types. Sponges, fruit flies, and humans belong to the same taxonomic group of metazoans, and although they differ in just a two-fold range in gene numbers, they differ markedly in terms of ways of interaction with the environment. Plants can achieve great sizes, but they lack by far the motility-based behavior of even the simplest (and much smaller) arthropod, while their behavior is even simpler than MC systems with architecturally much simpler body plans (e.g., swarms of bacteria). A lion and a plant are both considered organisms, but one cannot make two lions by cutting apart a lion and planting it in the soil. Everybody knows that a horse is more functionally integrated than a green alga. So what? The real challenge in explaining the different facets of multicellularity is that, as unanimously agreed by all approaches, all these systems mentioned above exhibit higher collaboration than conflict and they all seem to display a functional behavior at the global level. How then can we explain and distinguish between all these forms of multicellularity, and how can we relate their types and degrees of integration and the related morphological diversity to the different characters of MC systems, and consequently, to their status as individuals and/or organisms?

The problem is that even if most of the authors acknowledge the unicellular entity as the fundamental level of integrated (and therefore, organismal) organization, they also consider any MC group that persists and adapts in its environment—because of some kind of connectedness or interaction or/and communication between its unicellular constituents—as an entity that is also "integrated" in an organismal way. As we argue here, this is neither a safe result nor does it bear any explanatory power regarding the different characters of MC systems and of their biological organizations. Considering, as discussed before, that integration is a property that can be held to varying types and degrees, one needs to provide an analysis of the various forms of integration exhibited by different MC organizations as well as the tools for understanding the implication of these differences with respect to the proper characterization of MC systems.[5]

We begin in the section "The Integration of a Unicellular Organization" by arguing that unicellular organisms are entities that exhibit an integrated organization characterized by a mutual interdependence between its constitutive and interactive processes (which we call "constitutive–interactive closure"; Arnellos and Moreno, 2015). Next, in the "Three Types of Coordination of MC Interactions and the 'Constitutive–Interactive Closure' Principle" section, we discuss different types of coordination between unicellular constituents as they collectively interact with the environment, and we argue that not every form of multicellularity can support an analogous type of closure and interdependence between constitution and interaction as is found in cells. We then suggest that the required type of organismal integration in MC systems appears at the level of organizational complexity of Cnidaria. In the next section, "The Constitutive–Interactive Closure Principle Requires a Certain Type of Developmental Organization," we argue that this is not a contingent (or merely an empirical) issue, and we suggest that, on the contrary, such a special type of integration should be considered as the result of the nature and the specificities of a genuinely epithelial developmental organization exhibited by Eumetazoa. Next, in the "Some Notes on Plants" section, we briefly discuss the relationship between developmental and interactive aspects in plants, and its implications with respect to their capacity for "whole-plant" responses to the environment. We conclude by discussing the organismal status of different MC organizations with respect to their type of integration.

The Integration of a Unicellular Organization

Generically speaking, "integration" is a degree of connectedness; it is a function of numbers and intensity of any kind of interaction (relationship) between two (or more) "entities" that results in the correlation between their behaviors. The correlations can range from relatively simple ones, as these produced by linear cohesive forces, to more complex ones, as those produced by nonlinear dynamical interactions. The underlying relationship can be realized through solid links such as chemical bonds, through various kinds of force fields, through transmitted signals, and so on.[6] In this general respect, a piece of iron is "integrated" in the same sense—though to a different degree—as is an *amoeba* (Campbell, 1958; Mishler and Brandon, 1987).

Now, in the biological domain, an entity is causally integrated when the active interactions among its heterogeneous components are of such degree and type that they work in coordination with each other so as to maintain the whole. The unicellular entity exhibits such an integrated organization. In this respect, a cell is a *biological individual*; that is, a biological entity composed from diverse and distinguishable parts (membranes, catalysts, energy currencies, etc.) that are integrated into a functional organic unit (Wilson, 1999). In such a functional unit there will always be a sufficient *degree of connection* between its coordinated components to causally integrate them. However, the *type of coordination* among the different components of a cell has implications for the constitutive and operational characteristics of its integrated organization. Particularly, the cellular organization achieves its maintenance and reproduction by reinforcing the conditions for both the stability and the endogenous production of its components. What then is the type of coordination among the components of a cell that leads to the emergence of a cohesive self-maintaining and self-reproducing organization?

The construction and maintenance of a cell involves a very complex set of physico-chemical processes (metabolism) that continuously regenerate (by DNA-regulated protein biosynthesis) the decaying constitutive structures and their interrelations (i.e., the cellular organization) using an energetic gradient produced via the regulation of the maintenance of the necessary physicochemical conditions in the system through its specific interactions with the environment. Given that the appropriate physicochemical conditions are not always immediately available, robust self-construction and maintenance not only involve the control of metabolic processes (for instance, through modulation of the constitutive organization—e.g., the lac operon mechanism) but also the control of production, main-tenance, and modulation of the boundary conditions through the regulation of the system's interaction with the environment.

A characteristic example of such interaction is the chemotaxis of *Escherichia coli* bac-teria. *E. coli* is equipped with flagella that allow it to move following concentrations of sucrose (Berg, 2004). When the bacterium is close to a sugar gradient, certain proteins on its membrane detect sugar molecules and trigger metabolic pathways that change the movement of its flagellum, so that the bacterium swims up the sugar gradient (instead of performing the usual tumbling movement). Swimming and tumbling are two different interactions that are both *functional* in the sense that they contribute to the bacterium's self-maintenance in different conditions, and the bacterium can usually switch between them accordingly. The effect of these interactive functions—materialized as the changes produced between the system and its environment (e.g., more sugar) becomes input for the coordination of the interactions and consequently for the modulation of the mainte-nance of the whole bacterial organization. More specifically, a specialized subsystem regulates the metabolic regime, which in turn produces changes in motility, which modify the boundary conditions ensuring the maintenance of the metabolism. This regulation results in the coordination between membrane receptors and motor mechanisms, mediated by metabolic pathways in the bacterium. Indeed, the TCST subsystem of the bacterium operates as a *regulatory subsystem that functionally and actively coordinates the interac-tion* (Stock et al. 2000; Bijlsma and Groisman 2003). Chemotaxis is then a form of higher order control of bacterial metabolism; the regulatory control of chemotaxis is organized so as to operate over the basic metabolic functioning of the bacterium, and with the dynam-ics of the respective regulative processes exhibiting a degree of *decoupling* from those of the basic metabolic organization—otherwise the maintenance of the bacterial organization could be disrupted.[7]

Accordingly, a chemotactic interaction (and its regulation) is working well as long as the selected function (that results in a concentration gradient of metabolites across the bacterial membrane) satisfies the endogenously generated *norm* associated with the maintenance of the bacterial organization, that is, that the constitutive processes must occur as they do in order for the bacterium itself to exist (Barandiaran and Moreno, 2008). In more general terms, the set of constitutive–metabolic processes constructs and maintains the unicellular organization, and consequently, it constructs and maintains the regulatory subsystem that coordinates the organism's interaction with the environment so that the whole organization is recursively maintained.[8] As a result, in unicellular entities interactions are expressed as coordinated couplings with the environment, which are endogenously issued and regulated at the level of the cell (instead of only by a part of it) by the need to maintain the dynamic conditions necessary for the satisfaction of an also endogenously generated (constructive–metabolic) norm.

The endogenous regulation performed at the cell level that dynamically coordinates the constitutive with the interactive aspects entails *a particular type of integration* that characterizes unicellular systems. This type of integration entails that unicellular systems show a *functional* and *reciprocal* relationship between the interactive processes (and the *regulatory* subsystem that coordinates them) and the overall constructive–metabolic organization supporting materially and energetically this subsystem to the point that ultimately, it is not really possible to separate the bacterium's "doing" from its "being." In other words, the "*constitutive dimension*," which ensures both the ongoing bacterial construction and maintenance (including the regulatory subsystem that controls the boundary conditions of the former by coordinating the bacterial displacements in the environment), is ultimately dependent on the "*interactive dimension*," that is, the coordinated movements toward higher concentration of sugar. And in turn, the latter is governed/coordinated by the former (specifically, by the regulatory subsystem). As also discussed by Ruiz-Mirazo and Moreno (2012), this type of closure and mutual interdependence between constitution and interaction—via the endogenous production of regulation for both self-construction and self-maintenance through functional interactions with the environment—is characteristic of a basically autonomous biological organization, and in this respect, a unicellular entity is considered an organism.

Three Types of Coordination of MC Interactions and the "Constitutive–Interactive Closure" Principle

Whereas in cells interactions are attributable to the whole organism, in MC systems things are not that straightforward regarding their interactive dimension. The challenge regarding the integrated nature of MC interaction comes from the fact that in all cases the collective participation of distant cells (of various types) of the MC body is ensured, so that the interaction is expressed in a compact and synchronized way that is externally observed as an adaptive behavior coordinated by the whole system. However, as we recently argued in detail (Arnellos and Moreno, 2015), not all MC interactions exhibit the same type of coordination. There are three types of coordination of MC actions, which also seem to require different organizational relations with the constitutive (i.e., developmental and operational–metabolic) characteristics of an MC system. The first type of sensorimotor coordination characterizes the motility-based interactions realized by MC systems resulting from the aggregation of their constituents (e.g., *E. coli* biofilms, swarms of myxobacteria, or slime molds). Coordinated swarming is a mere result of self-organizing chemotactic movements, and the collective interaction is the emergent result of numerous local ones. In other words, coordination in swarms is realized equally by each individual cell; hence, swarming exhibits no central issuance attributable to the MC system as a whole. The fact that, as also discussed before, individual chemotactic interactions are strongly implemented in a metabolism-dependent way is also the reason why environmental factors seem to be the main drivers for the formation of the MC structures. In the case of *M. xanthus*, for instance, aspects such as sensed proximity to prey for nutrients, access to oxygen for respiration, and direct contact with the agar surface for movement are the sources that can trigger the coordination of cells (Berleman and Kirby, 2009). However, because of the high dependence on environmental factors, this type of sensorimotor coordination between cells is energetically incompatible with the development of the MC bodies. As a matter of fact, cells (e.g., of *M. xanthus* or *D. discoideum*) cannot at the same time be motile and participate in the construction and maintenance of their fruiting bodies.

In the cases of the second type of coordination, the absence of local interactions between the somatic cells is structurally compensated for by structural constraints imposed during development (e.g., the immersion of all somatic cells in the extracellular matrix [ECM] in the green alga *Volvox carteri*). This results in the enhancement of the spatiotemporal coordination between the behaviors of individual cells. In fact, phototactic swimming of *V. carteri* is a more immediate and precise interaction than swarming. However, such cases, similarly to the first type of coordination, lack the capacity for plastic and diverse actions. Any plasticity the alga could demonstrate in its phototactic swimming would be a direct function of the beat frequency of the flagella of individual cells. However, somatic cells are terminally differentiated. Also, despite the fact that the type of coordination between the somatic cells for phototactic swimming is compatible with the development and maintenance of the MC system (contrary to the case of aggregative constituents), there is again a strong dependence between motility and metabolism (Solari et al., 2006). Therefore, the way *Volvox* is integrated precludes any further diversification of its movements. Actually, *V. carteri* swims just as unicellular *Chlamydomonas* does.

The third type of coordination is characterized by a regulation as the one achieved by a nervous system (NS). Since the very beginning of its evolution, neural organization appeared as an extended network capable of producing recurrent dynamics of modulatory patterns that ensure viable sensorimotor loops with the environment. Unlike chemical signals circulating within the body, which directly interact with metabolic processes, electrical interchanges among neurons provide the possibility for recurrent interactions within the dynamic domain of the NS, in a strongly (compared to the other two types of coordination) decoupled way from metabolic dynamics (Barandiaran and Moreno, 2008). In addition, neural signals can bypass tissues and target only specific neural cells. Therefore, through signal(s) combination and regeneration, the NS provides the capacity for fast and plastic connection between sensors and effectors for the successful coordination of interactions with the environment. This type of sensorimotor coordination results in the versatile and sophisticated behavior demonstrated by the majority of metazoans. All animal species (except the Placozoa and the Porifera) can interact with their environments through a wide range of combinations of different behaviors characterized by richness in terms of the plasticity and velocity of the related MC movements.

For instance, jellyfish (phylum Cnidaria—invertebrate aquatic species) are capable of combining different parts of their bodies so as to swim softly for feeding and fast for avoiding predation, to halt and change their direction when they touch an obstacle. Some species are capable of efficient orientation, daily migration, and even of active pursuance of a prey. So, jellyfish have coordinated their motility in a way that supports the production of both diverse and precise interactions that can be appropriately and timely altered and shifted. But how is this possible, and specifically, how is the collective participation of distant cells (and of different types) ensured in the motility of the MC body so that the interaction is expressed in diverse but always compact and synchronized ways? This is possible mainly due to the jellyfish's epithelial nature.[9] Epithelialization allows distant cells to stay together by forming various sustainable effector structures capable of executing locally coordinated contractions (e.g., the bell, the tentacles, the manubrium, etc.) through the conduction of a symmetrically propagated electrical potential. The main problem of epithelial conduction in coordinating and synchronizing sheets of epitheliomuscular[10] effector cells is the lack of directional and selectively targeted propagation of impulses. In jellyfish, this limitation is overcome by nerve nets,[11] the evolution of which should be understood (as recently suggested by Keijzer et al., 2013) as the way to facilitate more rapid, and both more specific and plastic communication over longer distances of conducting cells within or connected to the epithelium. In other words, nerve nets serve to coordinate the sensory structures internalized within the epithelium with the self-organized contractile activity of effector epitheliomuscular tissues, thus creating a primitive neuromuscular structure through which sensorial input is communicated to motor–effector output in a way that allows the modulation of those contractions so as to produce different movements.[12] Therefore, behaviors like swimming, fast escaping, eating, attacking, and so forth, and their various combinations, are complex forms of sensorimotor coordination issued from the combination of signals operating both locally and globally, thereby regulating the coordination of a number of functional possibilities in order to achieve globally coherent interaction.

Beyond the globally coherent and plastic coordination of the interactions of the MC system, the NS of Cnidaria also bioregulates the whole constitutive organization via a neuroendocrine-like activity (Tarrant, 2005). Cnidarians have neither specialized endocrine organs nor a system of body fluids to distribute the secreted hormones. However, they achieve physiological regulation thanks to neuroendocrine cells synthesizing peptide-signaling molecules (acting as neurotransmitters or/and neurohormones)[13] whose circulation occurs primarily through diffusion. Neuropeptides like the FMRFamide and the GLWamide families are abundant in Cnidaria, and they play important roles in regulating a variety of developmental and physiological processes, such as the induction or inhibition of neural differentiation, neurogenesis, and oocyte maturation, the pumping activity of the body column, the tentacular movement leading to capture and ingestion of prey, morphogenesis and spawning, and the migration, settlement and metamorphosis of the Cnidarian planula larvae (see Takahashi and Hatta, 2011; Takahashi and Takeda, 2015, for reviews).

Therefore, the NS of Cnidaria not only regulates contractile epitheliomuscular tissues generating sensorimotor interactions but also operates as a bioregulator of intercellular metabolic processes so as to ensure development, growth, and global homeostasis. In a complete contrast to the other two cases, this third type of coordination is realized—as in unicellular organizations—through the endogenous production of *regulation* at the level of the whole system in a way that it also dynamically coordinates the constitutive with the interactive aspects, thus entailing *a particular type of integration* (analogous to the one exhibited by unicellular entities). While the constitutive organization is necessary for the development and maintenance of the regulatory subsystem of interaction (the NS), the latter in turn, firstly, operates as an active coordinator of interaction—decoupled from the constitutive–metabolic dynamics—and secondly, it continues to operate as a regulator of the whole constitutive organization. It is in this respect, as we have recently argued and suggested in detail, that such MC organizations exhibit the *constitutive–interactive closure principle*; that is, a closure and mutual interdependence between the constitutive and interactive processes of an MC organization so that it becomes an entity whose interactions represent its identity as an integrated whole (Arnellos and Moreno, 2015).

The Constitutive–Interactive Closure Principle Requires a Certain Type of Developmental Organization

The appearance of the NS with its wide regulatory capacities played a major role in the emergence of not only integrated but also complex (diversely and plastically executed) interactions with the environment. A necessary requirement for any form of multicellularity to allow such behaviors is that the regulatory subsystem of interactions be integrated with the constitutive organization in a way that the latter can support the material and energetic requirements of the former as well as the metabolic and biomechanical requirements of the behaviors it regulates. As a matter of fact, this specialized regulatory subsystem (the NS) is functional only in the context of the eumetazoan[14] organization and of the related body plans.

Cnidaria, for instance—which are considered the most basal Eumetazoa—are fully epithelial diploblastic animals.[15] Their neuromuscular structure is integrated in a body plan with a set of primitive and differentiated organs that provide the MC animal with the metabolic and biomechanical requirements for its behavior. Even though Cnidaria do not possess distinct organs (they have no head and no central NS, and they do not have any discrete gas exchange, secretory, or circulatory systems), there is a sufficient complexity in the body plan for the support of functionally rich behavior. In particular, the Cnidarian body plan is characterized by a very low (but necessary) degree of functional internalization[16] in the sense that the nerve cells are internalized as baso- and subepithelial nerve cells between the two epithelial layers. Due to the belt-like apical junctions the respective nerve nets are secluded from the environmental water, thus lying in an internal milieu (Koizumi, 2002). The same happens with the muscle cells, which are the most primitive in Eumetazoa, and which also lie under the epithelial cells, thus enjoying an internal environment gradually secluded from external physicochemical disturbances. Additionally, the partial internalization of the epithelial digestive cavity from the external environment gives Cnidaria the advantage in feeding on larger prey, as they can release and concentrate enzymes within an enclosed space. And it is through this epithelial gastrovascular cavity that lines the gut that Cnidaria partially carries digested food through their body. Actually, this central cavity could also be considered as a simple internal organ for digestion and partial transport of materials. Last but not least, the mesoglea is an oxygen, and for some species also a germ cells storage, and more importantly, it provides support in motility by playing the role of a primitive hydro-gelatinous skeleton supporting unified body displacements.

To sum up, the appearance of epithelial tissues, nerve nets, and a metabolically related functional differentiation, implemented through primitive organs internalized into a specific body plan, played a major role in the emergence of complex and integrated action in Cnidaria. Actually, functionally differentiated subsystems and parts are constitutively integrated in all types of eumetazoan body plans in a way that such MC organizations achieve integrated interactions. Then, the question to be asked is why this is the case—in other words, why do eumetazoan body plans fulfill the constitutive–interactive closure principle. And in a complementary way, is this type of closure and the fact that it appears at the organizational level of Cnidaria a contingent aspect? In the rest of the section we shall argue it is not, since the fulfillment of the constitutive–interactive closure strongly depends on the way an MC organization develops.

What characterizes the development of Eumetazoa are axial differentiation, cell-type specification, and regional tissue patterning. These are the result of intrinsic developmental programs comprising several developmental modules, which in turn consist of intercellular signaling mechanisms that drive, harness, and mobilize self-organizing physical processes, thereby mediating morphogenesis and cell-type differentiation on tissues and aggregates of cells (see, e.g., Davidson and Erwin, 2006; Newman, 2012; Newman and Müller, 2010).

Although Cnidaria have much simpler body plans than the bilateria, their genomes contain most of the molecular developmental toolkit that bilateria use to build their body plans (see Erwin, 2009, for details). Cnidaria also exhibit developmental similarities and/or homologies with bilaterian embryos such as the echinoderms (Martindale and Hejnol, 2009). Furthermore, in Cnidaria we find all virtually possible modes of gastrulation, an aspect that indicates a capacity for germ layer specification and formation. In general, Cnidaria share many common developmental characteristics with other Eumetazoa (trip-loblastic) animals with much more diverse cell types and complex morphologies. Most importantly, Cnidaria share with the other Eumetazoa a great part of a special developmental-genetic toolkit, which in principle enabled them to build morphologies homologous to body plans with axes differentiation, coelom-like cavities, and primitive organs (e.g., gut, pharynx, tentacles, and sense organs), and of course, a neuromuscular structure for an overall diverse behavior as a response to sensory stimulation. It is important to note that a complete epithelial tissue appears for the first time in the development of Cnidaria (Tyler, 2003; Magie and Martindale, 2008). So, besides the already mentioned important role of epithelia in locally coordinated contractility, at this point we will argue with respect to the key importance of epithelia also for understanding the specificity *of the developmental process.*

First, epithelial tissues provide the way to organize cellular movements so that the specific development of Eumetazoa is effectively driven and stabilized. Epithelia form adherent junctions that are mediated by small amounts of extracellular components and a number of integral plasma membrane proteins located on the contact surfaces and linked to cytoplasmic proteins via specific peripheral proteins (Capaldo, Farkas, and Nusrat, 2014). In this way, epithelial cells establish the condition for the implementation of complex intercellular signaling through the ECM (Pérez-Pomares and Muñoz-Chápuli, 2002; Tyler, 2003). In turn, the ECM plays a major role in intercellular signaling and consequently in development. Parts of its macromolecular complex, which is secreted by the epithelial cells, remain in communication with these cells via integrins that span the cell membranes. Hence, ECM provides a microenvironment, which through the basal lamina mediates cell movements and migration, cell differentiation, and growth, and thus tissue specificity during development and organogenesis by signaling directly to the cell nucleus (Rozario and DeSimone, 2010).

In such a highly protected and organized internal space, other cells are completely withdrawn from the environment and are internalized into the mesenchyme (initially appearing as a undifferentiated loose connective tissue).[17] In contrast to epithelial cells, mesenchymal cells are able to migrate through the ECM in other embryonic regions. They interact with the ECM in a three-dimensional intercellular environment, where they receive changing information from it, and secrete substances that modify it accordingly, sometimes irreversibly (Pérez-Pomares and Muñoz-Chápuli, 2002). The interactions of mesenchymal with epithelial cells through the ECM are modulated and controlled by the epigenetic regulatory network of intercellular signals and drive the development of organs in the MC organism (Lim and Thiery, 2012). However, for this to happen, a complete epithelium must have been previously formed. The role of epithelia in the endogenous production and further deployment of the epigenetic signaling network met in Eumetazoa is obvious from the initiation of gastrulation. The regulatory epigenetic network is deployed within an epithelial organization that provides the conditions for intercellular signaling that will drive morphogenetic movements, which in turn alter the spatiotemporal pattern of signaling in an interdependent way with these and other cellular movements (Newman, 2012; Miller and Davidson, 2013), thus resulting in an MC organization that achieves globally regulated development and constitution (Arnellos et al., 2014).

Second, epithelia allow for the physiological compartmentalization of MC bodies in discrete compartments with their own intercellular milieu sealed from external environments. The diffusion of molecules across the epithelial layer is prevented and the physiological function of the tissue is guaranteed, thus providing a specialized internal environment for the emergence of a highly elaborated intercellular signaling and of dynamic interactions between the tissues that results in the formation of complex organs (Magie and Martindale, 2008). Epithelia, then, seem to be the most important building block for creating and then regulating the conditions of the internal local milieu in which the development of MC bodies takes place, and under which its completion will be successful (Cereijido et al., 2004).

Conversely, Porifera and Placozoa, lacking a complete epithelium,[18] and unable to differentiate muscle and neural cells,[19] have a different type of development characterized by limited morphogenetic opportunities.[20] These MC systems are essentially sessile.[21] The explanation of the sharp difference between Porifera and Eumetazoa is quite challenging, especially if we consider that the genetic toolkit of sponges correlates with several important aspects of all metazoan body plans, and that sponges and Eumetazoa share many common pathways related to morphogenesis, patterning of distinct cell layers, and cell-type specification in their developmental-genetic toolkit (Adamska et al., 2011; Ereskovsky et al., 2013). Independently of these similarities, from the developmental perspective the key point, as we have argued, is that Porifera lack a complete epithelium, and consequently lack the relevant type of intercellular milieu and of the endogenously produced developmental regulation. For instance, although sponges possess several transcription factors involved in determining and differentiating muscles and nerves, they do not develop a neuromuscular system (Larroux et al., 2008).[22] In addition, sponge cell layers are found not to be homologous to eumetazoan germ layers, as they do not undergo progressive fate determination. This means that the existence of a regulatory network guiding progressive germ layer determination (through gastrulation) and the related capacity for diverse cellular differentiation and integration are characteristics of eumetazoan organizations (Nakanishi et al., 2014). This is in accordance with the fact that the highly radical expansion of rather complex signaling networks (all of them playing primary roles in the regulation of cell-type specification, axial pattern formation, and in gastrulation) was initiated at the level of organizational complexity exhibited in Cnidaria (Erwin, 2009). This expansion was concurrent with the evolution of Eumetazoa MC organizations and the diversification of their body plans.[23]

To sum up, the epithelial diploblastic organization seems to exhibit the minimal requirements for an MC organization that not only allows for the kind of development that leads to the generation of neural and muscle tissues and of organs; genuinely epithelial-based developmental organizations provide also an intercellular milieu in which cells stay together and are sealed from the environment in such a way that are able to communicate with each other so as to regulate, control, and modulate their further development as a whole, while they simultaneously interact as a whole with the environment. This is why in such MC organizations *action is at the same time distinct (decoupled) but also deeply and intimately intertwined with both construction–development and maintenance.* More specifically, an epithelial organization is required for both the endogenous production of developmental and interactive regulation at the level of the whole, or in other words, for both fully fledged organismal constitution and interaction. This is why in Cnidaria—as in all Eumetazoa—the overall MC organization exhibits a particular type and a very high degree of integration between *behavior, maintenance, development and growth, which is realized by the NS, acting as a global regulatory center within the developmental context of a genuinely epithelial (eumetazoan) organization.* This form of integration in Eumetazoa is the result of the nature of their developmental organization, due to which their constitutive and interactive dimensions are so deeply interwoven and decoupled. As genuinely epithelial organizations, since their very first developmental stages, Eumetazoa thus demonstrate an entangled relation between the functional complexity of their constitutive and interactive dimension.

Some Notes on Plants[24]

Explaining the type of integration between constitution and interaction in plants is even more challenging than in the case of sponges since, in the absence of epithelial tissues, plants—being the other primary group of multicellular eukaryotes—have evolved their developmental strategies independent of animals (Meyerowitz, 2002), while they also adapt to their environments in a different way. To a large degree, these differences depend on structural and functional differences at the cellular level.[25] Plant cells are much larger and surrounded by a rigid cellulosic wall, due to which they are immobile and cannot migrate in the plant body. They have a very different osmotic response, and they acquire turgidity due to absorption of fluids, which is the reason for the mechanical rigidity of the young tissues. While animal cells have to nourish themselves with a prey that they ingest by phagocytosis, plant cells are self-sufficient in an energetic sense by creating food inside them through photosynthesis. Since plants are both photosynthetic and fixed in place, they neither have nor can actively search for concentrated sources of energy. On the contrary, they must satisfy themselves with uncertain and sparse energy sources, whether luminous or not, and they do so by taking advantage of the rigid cell walls and their combined mechanical properties to develop large and wide surfaces for capturing that energy (Hallé, 2002).

Plants adapt to heterogeneous environments by modifying shoot and root architecture to allow optimal nutrient, water, and light capture, and they are continuously rebalancing the allocation of resources between roots and shoots in order to favor the growth of their growing parts placed in the resource-rich positions (Hutchings and de Kroon, 1994). Considering the sectorial nature of plants and its plasticity, plants adapt to unpredictable environments by changing their functional units through vascular reorientation during growth and development. A plant can then be seen as a collection of *semi-independent functional modules (physiological subunits)*[26] that engage in a within-plant competition for resources. And they do so on the basis of the competition between shoots for changing vascular contacts with the roots (Sachs et al., 1993; but see also Oborny, 2003, and Leyser, 2011). As argued and explained in detail by Sachs et al. (1993), the control of vascular differentiation is sectorial; the development of a given shoot enhances the formation of only those vascular tissues that connect it with the roots. Contacts in the immediate neighborhood, ones that have other orientations and connect with other shoots, are not enhanced. In this respect, the growth of individual shoot branches is not only and directly determined by their local light conditions or photosynthetic production, but also by their relative competitive ability for water and mineral nutrients from the root system. Overall, the most successful shoots will get better vascular contacts with the roots and thus increase their potential for growth and reproduction. And the competition reinforces itself. The better the contacts with the roots, the more rapidly a shoot develops and the more contacts with the root it monopolizes. It is according to this perspective that plants are considered as colonies or "metapopulations" with competing populations of redundant organs (White, 1979; Sachs et al., 1993; Leyser, 2011).[27]

Therefore, in plants, as modular organisms, growth and development, and the respective environmentally induced changes in the expression of traits (i.e., phenotypic plasticity) take place at the level of the semi-independent functional modules. However, semi-independency depends on the degree of integration of the responses. Indeed, in realistic plants the growth of an individual shoot or root part depends not only on the local light or nutrient condition experienced by itself but also on the environmental conditions experienced by all other parts of the same plant. Also, competition between meristems (shoots) for resources (space, light, carbon fixed by photosynthesis, minerals in solution) produces situations in which change in part of a plant could affect other parts at considerable distance (see Karban, 2008, for a review). However, contrary to animals, whose nervous and endocrine system can produce signals acting as integrators for multiple inputs in their development and behavior (see the "Three Types of Coordination of MC Interactions and the 'Constitutive–Interactive Closure' Principle" section), plants lack any controller for morphogenesis or interaction with the environment. There is no NS or any other morphological structure that could coordinate plant body parts (Novoplansky, 2002).

Adaptation of the whole plant to the environment depends, then, on the type and degree of integration between the individual semi-independent functional modules. This integration is achieved via plant hormones (e.g., auxin), which can modify the pattern of resource transport within the plant on a longer time scale than nutritious resources (Oborny, 2003), and whose self-organizing pattern of transport network allows the balanced growth of the entire shoot system under the proper conditions (Leyser, this volume). However, given that auxin transport routes ultimately trigger vascular differentiation, a longer term positive feedback is inherent in the system as the branches that export the most auxin will also develop the most vascular connectivity with the stem and thus attract the most water, nutrients, and so on (Oborny, 2003). Therefore, local responses may be modified through integration and module interactions in various ways. As de Kroon et al. (2005) explain, whole-plant plasticity is the sum of all environmentally induced modular responses, plus all interaction effects that are due to communication and integration of modules. This is how, in the absence of any integrative structure, plants achieve adaptive responses in their environments.

However, as also explained by de Kroon et al. (2005), firstly, the relevant environmental variation occurs at the scale of the semi-independent module, and not all modules perceive the same variation, secondly, even if they would do so, not all modules would respond uniformly due to the physiology of the response, and thirdly, any "whole-plant" response would be a net effect (due to the self-organizing operation of auxin) of heterogeneous ones. Therefore, whole-plant interactions with the environment seem unlikely if not impossible given the functional semi-independence of plant structures.

Conclusions

The basic form of organization of biological organisms is the cell. Whereas in a unicellular organization there is a strong form of integration between its constitutive (construction and metabolic maintenance) and interactive processes (the functional actions that the organism triggers in the environment), in an MC organization this relation can take very different forms. Therefore, contrary to the commonalities being manifested from the abstract treatment of the various approaches based on natural selection and adaptation, not all forms of multicellularity are the same. While the selective reasons approach will abstractly pick up all those MC systems that exhibit an integration that permits selection on the reproducible-functional individual, it does so in a horizontal and uniform way that hides the different types and degrees of multicellular integration.

In addition, the functional characteristics and properties of the unicellular organization are analogously but not symmetrically exported at the organizational level of MC organisms. According to the discussion in the section "The Integration of a Unicellular Organization," any unicellular action (e.g., chemotaxis)—being a form of regulatory control of metabolism—should be considered as the interactive means of the unicellular organism to display adaptation (based on modulation of metabolic processes). On the other hand, although a secondary process with respect to metabolism, chemotaxis is ultimately a biochemical extension of the self-reproducing and self-maintaining processes of the unicellular organism. Notwithstanding the operational decoupling between regulatory and metabolic processes (Bich et al., 2015), then, in unicellular systems the underlying organization of interaction and of morphological change is basically the same, since both are of biochemical nature, no matter how chemically complex and genetically specified they might be.

In MC systems, interactions with the environment come with high energetic expenses that have to be covered, especially as body size is increased. As we discussed in the section "Three Types of Coordination of MC Interactions and the 'Constitutive–Interactive Closure' Principle," not all different ways in which the MC systems deal with this problem guarantee the fulfillment of the constitutive–interactive closure principle. And in the case of an MC organization's achieving such form of closure through an NS, the organizational characteristics and properties of this particular type of multicellularity are implemented differently than in the cellular organization. Specifically, interaction and morphological change partially share the same underlying epithelial organization, and the regulatory subsystem is neither mediated by purely metabolic means nor does it directly regulate the transition between metabolic pathways. On the contrary, the NS is organized to operate as both a bioregulator of the interactive level, by functionally modulating muscles and other organs involved in sensorimotor coordination, and as a controller of intercellular metabolic processes in order to ensure proper homeostasis, development, and growth. It is in this respect that the NS operates as a *regulatory center* that integrates all the levels of an MC organization, resulting in an even more interwoven *constitutive– interactive closure* than the one exhibited by a unicellular organization (Arnellos and Moreno, 2015).

As we explained, it takes an endogenously regulated developmental organization to develop and maintain an endogenously regulated interactive organization, while the latter is also needed to maintain but also to develop the former. We consider this type of multicellularity as *strongly organismal*, since the unicellular constituents are engaging in a type of collaboration that is not expressed just as mere coordination of an adaptive group, but on the contrary, it is expressed as the identity of a new whole that achieves both organismal constitution and interaction (Arnellos et al., 2014; Arnellos and Moreno, 2015).

In all other cases, things are different. Of course, in almost all types of multicellularity there is also some collective action that induces changes in the environment that contribute to the maintenance of their constitutive identity. However, in these other cases, the MC entity can be developed independently of the interactive activity (or even be reversed toward a previous stage where the parts live autonomously—e.g., bacterial colonies). In a case where the MC system is externally provided with what it requires (energy, food, etc.), the constitutive identity can be maintained even without the activity of the system (e.g., isolated germ cells in *V. carteri* will grow and divide under euphotic conditions; Koufopanou and Bell, 1993). Furthermore, because of its colonial nature, the interactive dimension of *V. carteri* is not fully integrated into the constitutive dynamics. Swimming acts for the development of the new generation, which, on the contrary, contributes neither to the development nor to the maintenance of the source of the swimming. In general, the integration achieved by bacterial and early eukaryotic multicellularity supports a very weak interdependence between constitution and interaction.

The case of plants has its own particularity. Being sessile signifies also the necessity to confront environmental adversity and precariousness instead of fleeing from it, as done so often by animals. Whereas animals are postembryonically adapted to their environments through reversible behaviors and the respective changes in their physiology, plants adapt to their environments through the regulation of postembryonic growth and the related developmental events. In contrast to animals, plant development is a continuum, during which tissue establishment and organ formation are intimately and continuously connected to environmental adaptation. Hence, plants do not show a clear distinction between constitutive and interactive processes, since in plants it is the very development and growth that adapt to environmental changes. Or, in other words, in plants, in a more absolute way than in the unicellular case, interaction and morphological change *do* share the very same underlying organization. However, in contrast to unicellular organisms, the interactive dimension in plants—because of their colonial nature—is not symmetrically integrated into the constitutive dynamics. While in plants the constitutive norms are a property of genotypes, these norms are expressed at the level of semi-autonomous modular units, and this expression varies among modules (see also Tuomi and Vuorisalo, 1989). That is to say that the interactive dimension of plants does not demonstrate whole-plant actions, and the result of these actions does not satisfy an endogenously generated norm at the level of the whole plant.[28]

Only in Eumetazoa we see a particularly strong closure and mutual interdependence, and at the same time, an operational decoupling between the constitutive and the interactive dimension. Through its actions, the animal's interactive dimension not only allows for plastic and highly reversible behavior compared to the growth and maintenance of its soma (that in turn provide the animal with food and allow it to avoid predators and find sexual mates); it is at the same time so strongly embodied and causally entangled in the constitutive identity to the point that the one dimension cannot exist without the other outside of this selfsame system.

Acknowledgments

We are grateful to Stuart A. Newman and Karl J. Niklas for the invitation to participate and present these ideas in the KLI workshop "The Origins and Consequences of Multicellularity" in September 2014. AA would like to thank all the participants for the highly interesting discussion and the KLI for providing such a unique research environment. Alan Love is thanked for useful comments on an earlier draft.

Notes

1. Although the terms "individual" and "organisms" can mean different things, they are used interchangeably in biology. Most authors would agree that the set of biological individuals is greater than the one of organisms. However, the common conception is that an organism is an integrated biological entity, spatially separated from others, and made from interdependent parts that are integrated so that they work in coordination with each other for the proper function of the organized whole (Wilson, 1999).

2. Godfrey-Smith is clear in that the notion of integration (which he takes as a summary of features such as mutual dependence and loss of the autonomy of the parts, the extent of division of labor, and the maintenance of the boundary between the collective and its environment) is "elusive," and he accordingly aims only in coarse-grained comparisons.

3. In the selective reasons approach, what is mainly emphasized is the capacity of a MC organization to be reproducible and to be maintained in its environment. For individuality, what is considered as being reproduced is a causal relation between the parts of the system, sufficient for its maintenance. In this respect, exportation of fitness—independently of genetic composition and of the specificities of the processes of reproduction—would suffice for the job. This is what we mean by "reproducible-functional individual."

4. Actually, an increase in size is more likely to be accompanied by an increase in cell types (Bell and Mooers, 1997; Bonner, 2003).

5. We are putting forward an organizational perspective according to which the biological organization of MC systems—that is, the current causal relation between their parts—is emphasized compared to their history. The historical dimension is equally important since biological complexity requires an evolutionary process of accumulation and preservation. However, a theory of biological organization is fundamental for understanding the appearance and operation of the related evolutionary mechanisms because it is at the organizational level that the biological domain's organized complexity is fundamentally expressed. See Love (this volume) for an inclusive discussion on the necessity of a conceptual integration of the two approaches.

6. In any case, the interactions among the components of an entity are many or strong or both compared to the interactions of the entity with external components (see McShea and Venit, 2001, for a detailed discussion).

7. In metabolism-independent chemotaxis, although the TCST system is metabolically mediated, the functional operation of the regulatory subsystem is not specified or determined directly by the metabolism. Particularly, in case of high levels of repellent chemicals, the response regulator (CheY) will be phosphorylated (CheY-P) and will bind to the flagellar switch protein thus inducing tumbling behavior. High levels of attractants will significantly decrease the concentrations of the CheY-P levels, thus causing lower tumbling frequency that prolongs swimming. Even in cases of metabolism-based chemotaxis (where sensory behaviors are more directly integrated with metabolic activities than in the more standard metabolism-independent case since the signal for behavior is metabolism-dependent—e.g., changes in metabolism that result in oxygen concentration), flagellar rotation is *indirectly* influenced through the state of the metabolic intermediary of the electron-transport system by way of the chemotactic TCST (Alexandre, 2010). Overall, *E. coli* chemotaxis is regulated by the TCST system as the

selective choice of a subset of particular metabolic pathways among the available repertoire (see also van Duijn et al., 2006, for a relevant discussion).

8. Extensive research has shown that all bacterial chemosensory systems are variations on the chemotaxis TCST system. Variation and diversity of the standard two-component system are predominant in all motile Bacteria and Archaea. In general, the hallmark feature for all specialized two-component chemosensory systems is temporal regulation of signal transduction, based on which diverse bacterial organisms benefit for the control of their physiology and/or development (Kirby, 2009).

9. An epithelium is defined as a sheet of polarized cells that are joined by belt-like junctions around their apical margins and with an ECM being present only apically and basally. The formation of an epithelium is characteristic condition of MC animals, which happens early in embryonic development, and during which somatic cells are held together in a restricted space after the first divisions (Tyler, 2003).

10. Two types of cells with contractile properties are known across MC animals: true muscle cells and epitheliomuscular cells. Epitheliomuscular cells are contained within a tissue layer and are anchored to the ECM. All muscular structures described so far in Cnidaria are epitheliomuscular (Burton, 2008).

11. Cnidarians are the most primitive animals to possess an NS realized as a nerve net distributed radially around the body of the animal (Satterlie, 2002).

12. In this type of coordination, not only are sensors differentiated from effectors (contrary to the other two types), but their coordination exhibits an interesting plasticity; the several epitheliomuscular components can produce different output as a result of differing modulatory input (e.g., the contractions of the bell in normal swimming and escaping).

13. The role of neuroendocrine cells in Cnidaria is played by both sensory cells integrated into the epidermis and by subepidermal ganglion cells, where endocrine cells are epidermal epithelial cells and gastrodermal neurons. Almost all neurotransmitters, neurohormones, and nonneuronal hormones are present in Cnidaria. Most signaling molecules in Cnidaria are peptides, but there are also small nonpeptide regulators (see Leitz, 2001; Kass-Simon and Pierobon, 2007, for detailed reviews).

14. Eumetazoa is the clade comprising all animal groups except the Porifera (sponges) and Placozoa. Notwithstanding their morphological diversity, Eumetazoa are characterized by some common and fundamental body plan–related features such as a type of symmetry, diploblasty or triploblasty, coelom-like cavities or true coeloms, mouth–anus (as a single opening or separately being connected via a through gut), a neuromuscular system (from myoepithelial and nerve nets to proper muscles and centralized NS), skeletons (from pseudo-skeletons to various kinds of skeletons), various organs (from very primitive sense organs to highly complex internal organs).

15. Cnidaria are essentially composed of two epithelia: an outer epithelial layer (the ectoderm or epidermis), which contains the cnidocytes, and the endoderm or gastrodermis forming the coelenteron, which is the only body cavity—an inner digestive epithelial cavity that lines the gut and that is protected from the external environment (Tyler, 2003).

16. A common characteristic of the evolutionary developments that accompanied the explosion of behaviors in MC animals and the increase of their independency from the environment is the formation of body cavities and, as such, the creation of extra internal space for the development of organs and the internalization of physiological functions. See Rosslenbroich (2014) for a detailed discussion.

17. Mesenchymal cellular morphologies possess only spot-like junctions, they establish only transient contacts with other cells, and they do not share any type of collective organization (Tyler, 2003).

18. A complete epithelium (such as the one found in Cnidaria) has adherens junctions for structural support, septate junctions for sealing function and for controlling tissue homeostasis, gap junctions for communication between epithelial cells, and a basal lamina—a layer of collagen-rich ECM (see Magie and Martindale, 2008, for details). Sponges do not present a complete "epithelialization." First of all, they lack gap junctions. Septate

junctions are clearly identifiable only in the group of calcareous sponges (Leys and Hill, 2012), and the epithelia in desmosponges have been shown to have sealing functions (Adams et al., 2010), although septate junctional proteins have not been identified in the genome of the sponge *Amphimedon queenslandica* (Srivastava et al., 2010). Few sponges have structures similar to basal lamina, and homoscleromorphs are the only ones out of the four sponge groups to have a eumetazoan-like basal lamina (Boute et al., 1996). In general, proteins related with the formation of septate and gap junctions and of basal lamina appear to be largely eumetazoan innovations (Srivastava et al., 2010).

19. Keijzer (2015) argues there are good reasons to think that the evolution of the first NS took place as a functional and complementary division of myoepithelial cells.

20. Most sponge cells reside within the ECM to which they bind via integrins. Exhibiting mainly a mesenchymal organization, their cellular aggregations are subjected to morphogenetic capacities and dynamics different than those of epithelial tissues (see Newman et al., 2009, for details). The lack of epithelial-like mechanical integrity and elasticity in sponge tissues permit the continuous movement of individual cells so that sponges actively remodel their branched skeletal structures, thereby yielding a non–precisely determined anatomy in a body form that is shaped asymmetrically and plastically by its environment (Meroz-Fine et al., 2005).

21. Only some marine and freshwater species of sponges can move (very slowly) across the seabed through the cumulative amoeboid or crawling locomotion of the individual cells that compose their lower surfaces (Bond and Harris, 1988). Some tufted larvae (mainly in their parenchymella stage) show phototactic responses that are the result of independent responses (either photonegative or photopositive) of individual cells (operating both as sensors and effectors) in the ciliated posterior tuft of the larva (Maldonado et al., 2003).

22. On the contrary, ctenophores (i.e., comp jellies, which are also diploblastic epithelial animals like the Cnidaria) develop an NS and specify muscle cells despite the total absence of microRNAs and of important controllers of gene expression in other animals (see Moroz et al., 2014, for details).

23. The emergence of gap junctions (absent in epithelial organizations of sponges) is suggestive of the contribution of intercellular communication in the morphological diversity and elaboration of eumetazoan body plans (Abedin and King, 2010).

24. The case of plants deserves a separate treatment, but here we will try to briefly illuminate the relationship between constitutive–developmental and interactive aspects in plants, and its implications with respect to their capacity for whole-plant actions.

25. Both plants and animals are composed of eukaryotic cells, and they share several similarities at the level of fundamental molecular mechanisms, such as DNA replication, protein synthesis, membrane structure, and so forth. However, this does not mean that the cells are identical. On the contrary, there are several differences at the level of cellular integration.

26. Clonal plants achieve the highest degree of module independence, where each module develops a self-supporting shoot and root system, and therefore, the genetic individual (genet) can get fragmented into independent physiological individuals (ramets). Branches of trees seem to be relatively autonomous in their use of carbohydrates within a growing season (Watson and Casper, 1984). They cannot be as completely independent as ramets of clonal plants since they all obviously depend on the same root system for water and nutrients and on the stem for physical support. However, they can be largely independent of other branches as regards water and nutrient fluxes (Sprugel et al., 1991).

27. As explained in Sachs et al. (1993), this definition applies even to plants with one central trunk and one connection between shoot and root systems since they also have many redundant organs that carry out similar physiological roles in a semi-independent way (see Sachs et al., 1993, p. 765 for details).

28. This is not the same as suggesting that plants are less or more fully integrated than animals, which, as also mentioned by Leyser (2015), does not add much to the discussion.

References

Abedin, M., & King, N. (2008). The premetazoan ancestry of cadherins. *Science*, 319, 946–948.

Abedin, M., & King, N. (2010). Diverse evolutionary paths to cell adhesion. *Trends in Cell Biology*, 20, 734–742.

Adams, E. D. M., Goss, G., & Leys, S. (2010). Freshwater sponges have functional, sealing epithelia with high transepithelial resistance and negative transepithelial potential. *PLoS One*, 5, e15040.

Adamska, M., Degnan, B. M., Green, K., & Zwafink, C. (2011). What sponges can tell us about the evolution of developmental processes. *Zoology (Jena, Germany)*, 114, 1–10.

Alexandre, G. (2010). Coupling metabolism and chemotaxis-dependent behaviours by energy taxis receptors. *Microbiology*, 156, 2283–2293.

Arnellos, A., & Moreno, A. (2015). Multicellular agency: An organizational view. *Biology & Philosophy*, 30(3), 333–357. doi:.10.1007/s10539-015-9484-0

Arnellos, A., Moreno, A., & Ruiz-Mirazo, K. (2014). Organizational requirements for multicellular autonomy: Insights from a comparative case study. *Biology & Philosophy*, 29(6), 851–884.

Arnellos, A., Ruiz-Mirazo, K., & Moreno, A. (2013). Autonomy as a property that characterises organisms among other multicellular systems. *Contrastes (Murcia, Spain)*, 18, 357–372.

Barandiaran, X., & Moreno, A. (2008). Adaptivity: From metabolism to behavior. *Adaptive Behavior*, 16, 325–344.

Bell, G., & Mooers, A. O. (1997). Size and complexity among multicellular organisms. *Biological Journal of the Linnean Society. Linnean Society of London*, 60, 345–363.

Berg, H. C. (2004). *E. coli in motion*. New York: Springer-Verlag.

Berleman, J. E., & Kirby, J. R. (2009). Deciphering the hunting strategy of a bacterial wolfpack. *FEMS Microbiology Reviews*, 33, 942–957.

Bich, L., Mossio, M., Ruiz-Mirazo, K., & Moreno, A. (2015). Biological regulation: Controlling the system from within. *Biology and Philosophy*. doi: 10.1007/s10539-015-9497-8

Bijlsma, J. J., & Groisman, E. A. (2003). Making informed decisions: Regulatory interactions between two-component systems. *Trends in Microbiology*, 11, 359–366.

Bond, C., & Harris, A. K. (1988). Locomotion of sponges and its physical mechanism. *Journal of Experimental Zoology*, 246, 271–284.

Bonner, J. T. (1988). *The evolution of complexity by means of natural selection*. Princeton, NJ: Princeton University Press.

Bonner, J. T. (1999). The origins of multicellularity. *Integrative Biology*, 1, 27–36.

Bonner, J. T. (2003). On the origin of differentiation. *Journal of Biosciences*, 28, 523–528.

Boute, N., Exposito, J. Y., Boury-Esnault, N., Vacelet, J., Nor, N., Miyazaki, K., et al. (1996). Type IV collagen in sponges, the missing link in basement membrane ubiquity. *Biologie Cellulaire*, 88, 37–44.

Burton, P. M. (2008). Insights from diploblasts: The evolution of mesoderm and muscle. *Journal of Experimental Zoology. Part B, Molecular and Developmental Evolution*, 310, 5–14.

Buss, L. W. (1987). *The evolution of individuality*. Princeton, NJ: Princeton University Press.

Campbell, D. T. (1958). Common fate, similarity, and other indices of the status of aggregates of persons as social entities. *Behavioral Science*, 3, 14–25.

Capaldo, C. T., Farkas, A. E., & Nusrat, A. (2014). Epithelial adhesive junctions. *F1000prime Reports*, 6, 1. doi:.10.12703/P6-1

Carroll, S. B. (2001). Chance and necessity: The evolution of morphological complexity and diversity. *Nature*, 409, 1102–1109.

Cereijido, M., Contreras, R. G., & Shoshani, L. (2004). Cell adhesion, polarity, and epithelia in the dawn of metazoans. *Physiological Reviews*, 84, 1229–1262.

Davidson, E. H., & Erwin, D. H. (2006). Gene regulatory networks and the evolution of animal body plans. *Science*, 311, 796–800.

de Kroon, H., Huber, H., Stuefer, J. F., & van Groenendael, J. M. (2005). A modular concept of phenotypic plasticity in plants. *New Phytologist*, 166, 73–82.

Dupré, J., & O'Malley, M. A. (2009). Varieties of living things: Life at the intersection of lineage and metabolism. *Philosophy and Theory in Biology*, 1, e003.

Ereskovsky, A. V., Renard, E., & Borchiellini, C. (2013). Cellular and molecular processes leading to embryo formation in sponges: Evidences for high conservation of processes throughout animal evolution. *Development Genes and Evolution*, 223, 5–22.

Erwin, D. H. (2009). Early origin of the bilaterian developmental toolkit. *Philosophical Transactions of the Royal Society of London. Series B, Biological Sciences*, 364, 2253–2261.

Folse, H. J., III, & Roughgarden, J. (2010). What is an individual organism? A multilevel selection perspective. *Quarterly Review of Biology*, 85, 447–472.

Gilbert, S. F., Sapp, J., & Tauber, A. I. (2012). A symbiotic view of life: We have never been individuals. *Quarterly Review of Biology*, 87, 325–341.

Godfrey-Smith, P. (2009). *Darwinian populations and natural selection*. Oxford, UK: Oxford University Press.

Grosberg, R. K., & Strathmann, R. R. (2007). The evolution of multicellularity: A minor major transition? *Annual Review of Ecology Evolution and Systematics*, 38, 621–654.

Hallé, F. (2002). *In praise of plants*. Cambridge, UK: Timber Press.

Herron, M. D., Rashidi, A., Shelton, D. E., & Driscoll, W. W. (2013). Cellular differentiation and individuality in the 'minor' multicellular taxa. *Biological Reviews of the Cambridge Philosophical Society*, 88(4), 844–861. doi:.10.1111/brv.12031

Hutchings, M. J., & de Kroon, H. (1994). Foraging in plants: The role of morphological plasticity in resource acquisition. *Advances in Ecological Research*, 25, 159–238.

Kaiser, D. (2001). Building a multicellular organism. *Annual Review of Genetics*, 35, 103–123.

Karban, R. (2008). Plant behaviour and communication. *Ecology Letters*, 11, 727–739.

Kass-Simon, G., & Pierobon, P. (2007). Cnidarian chemical neurotransmission, an updated overview. *Comparative Biochemistry and Physiology. Part A, Molecular & Integrative Physiology*, 146, 9–25.

Keijzer, F. (2015). Moving and sensing without input and output: Early nervous systems and the origins of the animal sensorimotor organization. *Biology & Philosophy*. doi:.10.1007/s10539-015-9483-1

Keijzer, F., van Duijn, M., & Lyon, P. (2013). What nervous systems do: Early evolution, input–output, and the skin brain thesis. *Adaptive Behavior*, 21, 67–85.

King, N. (2004). The unicellular ancestry of animal development. *Developmental Cell*, 7, 313–325.

Kirby, J. R. (2009). Chemotaxis-like regulatory systems: Unique roles in diverse bacteria. *Annual Review of Microbiology*, 63, 45–59.

Koizumi, O. (2002). Developmental neurobiology of hydra, a model animal of cnidarians. *Canadian Journal of Zoology*, 80, 1678–1689.

Koufopanou, V., & Bell, G. (1993). Soma and germ: An experimental approach using *Volvox*. *Proceedings of the Royal Society of London. Series B, Biological Sciences*, 254, 107–113.

Larroux, C., Luke, G. N., Koopman, P., Rokhsar, D., Shimeld, S. M., & Degnan, B. M. (2008). Genesis and expansion of metazoan transcription factor gene classes. *Molecular Biology and Evolution*, 25, 980–996. doi:.10.1093/molbev/msn047

Leitz, T. (2001). Endocrinology of the Cnidaria: State of the art. *Zoology—Analysis of Complex Systems*, 103, 202–221.

Leys, S. P., & Hill, A. (2012). The physiology and molecular biology of sponge tissues. *Advances in Marine Biology*, 62, 1–56.

Leyser, O. (2011). Auxin, self-organisation, and the colonial nature of plants. *Current Biology*, 21, R331–R337.

Lim, J., & Thiery, J. P. (2012). Epithelial–mesenchymal transitions: Insights from development. *Development*, 139, 3471–3486.

Magie, C. R., & Martindale, M. Q. (2008). Cell–cell adhesion in the Cnidaria: Insights into the evolution of tissue morphogenesis. *Biological Bulletin*, 214, 218–232.

Maldonado, M., Durfort, M., McCarthy, D. A., & Young, C. M. (2003). The cellular basis of photobehavior in the tufted parenchymella larva of demosponges. *Marine Biology (Berlin)*, 143, 427–441.

Martindale, M. Q., & Hejnol, A. (2009). A developmental perspective: Changes in the position of the blastopore during bilaterian evolution. *Developmental Cell*, 17, 162–174. doi:.10.1016/j.devcel.2009.07.024

Maynard Smith, J., & Szathmáry, E. (1995). *The major transitions in evolution*. New York: Freeman.

McShea, D. W., & Venit, E. P. (2001). What is a part? In G. P. Wagner (Ed.), *The character concept in evolutionary biology* (pp. 259–284). San Diego, CA: Academic Press.

Medina, M., Collins, A. G., Taylor, J. W., Valentine, J. W., Lipps, J. H., Amaral-Zettler, L., & Sogin, M. L. (2003). Phylogeny of Opisthokonta and the evolution of multicellularity and complexity in Fungi and Metazoa. *International Journal of Astrobiology*, 2, 203–211.

Meroz-Fine, E., Shefer, S., & Ilan, M. (2005). Changes in morphology and physiology of an East Mediterranean sponge in different habitats. *Marine Biology*, 147, 243–250.

Meyerowitz, E. M. (2002). Plants compared to animals: The broadest comparative study of development. *Science*, 295, 1482–1485.

Michod, R. E. (2007). Evolution of individuality during the transition from unicellular to multicellular life. *Proceedings of the National Academy of Sciences of the United States of America*, 104, 8613–8618.

Miller, C. J., & Davidson, L. A. (2013). The interplay between cell signaling and mechanics in developmental processes. *Nature Reviews. Genetics*, 14, 733–744.

Mishler, B. D., & Brandon, R. N. (1987). Individuality, pluralism, and the phylogenetic species concept. *Biology & Philosophy*, 2, 397–414.

Moreno, A., Umerez, J., & Ibáñez, J. (1997). Cognition and life: The autonomy of cognition. *Brain and Cognition*, 34, 107–129.

Moroz, L. L., Kocot, K. M., Citarella, M. R., Dosung, S., Norekian, T. P., Povolotskaya, I. S., et al. (2014). The ctenophore genome and the evolutionary origins of neural systems. *Nature*, 510, 109–114. doi:.10.1038/nature13400

Nakanishi, N., Sogabe, S., & Degnan, B. M. (2014). Evolutionary origin of gastrulation: Insights from sponge development. *BMC Biology*, 12, 26.

Newman, S. A. (2012). Physico-genetic determinants in the evolution of development. *Science*, 338, 217–219.

Newman, S. A., & Bhat, R. (2009). Dynamical patterning modules: A "pattern language" for development and evolution of multicellular form. *International Journal of Developmental Biology*, 53, 693–705.

Newman, S. A., Bhat, R., & Mezentseva, N. V. (2009). Cell state switching factors and dynamical patterning modules: Complementary mediators of plasticity in development and evolution. *Journal of Biosciences*, 34, 553–572.

Newman, S. A., & Müller, G. B. (2010). Morphological evolution: Epigenetic mechanisms. In *Encyclopedia of life sciences (ELS)*. Chichester: John Wiley & Sons. doi: 10.1002/9780470015902.a0002100.pub2

Niklas, K. J., & Newman, S. A. (2013). The origins of multicellular organisms. *Evolution & Development*, 15, 41–52. doi:.10.1111/ede.12013

Novoplansky, A. (2002). Developmental plasticity in plants: Implications of non-cognitive behavior. *Evolutionary Ecology*, 16, 177–188.

Oborny, B. (2003). External and internal control in plant development. *Complexity*, 9(3), 22–28.

Pérez-Pomares, J. M., & Muñoz-Chápuli, R. (2002). Epithelial–mesenchymal transitions: A mesodermal cell strategy for evolutive innovation in metazoans. *Anatomical Record*, 268, 343–351.

Queller, D. C., & Strassmann, J. E. (2009). Beyond society: The evolution of organismality. *Philosophical Transactions of the Royal Society of London. Series B, Biological Sciences*, 364, 3143–3155.

Rokas, A. (2008 a). The origins of multicellularity and the early history of the genetic toolkit for animal development. *Annual Review of Genetics*, 42, 235–251.

Rokas, A. (2008 b). The molecular origins of multicellular transitions. *Current Opinion in Genetics & Development*, 18, 472–478.

Rosslenbroich, B. (2014). *On the origin of autonomy: A new look at the major transitions in evolution*. Cham, Switzerland: Springer.

Rozario, T., & DeSimone, D. W. (2010). The extracellular matrix in development and morphogenesis: A dynamic view. *Developmental Biology*, 341, 126–140.

Ruiz-Mirazo, K., Etxeberria, A., Moreno, A., & Ibáñez, J. (2000). Organisms and their place in biology. *Theory in Biosciences*, 119, 43–67.

Ruiz-Mirazo, K., & Moreno, A. (2012). Autonomy in evolution: From minimal to complex life. *Synthese*, 185, 21–52.

Sachs, T., Novoplansky, A., & Cohen, D. (1993). Plants as competing populations of redundant organs. *Plant, Cell & Environment*, 16, 765–770.

Satterlie, R. A. (2002). Neuronal control of swimming in jellyfish: A comparative story. *Canadian Journal of Zoology*, 80, 1654–1669.

Solari, C., Ganguly, S., Kessler, J. O., Michod, R. E., & Goldstein, R. E. (2006). Multicellularity and the functional interdependence of motility and molecular transport. *Proceedings of the National Academy of Sciences of the United States of America*, 103(5), 1353–1358.

Sprugel, D. G., Hinckley, T. M., & Schaap, W. (1991). The theory and practice of branch autonomy. *Annual Review of Ecology and Systematics*, 22, 309–334.

Srivastava, M., Simakov, O., Chapman, J., Fahey, B., Gauthier, M., Mitros, T., et al. (2010). The *Amphimedon queenslandica* genome and the evolution of animal complexity. *Nature*, 466, 720–727.

Stock, A. M., Robinson, V. L., & Goudreau, P. N. (2000). Two-component signal transduction. *Annual Review of Biochemistry*, 69, 183–215.

Takahashi, T., & Hatta, M. (2011). The importance of GLWamide neuropeptides in Cnidarian development and physiology. *Journal of Amino Acids*, 2011. doi:.10.4061/2011/424501

Takahashi, T., & Takeda, M. (2015). Insight into the molecular and functional diversity of cnidarian neuropeptides. *International Journal of Molecular Sciences*, 16, 2610–2625. doi:.10.3390/ijms16022610

Tarrant, A. M. (2005). Endocrine-like signaling in cnidarians: Current understanding and implications for ecophysiology. *Integrative and Comparative Biology*, 45, 201–214.

Tuomi, J., & Vuorisalo, T. (1989). Hierarchical selection in modular organisms. *Trends in Ecology & Evolution*, 4, 209–213.

Tyler, S. (2003). Epithelium—The primary building block for metazoan complexity. *Integrative and Comparative Biology*, 43, 55–63.

Van Duijn, M., Keijzer, F., & Franjen, D. (2006). Principles of minimal cognition: Casting cognition as sensorimotor coordination. *Adaptive Behavior*, 14, 157–170.

Watson, M. A., & Casper, B. B. (1984). Morphogenetic constraints on patterns of carbon distribution in plants. *Annual Review of Ecology and Systematics*, 15, 233–258.

White, J. (1979). The plant as a metapopulation. *Annual Review of Ecology and Systematics*, 10, 109–145.

Wilson, J. (1999). *Biological individuality: The individuation and persistence of living entities*. Cambridge, UK: Cambridge University Press.

14 Explaining the Origins of Multicellularity: Between Evolutionary Dynamics and Developmental Mechanisms

Alan C. Love

The Origin and Evolution of Multicellularity: A Complex Problem

John Tyler Bonner was a lone voice in the wilderness for many years as he actively pursued questions surrounding the origin and evolution of multicellularity, motivated in part by his experimental system: slime molds (Bonner, 2000; Nanjundiah, this volume). However, there has been a veritable renaissance of scientific work on multicellularity in recent years. The sources for this renewed interest and investment hail from several domains. One source derives from multilevel selection theory and formulating models of fitness transitions that occur when a collective becomes a new individual (Michod, 1999; Michod and Nedelcu, 2003). Another source is experimental investigations into the evolution of development in animals (Rokas, 2008b; Leininger et al., 2014; Ruiz-Trillo, this volume) and plants (Niklas, 2014), with special attention to how their developmental genetic toolkit facilitated the emergence of multicellular complexity. A third source draws on paleontology and phylogenetic systematics where patterns of origination across different lineages can be compared and contrasted (Grosberg and Strathmann, 2007; Knoll, 2011). Substantial contributions also derive from combinations of these approaches, including experimental investigations that utilize considerations from multilevel selection theory (Ratcliff et al., 2012; Conlin and Ratcliff, this volume), developmental genetic investigations of novel taxa that are phylogenetically illuminating (Suga and Ruiz-Trillo, 2013), and model clades with high phylogenetic resolution where specific instantiations of fitness transitions from a collective to a new individual can be mapped in detail (Herron, 2009). Major advances in these domains have moved the origins of multicellularity from a relatively speculative area of evolutionary biology into a poster child for the combined power of theoretical, experimental, and phylogenetic investigative methodologies.

Philosophers have tracked many of these intellectual developments and, in some cases, have even contributed to them. For example, Samir Okasha provided an illuminating analysis of different approaches to multilevel selection theory through the lens of the Price equation that clarifies what it means for selection to operate at a given hierarchical level (Okasha, 2006). Philosophers have analyzed concepts of individuality and organization that facilitate thinking about the origins of multicellularity and other major transitions in the history of life (e.g., Bouchard and Huneman, 2013; Calcott and Sterelny, 2011; Arnellos et al., 2014; Arnellos and Moreno, this volume). This mode of philosophical work can be labeled "theoretical collaboration" because the philosopher is offering theoretical contributions in collaboration with scientists via a similar mode of analysis, sometimes with mathematical models and sometimes without. Peter Godfrey-Smith exemplifies this type of work, both in terms of mathematical modeling and analyzing key concepts such as individuality and reproduction (Kerr and Godfrey-Smith, 2002; Godfrey-Smith and Kerr, 2013; Godfrey-Smith, 2009). The philosopher as theoretical collaborator operates most fruitfully in the realm of abstract theorizing, but this is only one domain where contemporary discussions surrounding multicellularity have gained traction. Much less attention has been given to experimental and phylogenetic investigations of the evolution of development where abstract theorizing plays a less prominent role in advancing inquiry (Rokas, 2008a; Ruiz-Trillo, this volume).

One interpretation of this situation is that philosophers cannot contribute to these types of investigations and therefore conceptual work should be concentrated in the realm of abstract theorizing. This interpretation is neither necessary nor warranted. The reasoning practices found in experimental, phylogenetic, and paleontological investigations surrounding the origin and evolution of multicellularity can be scrutinized and evaluated with respect to their suitability for achieving the stated explanatory aims of biologists. This can take at least two forms: reasoning explication and problem clarification (Love, 2008b).

Reasoning explication includes reconstructing different kinds of reasoning used in scientific investigation, such as reductionist research strategies, to identify their characteristic strengths and latent biases (Love, 2010b). Problem clarification involves clarifying the structure of problems and their interrelations across biological disciplines, such as how different approaches contribute to multidisciplinary explanations of complex phenomena (Love, 2008a). Although these forms of philosophical analysis differ from theoretical collaboration, they are oriented to be of potential value to working scientists: "The philosopher of science would be analyzing or criticizing an activity in terms of how well it served the ends of the scientist, and in each case, the activity itself and the analysis of it further these ends" (Wimsatt, 2007, p. 244). We can label this complementary type of philosophical work "conceptual elaboration" because it concentrates on elaborating the reasoning practices of scientists through clarification or by making them explicit. It is not directly collaborative, in part because it proceeds via modes of analysis that differ from what scientists routinely work with, but it is of potential value to ongoing biological inquiry. Conceptual elaboration can provide an evaluation of scientific practices from a perspective that indicates more or less fruitful lines of inquiry, modeling strategies, and data gathering or suggests ways to construct theoretical frameworks, interpret data, and understand criteria of adequacy for good explanations (Love, 2008b).

My goal in this chapter is to engage in conceptual elaboration in the form of problem clarification with respect to research on the origin of multicellularity. This strategy has the potential to clarify aspects of the complex architecture of the problem of the origin of multicellularity and thereby make explicit some of the explanatory standards of different communities of biologists pursuing these questions. I begin by reviewing a helpful analysis of two different theoretical perspectives on modularity that identifies a conceptual fault line between evolutionary dynamics and developmental mechanisms (Winther, 2005). This fault line marks a key bifurcation in the structure of the problem agenda of the origin of multicellularity. Once this is made explicit, a characterization of several dimensions of the problem agenda facilitates the identification of different criteria of adequacy that govern these distinct conceptualizations of research questions pertaining to multicellularity. Researchers differ in how they engage in abstraction and generalization as a consequence of addressing different kinds of research questions. Recognizing these differences opens up a space for understanding how the two approaches do not involve competing hypotheses but frequently lead to investigative communities that exhibit little interaction. This recognition also suggests ways one might formulate more integrated explanatory models to yield a deeper understanding of the phenomena in view, such as by identifying epistemological assumptions embedded in the approaches through linking answers to research questions across the fault line between evolutionary dynamics and developmental mechanisms.

Competition (Evolutionary Dynamics) and Integration (Developmental Mechanisms)

The study of modularity has experienced an explosion of activity over the past two decades (see, e.g., Schlosser and Wagner, 2004). Modules of different kinds are evolutionarily relevant because of their potential to facilitate the independent evolution of traits while maintaining organismal integrity (Kirschner and Gerhart, 1998). Two different perspectives on modularity have been characterized: integration and competition (Winther, 2005). The former focuses on how developmental or physiological parts are combined to yield structural or functional outcomes, such as organs composed of multiple tissues or circulatory components that operate sequentially to move blood through an organism to provide nutrients to different regions of the body. The latter focuses on selection processes that occur between parts located at different hierarchical levels and result in the differential survival and proliferation of those parts.

These two perspectives encourage contrasting approaches to research on modularity (Winther, 2001). An integration perspective seeks to understand how modules are combined into a more unified whole, such as an organism, and how these modules might be developmentally modified to yield phenotypic change in evolution. A competition perspective seeks to understand how selective forces among modules that exhibit variable properties change their frequencies over time due to differential fitness, and how these modules remain stable against subversion from component parts, such as via mechanisms for suppressing cheating (Grosberg and Strathmann, 2007). The integration of modules is a predominant theme in evolutionary developmental biology, whereas competition among modules is the dominant theme among evolutionary population biologists. These distinct perspectives yield divergent interpretations of the phenomenon of modularity. The assignment of fitness values to modules is standard for the competition perspective, but fitness is not a property of modules on the integration perspective. Cooperation among modules is (potentially) a fitness-increasing individual strategy on the competition perspective; the complex combinatorial outcomes of modules in physiology or development are not conceptualized as cooperative on the integration perspective. Cancer is a pathological side effect of modularity on the integration perspective but a cheating strategy that escapes detection in the competition perspective.

All of these and other divergent interpretations are linked to differences in methodology, modeling practices, and theory formulation between the two perspectives. Winther offers three interpretations of how the research questions posed within each perspective might be related (cf. Winther, 2001): (1) they address different questions (differential proliferation of modules vs. integrative mechanisms that coalesce modules); (2) they address the same questions but concern different episodes or time scales (evolutionary transitions vs. mechanisms of stability for multicellular individuality); or (3) they address the same questions and offer competing answers, which means they are in conflict rather than complementary. Of these three options, (1) is more plausible than (2) or (3), but a difficulty with discriminating among this set of alternatives is that we have no resources for understanding what it means for questions to be different, or how relationships obtain among different questions pertaining to modularity. A richer, *erotetic* ontology is needed to discern how differences in questions might translate into divergent interpretations of modularity and foster distinct approaches in how to study it.[1]

The distinction between integration and competition, or what I will label *developmental mechanisms* and *evolutionary dynamics*, is potentially relevant for clarifying the problem structure for the origin of multicellularity. Individual cells are a type of module, and research oriented around their fitness values in a collective is distinct from research oriented around how specific cells combine into multicellular structures in different lineages. The gap between developmental mechanisms and evolutionary dynamics is manifested in distinct interpretations, different methods and modeling strategies, and explanatory theories formulated to account for the origins of multicellularity. We should expect to find these kinds of differences if the fault line of evolutionary dynamics versus developmental mechanisms captures an important distinction in biological reasoning. However, as noted, despite the plausibility that these perspectives address different questions, we require more details about what it means to have different research questions about the origin and evolution of multicellularity and some grasp of how they are related to one another. In particular, this is critical for understanding the criteria of adequacy used to evaluate purported answers to these questions.

The Problem Agenda of the Origin of Multicellularity

The first step in developing an erotetic ontology that is sufficient to capture what it means for there to be different kinds of questions about the origin of multicellularity is to distinguish *problem agendas* and *research questions*. A problem agenda is a broad domain of interrelated questions, such as adaptation, speciation, homology, or cellular differentiation (Love, 2010a). It is complex in the sense of having many component parts of different kinds in dynamic interaction. We can distinguish three strands to this erotetic complexity: (1) "many component parts"—multiple subproblems or research questions with hierarchical relations; (2) "of different kinds"—a heterogeneity of subproblems or question kinds; and (3) "in dynamic interaction"—historical development and relationships among the component questions of different kinds (Love, 2014; Brigandt and Love, 2012). Thus, the complex structure of a problem agenda refers to these different strands and problem clarification concerns characterizing them explicitly. Once made explicit, this structure assists in the isolation and explication of criteria of adequacy for addressing aspects of the problem agenda, such as how and where different disciplinary contributions are required for answering particular research questions. In this respect, a problem agenda is a kind of list of what needs to be addressed—a catalog of multiple, interrelated questions (Love, 2008a). By contrast, a research question is one component part, which can be one kind rather than another (e.g., empirical, conceptual, or theoretical), and bears relationships to other research questions within the problem agenda.

The origins of multicellularity can be interpreted as a problem agenda and analyzed profitably along these lines. For example, a research question about the conditions under which a collaborative division of labor between germ cells and somatic cells obtains in a particular model of evolutionary transitions in individuality does not exhaust the research questions that might be tackled in the problem agenda. This theoretical question can be linked to allied conceptual questions (how do we assign fitness values? what counts as an individual?) and empirical questions (can we confirm the emergence of a collaborative division of labor using experimental evolution research?). These questions can exhibit hierarchical relations. An answer to the conceptual question of what counts as an individual has an impact on theoretical questions about how fitness becomes aligned among cellular constituents through mechanisms such as a unicellular bottleneck ("alignment of fitness") and then is exported to the composite individual through genuine interdependency with respect to a reproductive cycle ("export of fitness") (Folse and Roughgarden, 2010).

This partial structure to the problem agenda of the origin of multicellularity already exhibits the three strands of erotetic complexity: many component questions of different kinds in dynamic interaction. But additional complexity is present because there are other questions that appear less connected to those described thus far. For example, what components of the genetic toolkit were necessary for cellular aggregation in the transition to multicellularity for animals (Rokas, 2008a)? This is an empirical question that requires the investigation of concrete elements of genomes, but also relates to conceptual questions (what counts as cellular aggregation [Kaneko, this volume]?) and theoretical questions (on what basis is the phylogeny of metazoans reconstructed to infer the presence or absence of elements in the ancestral genetic toolkit [Ruiz-Trillo, this volume]?). Hierarchical relations are present because there are different types of cellular aggregation, as well as different genetic toolkit elements, which means answers to some types of research questions shape others.

The overarching advantage of teasing apart the various dimensions of erotetic structure in problem agendas is the ability to make explicit the criteria of adequacy for answering these questions. It is not surprising that an answer to a question about elements present in the metazoan toolkit is governed by standards that are distinct from those that apply to questions about the alignment or export of fitness. However, actually detailing these differences is crucial for identifying more or less fruitful lines of inquiry, using appropriate principles of data interpretation, and discerning whether an explanation meets the criteria or is in competition with other explanations. Two basic types of criteria are relevant: characterizational and explanatory.

Characterizational criteria of adequacy pick out requirements on descriptive detail and conceptual precision for questions in a problem agenda. For some questions, concrete detail about the relevant objects is required (e.g., specific genetic toolkit elements); for others, a high degree of abstraction is appropriate (e.g., models of fitness alignment). These requirements stem from investigative preconditions for attempting to answer these questions. We need a characterization of the specific toolkit element before we can ascertain whether it helps to explain the transition to multicellularity in metazoans or plants. This concrete detail is critical if an experimental procedure is required to isolate that element in different lineages. A high degree of abstraction is necessary if mathematical representation is required for modeling. We need to formulate variables and relations among them before we can determine whether alignment of fitness really occurs for particular parameter choices. Characterization has received less attention than explanatory theories in philosophical analysis, but it comprises an essential precondition for explanations and is ubiquitous in the sciences (Powers, under review).

Explanatory criteria of adequacy pick out requirements on the causal efficacy of material constituents and dynamical sufficiency of their joint interactions in answering particular research questions in a problem agenda (Love, 2008a). For most questions, this will involve disciplinary contributions that are positioned to ascertain whether this causal efficacy or dynamical sufficiency obtains. Developmental genetic methods and phylogenetic reconstruction are required to determine whether a genetic toolkit element is necessary for cellular aggregation or was present in an ancestral lineage. Mathematical models of population dynamics are required to determine whether the interactions among particles organized in a collective under ranges of fitness assignments favor transitions to a new, integrated unit. These criteria of adequacy vary across questions and encompass a variety of issues, including generalization (e.g., to what degree can we extrapolate from current experimental scrutiny of model organisms to past historical junctures?) and level of organization (e.g., do transitions from unicellular to multicellular individuals differ from other transitions to new types of individuals?). Most importantly, these criteria are not manifested identically across research questions, and not meeting the criteria with respect to particular research questions signals a deficiency in the explanatory account offered.

Before exploring how the distinction between evolutionary dynamics and developmental mechanisms permits us to further characterize the problem agenda of the origins of multicellularity, it is worthwhile to pause over the question of definitions. One reaction to the foregoing discussion is that we might be able to eliminate much of the structural complexity of the problem agenda by providing a precise definition of "multicellularity." With this in hand, questions about the origins of multicellularity could be more focused and we might stand a better chance of generating a unified explanatory theory. The difficulty is that differences in criteria of adequacy lead to different characterizations of multicellularity rather than a single definition. According to one criterion emphasizing cellular aggregation, multicellularity has evolved independently at least 25 times in different lineages (Grosberg and Strathmann, 2007). According to another set of criteria (cell–cell interconnection, communication, and cooperation), "complex" multicellularity has evolved 10 times (Niklas, 2014). Several years ago a research team claimed to experimentally evolve multicellularity (Ratcliff et al., 2012; Conlin and Ratcliff, this volume), but not everyone agreed because of differences in how multicellularity is characterized.

Debates over the meaning of a term like "multicellularity" signal that we are dealing with a problem agenda. The same phenomenon appears in problem agendas such as adaptation, speciation, novelty, and homology (Brigandt and Love, 2012). These definitional debates do not prevent ongoing, fruitful inquiry, and, in some ways, they encourage it. One reason that this difficulty with definitions appears to demand correction is an assumption that concepts only serve the purpose of categorization. This suggests that the primary concern is unambiguous classification, a kind of essentialism about the role played by concepts in scientific reasoning. However, concepts play many roles besides classification, and in the case of problem agenda terms like "homology" or "multicellularity" they serve to coordinate investigation and govern the choice of appropriate methods of inquiry (Brigandt and Love, 2012; Love, 2014). Once we shift attention to the role of a concept in marking out a structured array of questions (i.e., a problem agenda), definitional difficulties receive an illuminating interpretation. They arise from differences in the criteria of adequacy within a problem agenda due to its erotetic complexity. An appropriate philosophical response is to characterize that complexity in order to better understand this role in biological reasoning.

An Erotetic Fault Line: Evolutionary Dynamics versus Developmental Mechanisms

Within the problem agenda of the origin of multicellularity, there is a primary fault line that is represented in how arrays of questions cluster around evolutionary dynamics or developmental mechanisms (Rokas, 2008a). Evolutionary dynamics questions involve different frameworks for multilevel selection (e.g., MLS1, where the focal level is different cell types in a population that are subdivided into multicellular collectives, and MLS2, where the focal level is the multicellular collectives themselves), the significance of a division of labor (especially germ vs. soma), and the selective conditions for transitions of individuality. Developmental mechanism questions revolve around the structural and functional features of cellular differentiation or communication, the physical conditions of aggregation, and the lineage-specific acquisition and deployment of material resources and interactions that are critical for generating multicellular entities. We find many component questions of different kinds in relation to one another on both sides of the fault line. This is what makes an acknowledgment of the fault line so significant. It is an additional feature of erotetic complexity in the problem agenda of the origin of multicellularity. Theoretical research questions on the side of evolutionary dynamics may have different criteria of adequacy than empirical research questions (e.g., degree of abstraction), but theoretical questions from the side of developmental mechanisms can have distinct criteria of adequacy from theoretical questions that arise in the realm of evolutionary dynamics.

One manifestation of the fault line in the problem agenda is that each side presumes the other as a boundary condition in their models (cf. Sterelny, 2000). Consider cellular aggregation. For models from evolutionary dynamics, it is irrelevant how the aggregation occurs or what the unicellular ancestral lineage is and, accordingly, each is masked by an abstract placeholder:

We assume a spatially unstructured world of unicellular organisms (*I* for individuals) free to move about and capable of reproduction. With each division there is a probability that a mutation will produce a new type of cell (*g*).... Consider the possibility a *g* cell, arising by mutation from *I*, expresses a gene that encodes an adhesive polymer and thereby forms a group. (Libby and Rainey, 2013, p. 3)

From the perspective of developmental mechanisms, aggregation occurs differently in plants and animals because of distinct constituent proteins. It matters whether the aggregation is manifested via cadherins or pectins, and this makes a difference to the kinds of multicellularity that can obtain. It matters whether the ancestral lineage was a choanoflagellate (for metazoans; King et al., 2008) or some other algal species (for plants; Niklas and Newman, 2013). Whether or not a new cell type could originate by a single mutation is also debatable, as opposed to a cell that acquires a new capacity via multiple mutations that rearrange a gene regulatory network. The reverse situation holds as well. Developmental mechanism approaches do not assign fitness values to the cells being investigated, nor do they consider them as agents with different behavioral strategies (e.g., cooperation vs. defection). These features of direct relevance for evolutionary dynamics are simply boundary conditions for researchers focused on developmental mechanisms.

What do the different sides of the fault line in the problem agenda look like? On the side of evolutionary dynamics, many of the questions have been articulated explicitly: "To fully understand [evolutionary transitions in individuality], a set of interrelated questions must be answered" (Michod, 2005, p. 968):

The Fitness Question Heritable variation in fitness is the basis of natural selection. How do new levels of fitness heritability arise?

The Individuality Question How does a group become an individual?

The Major Transitions Question Units capable of independent replication before the transition only replicate as part of a larger whole after the transition. How does this larger whole evolve, and how is the replication of parts regulated?

The Cooperation Question Cooperation exports fitness from lower to higher levels. Under what conditions will cooperation evolve and be stable?

The Complexity Question How do new emergent properties arise at the higher level?

The Germ–soma Question Under what conditions will a group of individuals specialize in reproductive and vegetative functions?

The Life History Question As group size increases, when is it better to gain in one fitness component at a cost to another?

Although this list does not exhaust the research questions of evolutionary dynamics in the problem agenda of the origin of multicellularity, it amply illustrates the erotetic complexity and divergence in criteria of adequacy. There are different kinds of questions: theoretical (under what conditions will a group of individuals specialize in reproductive and vegetative functions?), empirical (what parameter changes make a difference in whether cooperation emerges in experimental populations?), and conceptual (what counts as an individual?). These are organized hierarchically in terms of abstraction and generalization. Rate of cell division or type of division (mitosis vs. fission) can be discriminated or treated abstractly as fecundity. Germ–soma differentiation is an especially important form of division of labor, but cellular specialization can be treated more abstractly. Issues of generalization are also present, including whether the MLS1 to MLS2 transition model is fully general (e.g., multinucleate to multicellular transitions; Niklas, Cobb, and Crawford, 2013), whether clonal growth facilitates transitions across all clades, and how frequently reversals to unicellularity occur. As expected, Michod's general framework of a "reorganization of fitness" is tailored to address all of these questions.

On the side of developmental mechanisms, we find an array of different questions with various interrelations pertaining to the origins of multicellularity. These include the following:

The Phylogenetic Question What genetic components were available in different lineages related to different developmental potentialities? Are different origins of multicellularity independent?

The Specificity Question How do particular molecules bring about specific developmental potentialities? What particular mechanisms are necessary and/or sufficient for the developmental potentialities underlying multicellularity (e.g., differentiation)?

The Control Question Are there particular elements that play consistent governing roles in the transition to multicellular developmental mechanisms?

The Homology Question To what degree are the developmental components or mechanisms shared across lineages?

The Physics Question What role do generic physical forces and dynamics have in bringing about developmental potentialities?

The erotetic complexity and divergence in criteria of adequacy are displayed unambiguously. There are theoretical questions (are particular morphological transitions ruled out by physical considerations?), empirical questions (what genetic toolkit components related to establishing polarity were present in the opisthokont common ancestor?), and conceptual questions (what is a life cycle? what is "complex" multicellularity?). Hierarchical relations in terms of abstraction are present: cellular adhesion can be treated more concretely as cadherins or integrins in animals or pectins and phlorotannins in plants (Niklas, 2014), we can distinguish cell adhesion from failure to separate after cell division (Rokas, 2008a; Ratcliff et al., 2012; Conlin and Ratcliff, this volume), and intercellular communication channels can be understood more or less specifically (e.g., plasmodesmata, desmosomes, gap junctions, or cytoplasmic bridges). The same is true for generalization. Were *T-box* genes present in the last common ancestor of metazoans, choanoflagellates, ichthyosporans, and capsasporans (Sebé-Pedrós et al., 2013)? Are specific types of cell–cell communication a prerequisite for multicellularity? Does the number or combination of patterning modules available in metazoans account for their multicellular diversity (Newman and Bhat, 2009)? Explanatory frameworks in developmental mechanism approaches are structured to address these different types of interrelated questions (Niklas, 2014; Niklas and Newman, 2013). In particular, there is a distinct concern about the consequences of multicellularity in specific clades, which cannot be addressed apart from attention to concrete features of particular lineages with rigorous phylogenetic comparisons. This may account for why the evolution of distinctive developmental features associated with multicellularity is the province of the developmental mechanisms approach, such as the origin of the animal egg or the phylotypic stage (Newman, 2011; Newman, this volume).

Epistemological Consequences of the Erotetic Fault Line

The value in making explicit the distinction between evolutionary dynamics and developmental mechanisms in the problem agenda of the origin of multicellularity is that it surfaces differences in criteria of adequacy, which can be utilized in ongoing research and clarify when particular explanatory models are competing or complementary.[2] Characterizational criteria of adequacy (i.e., requirements on descriptive detail and conceptual precision) for evolutionary dynamics differ from those for developmental mechanisms. The specificity of the latter (e.g., genes that permit cell–cell communication in metazoan or plant lineages) is often in contrast to the abstractness of the former (e.g., the conditions for the export of fitness from lower to higher levels). Explanatory criteria of adequacy (i.e., requirements on causal efficacy and dynamic sufficiency) also diverge across the fault line. What selective conditions encourage the transition from MLS1 to MLS2? What genetic components and physical conditions are necessary or sufficient to achieve stable cellular aggregates?

   We can translate these differences in criteria of adequacy into procedural recommendations for what counts as an answer to questions within each division of the problem agenda. In both cases, abstract modeling must be linked to empirical evidence though not necessarily in the same way. For evolutionary dynamics, it is not just that a division of labor between germ and soma happens, but that it happens because a subset of cells incur a disproportionately high mutagenic load due to increased metabolic activity (Goldsby et al., 2014) or due to a change in the expression of a particular transcriptional repressor in a lineage (König and Nedelcu, this volume). It is not just that there is selection for increased size, but that selection for increased size is driven by increased predation during the Proterozoic (Knoll and Lahr, this volume) or the ability to avoid complete decapitation (Leyser, this volume), thereby making abstract fitness assumptions attach to concrete ecological factors. For developmental mechanisms, it is not just that the existing genetic toolkit inherited from choanoflagellates (or some other lineage) contained genes necessary for cell adhesion that could yield cellular aggregation, but that it happened in metazoans because an increase in ambient calcium ions in coastal marine environments converted preexisting cadherins into homophilic cell attachment molecules (Fernàndez-Busquets et al., 2009).

   The differences across the fault line in the problem agenda go beyond what has been articulated thus far. Consider the radically divergent sets of concepts associated with evolutionary dynamics (genetic bottleneck, size-related advantage, division of labor, metabolic cooperation, trade-offs, genetic conflicts, cheaters/defectors, self/non-self recognition, maternal control, etc.) and developmental mechanisms (transcriptional regulation/ transcription factors, genetic toolkit, cell–cell signaling/communication, gene duplication, protein domain, cadherins, Wnt pathway, epithelia, etc.). Working on either side of the fault line involves different conceptual competencies. It also leads to differing commitments. For evolutionary dynamics, experimental organisms or clades (e.g., such as microbial models in experimental evolution or volvocine algae) play the role of testing model assumptions and parameters. As long as these systems are experimentally tractable or have enough known representatives, they can serve this function. Broader phylogenetic relations are irrelevant. If the principles are vetted in these systems, then they have broad applicability. For developmental mechanisms, phylogeny is central because the focus is on specific ancestral lineages with particular properties due to their genomic complement or physical circumstances. Paleontology matters, such as the first appearance of morphological forms (Chen et al., 2014; Knoll, 2014), and model organisms play the role of comparing and contrasting components and mechanisms in these specific lineages. Thus, depending on the side of the fault line where research questions get addressed, there are differing views about how to understand the role of experimentation and the nature of evidence. This can make communication across the fault line difficult, and it is no surprise that the communities working on either side have not frequently collaborated.

At this point a suggestive interpretation of the conceptual divide in the problem agenda is that it represents an instantiation of the difference between proximate and ultimate explanation. Evolutionary dynamics addresses "why" multicellularity happened, and developmental mechanisms addresses "how" multicellularity happened:

Ultimate explanations are concerned with the fitness consequences of a trait or behavior and whether it is (or is not) selected ... proximate explanations are concerned with the mechanisms that underpin the trait or behavior—that is, how it works. (Scott-Phillips, Dickins, and West, 2011, p. 38)

There are at least two reasons to resist this interpretation. First, the developmental mechanisms approach is evolutionary and diachronic. It is not merely concerned with "how" multicellularity works but rather is focused on the conditions of the origination of multicellularity through particular segments of phylogeny in deep time. Second, the evolutionary dynamics approach focuses on mechanisms that underlie transitions of individuality (i.e., how it works). It is not only concerned with fitness consequences but also with the kinds of traits involved in the origin of multicellularity (e.g., germ–soma differentiation). Both of these reasons stem from a broader worry that the proximate–ultimate distinction is a problematic way to conceptualize the causal relationships involved in development and evolution (Laland et al., 2011). It encourages an insulation of evolutionary dynamics research questions from those of developmental mechanisms and vice versa, which discourages seeking linkages between them. The rationale behind characterizing the erotetic complexity in the problem agenda and making its criteria of adequacy explicit is precisely the reverse.

Forging Integrated Explanatory Models across the Erotetic Fault Line

My overall diagnosis is that researchers investigating the problem agenda of the origin of multicellularity exhibit a conceptual divide between evolutionary dynamics and developmental mechanisms. This complicates the standard threefold erotetic complexity within the problem agenda—many component questions with hierarchical relations, questions of different kinds, and interactive relationships among questions—that already poses challenges for building integrated explanatory models that meet criteria of adequacy (Love, 2008a). The standard complexity obtains distinctively on both sides of the fault line. Each approach concentrates on different aspects of the complex biological phenomenon of evolution. In evolutionary dynamics, development is decoupled from inheritance and idealized or abstracted away; evolution is understood as functional transitions due to fitness differences. In developmental mechanisms, performance considerations are idealized or abstracted away; evolution is understood as structural transitions due to material properties of genetic and physical mechanisms.

Explicitly recognizing these differences through the lens of problem agendas demonstrates that these approaches are best understood as addressing different constellations of research questions rather than offering competing hypotheses. Given that these research questions are a part of the problem agenda of the origin of multicellularity, answering both types is necessary for building an adequate explanatory framework. An evolutionary dynamics or developmental mechanisms approach alone is inadequate. At the same time, this fault line nurtures a tendency for research within each approach to proceed in parallel without regular and substantive interaction across approaches. The formulation of integrated explanatory models for the origin and evolution of multicellularity requires overcoming this epistemic gap in the problem agenda. Fortunately, the clarification and characterization of the problem agenda achieved through conceptual elaboration positions us to see strategies that bridge the erotetic fault line and thereby secure a deeper understanding of the phenomena in view.

One strategy is to flag assumptions operating in each of the approaches and use this as a basis for linking answers to research questions across the erotetic fault line. Researchers in the evolutionary dynamics approach have expressed a concern that the transition from MLS1 to MLS2 is opaque within the multilevel selection framework but is central to understanding the origin of multicellularity (and any new level of individuality):

> The difficulty is that MLS theory fails to explain how the transition from MLS1 to MLS2 comes about.... Variation, heritability and reproduction are derived properties and their emergence at the group level requires an evolutionary explanation. (Libby and Rainey, 2013, p. 2)

However, what standards are used to determine the failure of explanation? What counts as an "evolutionary" explanation? For Libby and Rainey, this involves determining how you get group-level reproduction. But "how" you get group-level reproduction is treated abstractly. A more integrated explanatory framework across approaches might utilize different concrete answers to what group-level reproduction looks like in different taxa. This could yield more robust answers to the two-part sufficiency criterion for the origin of primitive multicellularity discussed by Libby and Rainey: that there must be a recognizable group state (what does this look like concretely?) with reproductive capacities (what does this look like concretely?). A similar type of analysis can be applied to claims that "the evidence so far suggests that cellular diversity can evolve easily when functionally called for by selection" (Grosberg and Strathmann, 2007, p. 644). What counts as "easily" does not (seemingly) depend on developmental details, but a more robust account of why it is easy could be derived from the study of specific developmental mechanisms. For example, the choanocytes of sponges combine multiple cellular functions and might serve as a distinctive platform for the origin of metazoan multicellularity via functional partitioning within life cycles containing colonial stages (Adamska, this volume).

We can pursue this same strategy by examining assumptions of developmental mechanisms approaches:

Examination of the DNA records of the earliest-branching animal phyla and their closest protist relatives has begun to shed light on the origins and assembly of this toolkit. Emerging data favor a model of gradual assembly, with components originating and diversifying at different time points prior to or shortly after the origin of animals (Rokas, 2008b, p. 235).

At least two assumptions are embedded in the viewpoint offered by Rokas: (1) a notion of "gradual assembly" and (2) a privileging of the metazoan model. An evolutionary dynamics approach might probe the notion of gradual assembly ("prior to," "shortly after") in such a way that the model increases in precision or more explicitly correlates toolkit assembly with the roles played by its members in the origin of multicellularity. By shifting to a more abstract notion of developmental role, a more general model of toolkit assembly relevant for explaining the origin of multicellularity in other lineages might be constructed (e.g., Dickinson, Nelson, and Weis, 2012). A similar type of analysis can be undertaken with respect to claims about complexity, which predominate in developmental mechanism approaches: "increases in organismal complexity can be associated with modifications of lineage-specific proteins rather than large-scale invention of protein-coding capacity" (Prochnik et al., 2010, p. 223). What is meant by "complexity"—is cell type number an adequate measure (Niklas, Cobb, and Dunker, 2014)—and what relative contributions do modifications of lineage-specific proteins or the invention of protein-coding capacity make to the evolution of complexity? Could it be that the availability of these and other routes underlies the ease with which multicellular transitions appear to have taken place in multiple lineages?

A host of benefits emerge from pursuing these types of strategies within the problem agenda of the origin of multicellularity. Consistent with conceptual elaboration via problem clarification, methodological and epistemological benefits can be derived within and across approaches. For evolutionary dynamics, making abstract selective advantages concrete through particular ecological drivers enhances explanatory power; for developmental mechanisms, shifting away from concrete toolkit elements to more abstract developmental roles can enhance explanatory generalizations across diverse taxa. For evolutionary dynamics, the use of niche construction models probes a wider possibility space for individuality transitions and makes natural connections with ontogenetic processes (Laland et al., 2008; Newman, this volume); for developmental mechanisms, diversifying outgroup comparisons can clarify different senses of multicellularity and the meanings of complexity. For evolutionary dynamics, experimental evolution methods can test selective regimes in the presence and absence of particular genetic toolkit members; for developmental mechanisms, the pursuit of generalizations across plants and animals can reveal novel contributors, such as transcription factors with intrinsically disordered domains (Niklas et al., 2014; Niklas and Dunker, this volume). The developmental mechanisms approach can identify models that better integrate genetic and generic (or physical) causal factors through the consideration of abstract developmental roles for components and their interactions (Newman and Bhat, 2009; Furusawa and Kaneko, 2002; Kaneko, this volume; Hernández-Hernández et al., 2012; Arias del Angel et al., this volume), as well as isolate neglected factors such as phenotypic plasticity (Nanjundiah, this volume; Newman, this volume). The evolutionary dynamics approach can reconceptualize individuality outside of the MLS1/MLS2 framework in light of lineage-specific phenomena (Niklas et al., 2013).

It is critical to stress that potential benefits of this kind accrue from comparing assumptions about questions not only across approaches but also within a particular approach. The privileging of metazoan models can be interrogated from within a developmental mechanisms approach through a contrast with plant lineages (Niklas and Newman, 2013). For example, the origin and diversification of molecular toolkit components for multicellular plants follow distinct trajectories in several different lineages. Different components play similar functional roles that are pertinent to multicellularity. For cellular adhesion, land plants use pectins, fungi use glycoproteins, and brown algae use phlorotannins and other molecules (Niklas, 2014). A similar point applies to distinct molecular components used for intercellular communication in plant lineages. This exposes that the lineage specificity which is distinctive of questions surrounding developmental mechanisms can beget a particular type of bias: focusing primarily on one lineage, such as metazoans, rather than others, such as land plants. When we are aware of this, the same research questions that motivate inquiry into the developmental mechanisms underlying the origin of multicellularity in different lineages can be compared and contrasted to characterize these biases and thereby advance inquiry.

One way to construe these various benefits is that what was previously implicit as a boundary condition in an approach becomes an explicit, exploratory tool to augment inquiry within and across approaches. This can only be executed when the implicit erotetic complexity within the problem agenda of the origin of multicellularity divided by the fault line of evolutionary dynamics and developmental mechanisms is clarified and character-ized. Payoff comes in diverse forms, from indicating fruitful strategies for modeling or gathering data to facilitating more integrated explanatory models based on an understand-ing of diverse criteria of adequacy in operation when offering answers to research ques-tions. Although there is much more to be done in this regard, such as scrutinizing the role of computational modeling and simulation for both approaches (Duran-Nebreda et al., this volume; Conlin and Ratcliff, this volume), the conceptual elaboration undertaken herein suggests that this mode of philosophical analysis can bear much fruit for ongoing theoreti-cal and experimental investigation. In conjunction with efforts in theoretical collaboration (Arnellos and Moreno, this volume; Godfrey-Smith and Kerr, 2013; Bouchard and Huneman, 2013), a strong case can be made for the continued involvement of philosophers in diverse aspects of inquiry into the origins of multicellularity as a means to generating increasingly integrated explanatory models of complex evolutionary phenomena.

Acknowledgments

I am grateful to Stuart A. Newman and Karl J. Niklas for the invitation to participate in the 2014 KLI workshop on "The Origins and Consequences of Multicellularity" and for their helpful feedback on an earlier version of the manuscript. Thanks are due to all of the participants because my analysis was shaped substantively through discussions during the workshop. The research and writing of this chapter was supported in part by a grant from the John Templeton Foundation (Integrating Generic and Genetic Explanations of Biologi-cal Phenomena; ID 46919).

Notes

1. "Erotetic" means "of or pertaining to questioning" and derives from the Greek word *erōtētikós*.

2. Although the explicit distinction between evolutionary dynamics and developmental mechanisms exposes differences in criteria of adequacy within the problem agenda of the origin of multicellularity, it does not ferret out all of the divisions that exist between these approaches. The epistemic value of the distinction in this context does not assume that it uniquely captures all differences between evolutionary dynamics and developmental mechanisms, either with respect to the origin of multicellularity or, more generally, for other problem agendas within evolutionary biology.

References

Arnellos, A., Moreno, A., & Ruiz-Mirazo, K. (2014). Organizational requirements for multicellular autonomy: Insights from a comparative case study. *Biology & Philosophy*, 29, 851–884.

Bonner, J. T. (2000). *First signals: The evolution of multicellular development*. Princeton, NJ: Princeton Uni-versity Press.

Bouchard, F., & Huneman, P. (Eds.). (2013). *From groups to individuals: Evolution and emerging individuality.* Cambridge, MA: MIT Press.

Brigandt, I., & Love, A. C. (2012). Conceptualizing evolutionary novelty: Moving beyond definitional debates. *Journal of Experimental Zoology. Part B, Molecular and Developmental Evolution, 318B, 417–427.*

Calcott, B., & Sterelny, K. (Eds.). (2011). *The major transitions in evolution revisited.* Cambridge, MA: MIT Press.

Chen, L., Xiao, S., Pang, K., Zhou, C., & Yuan, X. (2014). Cell differentiation and germ–soma separation in Ediacaran animal embryo-like fossils. *Nature, 516,* 238–241.

Dickinson, D. J., Nelson, W. J., & Weis, W. I. (2012). An epithelial tissue in *Dictyostelium* challenges the traditional origin of metazoan multicellularity. *BioEssays, 34,* 833–840.

Fernàndez-Busquets, X., Kornig, A., Bucior, I., Burger, M. M., & Anselmetti, D. (2009). Self-recognition and Ca^{2+}-dependent carbohydrate–carbohydrate cell adhesion provide clues to the Cambrian Explosion. *Molecular Biology and Evolution, 26,* 2551–2561.

Folse, H. J., Jr., & Roughgarden, J. (2010). What is an individual organism? A multilevel selection perspective. *Quarterly Review of Biology, 85,* 447–472.

Furusawa, C., & Kaneko, K. (2002). Origin of multicellular organisms as an inevitable consequence of dynamical systems. *Anatomical Record, 268,* 327–342.

Godfrey-Smith, P. (2009). *Darwinian populations and natural selection.* New York: Oxford University Press.

Godfrey-Smith, P., & Kerr, B. (2013). Gestalt-switching and the evolutionary transitions. *British Journal for the Philosophy of Science, 64,* 205–222.

Goldsby, H. J., Knoester, D. B., Ofria, C., & Kerr, B. (2014). The evolutionary origin of somatic cells under the dirty work hypothesis. *PLoS Biology, 12,* e1001858.

Grosberg, R. K., & Strathmann, R. R. (2007). The evolution of multicellularity: A minor major transition? *Annual Review of Ecology Evolution and Systematics, 38,* 621–654.

Hernández-Hernández, V., Niklas, K. J., Newman, S. A., & Benítez, M. (2012). Dynamical patterning modules in plant development and evolution. *International Journal of Developmental Biology, 56,* 661–674.

Herron, M. (2009). Many from one: Lessons from the volvocine algae on the evolution of multicellularity. *Communicative & Integrative Biology, 2,* 368–370.

Kerr, B., & Godfrey-Smith, P. (2002). Individualist and multi-level perspectives on selection in structured populations. *Biology & Philosophy, 17,* 477–517.

King, N., Westbrook, M. J., Young, S. L., Kuo, A., Abedin, M., Chapman, J., et al. (2008). The genome of the choanoflagellate *Monosiga brevicollis* and the origin of metazoans. *Nature, 451,* 783–788.

Kirschner, M., & Gerhart, J. (1998). Evolvability. *Proceedings of the National Academy of Sciences of the United States of America, 95,* 8420–8427.

Knoll, A. H. (2011). The multiple origins of complex multicellularity. *Annual Review of Earth and Planetary Sciences, 39,* 217–239.

Knoll, A. H. (2014). Paleobiological perspectives on early eukaryotic evolution. *Cold Spring Harbor Perspectives in Biology, 6,* a016121.

Laland, K. N., Odling-Smee, J., & Gilbert, S. F. (2008). EvoDevo and niche construction: Building bridges. *Journal of Experimental Zoology. Part B, Molecular and Developmental Evolution, 310B,* 549–566.

Laland, K. N., Sterelny, K., Odling-Smee, J., Hoppitt, W., & Uller, T. (2011). Cause and effect in biology revisited: Is Mayr's proximate–ultimate dichotomy still useful? *Science, 334,* 1512–1516.

Leininger, S., Adamski, M., Bergum, B., Guder, C., Liu, J., Laplante, M., et al. (2014). Developmental gene expression provides clues to relationships between sponge and eumetazoan body plans. *Nature Communications*, 5, 3905.

Libby, E., & Rainey, P. B. (2013). A conceptual framework for the evolutionary origins of multicellularity. *Physical Biology*, 10, 035001.

Love, A. C. (2008 a). Explaining evolutionary innovation and novelty: Criteria of explanatory adequacy and epistemological prerequisites. *Philosophy of Science*, 75, 874–886.

Love, A. C. (2008 b). From philosophy to science (to natural philosophy): Evolutionary developmental perspectives. *Quarterly Review of Biology*, 83, 65–76.

Love, A. C. (2010 a). Rethinking the structure of evolutionary theory for an extended synthesis. In M. Pigliucci & G. B. Müller (Eds.), *Evolution—The extended synthesis* (pp. 403–441). Cambridge, MA: MIT Press.

Love, A. C. (2010 b). Idealization in evolutionary developmental investigation: A tension between phenotypic plasticity and normal stages. *Philosophical Transactions of the Royal Society of London. Series B, Biological Sciences*, 365, 679–690.

Love, A. C. (2014). The erotetic organization of developmental biology. In A. Minelli & T. Pradeu (Eds.), *Towards a theory of development* (pp. 33–55). Oxford, UK: Oxford University Press.

Michod, R. E. (1999). *Darwinian dynamics: Evolutionary transitions in fitness and individuality*. Princeton, NJ: Princeton University Press.

Michod, R. E. (2005). On the transfer of fitness from the cell to the multicellular organism. *Biology & Philosophy*, 20, 967–987.

Michod, R., & Nedelcu, A. M. (2003). On the reorganization of fitness during evolutionary transitions in individuality. *Integrative and Comparative Biology*, 43, 64–73.

Newman, S. A. (2011). Animal egg as evolutionary innovation: A solution to the "embryonic hourglass" puzzle. *Journal of Experimental Zoology. Part B, Molecular and Developmental Evolution*, 316B, 467–483.

Newman, S. A., & Bhat, R. (2009). Dynamical patterning modules: A "pattern language" for development and evolution of multicellular form. *International Journal of Developmental Biology*, 53, 693–705.

Niklas, K. J. (2014). The evolutionary-developmental origins of multicellularity. *American Journal of Botany*, 101, 6–25.

Niklas, K. J., Cobb, E. D., & Crawford, D. R. (2013). The evo-devo of multinucleate cells, tissues, and organisms, and an alternative route to multicellularity. *Evolution & Development*, 15, 466–474.

Niklas, K. J., Cobb, E. D., & Dunker, A. K. (2014). The number of cell types, information content, and the evolution of complex multicellularity. *Acta Societatis Botanicorum Poloniae*, 83, 337–347.

Niklas, K. J., & Newman, S. A. (2013). The origins of multicellular organisms. *Evolution & Development*, 15, 41–52.

Okasha, S. (2006). *Evolution and the levels of selection*. New York: Oxford University Press.

Powers, J. (under review). Atrazine research and criteria of characterizational adequacy.

Prochnik, S. E., Umen, J., Nedelcu, A. M., Hallmann, A., Miller, S. M., Nishii, I., et al. (2010). Genomic analysis of organismal complexity in the multicellular green alga *Volvox carteri*. *Science*, 329, 223–226.

Ratcliff, W. C., Denison, R. F., Borrello, M., & Travisano, M. (2012). Experimental evolution of multicellularity. *Proceedings of the National Academy of Sciences of the United States of America*, 109, 1595–1600.

Rokas, A. (2008 a). The molecular origins of multicellular transitions. *Current Opinion in Genetics & Development*, 18, 472–478.

Rokas, A. (2008 b). The origins of multicellularity and the early history of the genetic toolkit for animal development. *Annual Review of Genetics*, 42, 235–251.

Schlosser, G., & Wagner, G. P. (Eds.). (2004). *Modularity in development and evolution*. Chicago: University of Chicago Press.

Scott-Phillips, T., Dickins, T. E., & West, S. A. (2011). Evolutionary theory and the ultimate–proximate distinction in the human behavioral sciences. *Perspectives on Psychological Science*, 6, 38–47.

Sebé-Pedrós, A., Ariza-Cosano, A., Weirauch, M. T., Leininger, S., Yang, A., Torruella, G., et al. (2013). Early evolution of the T-box transcription factor family. *Proceedings of the National Academy of Sciences of the United States of America*, 110, 16050–16055.

Sterelny, K. (2000). Development, evolution, and adaptation. *Philosophy of Science*, 67, S369–S387.

Suga, H., & Ruiz-Trillo, I. (2013). Development of ichthyosporeans sheds light on the origin of metazoan multicellularity. *Developmental Biology*, 377, 284–292.

Wimsatt, W. C. (2007). *Re-engineering philosophy for limited beings: Piecewise approximations to reality*. Cambridge, MA: Harvard University Press.

Winther, R. G. (2001). Varieties of modules: Kinds, levels, origins, and behaviors. *Journal of Experimental Zoology. Part B, Molecular and Developmental Evolution*, 291B, 116–129.

Winther, R. G. (2005). Evolutionary developmental biology meets levels of selection: Modular integration or competition, or both? In W. Callebaut & D. Rasskin-Gutman (Eds.), *Modularity: Understanding the development and evolution of natural complex systems* (pp. 61–97). Cambridge, MA: MIT Press.

Contributors

Maja Adamska, Sars International Centre for Marine Molecular Biology, University of Bergen

Juan Antonio Arias del Angel, Laboratorio Nacional de Ciencias de la Sostenibilidad, Instituto de Ecología, Universidad Nacional Autónoma de México, Mexico; C3, Centro de Ciencias de la Complejidad, Universidad Nacional Autónoma de México, Mexico

Argyris Arnellos, The KLI Institute

Eugenio Azpeitia, Virtual Plants INRIA team, UMR AGAP, Montpellier, France

Mariana Benítez, Laboratorio Nacional de Ciencias de la Sostenibilidad, Instituto de Ecología, Universidad Nacional Autónoma de México, Mexico; C3, Centro de Ciencias de la Complejidad, Universidad Nacional Autónoma de México, Mexico

Adriano Bonforti, ICREA-Complex Systems Lab, Universitat Pompeu Fabra

John Tyler Bonner, Department of Ecology and Evolutionary Biology, Princeton University

Peter L. Conlin, Department of Biology and BEACON Center for the Study of Evolution in Action, University of Washington

A. Keith Dunker, Center for Computational Biology and Bioinformatics, Indiana University

Salva Duran-Nebreda, ICREA-Complex Systems Lab, Universitat Pompeu Fabra

Ana E. Escalante, Laboratorio Nacional de Ciencias de la Sostenibilidad, Instituto de Ecología, Universidad Nacional Autónoma de México, Mexico

Valeria Hernández-Hernández, Laboratorio Nacional de Ciencias de la Sostenibilidad, Instituto de Ecología, Universidad Nacional Autónoma de México, Mexico; C3, Centro de Ciencias de la Complejidad, Universidad Nacional Autónoma de México, Mexico

Kunihiko Kaneko, Research Center for Complex Systems Biology, University of Tokyo

Andrew H. Knoll, Department of Organismic and Evolutionary Biology, Harvard University

Stephan G. König, Department of Biology, University of New Brunswick

Daniel J. G. Lahr, Department of Zoology, Institute of Biosciences, University of São Paulo

Ottoline Leyser, Sainsbury Laboratory, University of Cambridge

Alan C. Love, Department of Philosophy, Minnesota Center for Philosophy of Science, University of Minnesota

Raúl Montañez, ICREA-Complex Systems Lab, Universitat Pompeu Fabra

Emilio Mora Van Cauwelaert, C3, Centro de Ciencias de la Complejidad, Universidad Nacional Autónoma de México, Mexico

Alvaro Moreno, Department of Logic and Philosophy of Science, University of the Basque Country

Vidyanand Nanjundiah, Jawaharlal Nehru Centre for Advanced Scientific Research

Aurora M. Nedelcu, Department of Biology, University of New Brunswick

Stuart A. Newman, Department of Cell Biology & Anatomy, New York Medical College

Karl J. Niklas, School of Integrative Plant Biology, Cornell University

William C. Ratcliff, School of Biology, Georgia Institute of Technology

Iñaki Ruiz-Trillo, Departament de Genètica, Universitat de Barcelona

Ricard Solé, ICREA-Complex Systems Lab, Universitat Pompeu Fabra

Index

Printed in the United States
by Baker & Taylor Publisher Services